BOSTON STUDIES IN THE PHILOSOPHY OF SCIENCE

VOLUME LIII

THE STRUCTURE OF APPEARANCE

SYNTHESE LIBRARY

MONOGRAPHS ON EPISTEMOLOGY,

LOGIC, METHODOLOGY, PHILOSOPHY OF SCIENCE,

SOCIOLOGY OF SCIENCE AND OF KNOWLEDGE,

AND ON THE MATHEMATICAL METHODS OF

SOCIAL AND BEHAVIORAL SCIENCES

VOLUME 107

BOSTON STUDIES IN THE PHILOSOPHY OF SCIENCE

EDITED BY ROBERT S. COHEN AND MARX W. WARTOFSKY

VOLUME LIII

NELSON GOODMAN

THE STRUCTURE OF APPEARANCE

Third Edition
with an Introduction by

GEOFFREY HELLMAN

D. REIDEL PUBLISHING COMPANY

DORDRECHT-HOLLAND / BOSTON-U.S.A.

Library of Congress Cataloging in Publication Data

Goodman, Nelson.
The structure of appearance.

(Boston studies in the philosophy of science; v. 53)
(Synthese Library; v. 107)
Bibliography: p.
Includes index.
1. Phenomenology. 2. Structuralism. 3. System theory. 4. Science—Philosophy. I. Title. II. Series.
Q174.B67 vol. 53 [B829.5] 501s [142'.7] 77-24191
ISBN 90-277-0773-1
ISBN 90-277-0774-X pbk.

Published by D. Reidel Publishing Company,
P.O. Box 17, Dordrecht, Holland

Sold and distributed in the U.S.A., Canada, and Mexico
by D. Reidel Publishing Company, Inc.
Lincoln Building, 160 Old Derby Street, Hingham,
Mass. 02043, U.S.A.

Printed in The Netherlands

To my father

HENRY L. GOODMAN

1874–1941

EDITORIAL PREFACE

With this third edition of Nelson Goodman's *The Structure of Appearance*, we are pleased to make available once more one of the most influential and important works in the philosophy of our times. Professor Geoffrey Hellman's introduction gives a sustained analysis and appreciation of the major themes and the thrust of the book, as well as an account of the ways in which many of Goodman's problems and projects have been picked up and developed by others. Hellman also suggests how *The Structure of Appearance* introduces issues which Goodman later continues in his essays and in the *Languages of Art.* There remains the task of understanding Goodman's project as a whole; to see the deep continuities of his thought, as it ranges from logic to epistemology, to science and art; to see it therefore as a complex yet coherent theory of human cognition and practice. What we can only hope to suggest, in this note, is the broad significance of Goodman's apparently technical work for philosophers, scientists and humanists.

One may say of Nelson Goodman that his bite is worse than his bark. Behind what appears as a cool and methodical analysis of the conditions of the construction of systems, there lurks a radical and disturbing thesis: that the world is, in itself, no more one way than another, nor are we. It depends on the ways in which we take it, *and on what we do*. What we do, as human beings, is talk and think, make, act and interact. In effect, we construct our worlds by construing them, this way or that. The conditions on the construction of symbol systems are, by extension and interpretation, conditions on our construction of worlds, and of ourselves as part of the ways 'the world' is.

It would be impertinent and impetuous to impose on Goodman any grand philosophical programs. He is the model of deflationary analysis, both in his methodological nominalism and in his ontological relativism. Yet the Goodmanian bite is infectious, and suggests a much broader program than it is his style to admit. One form of rabid Goodmania would suggest a sort of dynamic and pluralistic Kantianism, in which *a prioris* are as plentiful as blackberries. Serious choices among them, however, are not. As Poincaré once put it, though, "Conventions, yes – arbitrary, no"[1]. The empirical – one may say, objective – pull of Goodman's constructionism is that, historically and cul-

[1] *Science and Hypotheses*, in *The Foundations of Science* (New York, 1929), p. 106.

turally speaking, we have chosen certain crucial constructions as more canonical than others. Goodman's argument is that it is open to us to discover, choose, invent others, because the constraints we imagine to be imposed on our choices are open to revision. By us.

Center for the Philosophy and History of Science,
Boston University
May 1977

ROBERT S. COHEN
MARX W. WARTOFSKY

TABLE OF CONTENTS

PART TWO / ON QUALITIES AND THE CONCRETE

FOREWORD TO THIS EDITION

This third edition incorporates many small changes as well as a rewriting of the final section of the first chapter, and a brief addition to the tenth. The new introduction by Geoffrey Hellman should facilitate access for many readers and help correct some persistent and prevalent misunderstandings. Moreover, *Problems and Projects*, a collection of my essays containing further discussion of some matters dealt with here, is now available.

Unfortunately, the hoped-for day when philosophy will be "discussed in terms of investigation rather than controversy, and philosophers, like scientists, be known by the topics they study rather than the views they hold" has not yet come. I can only repeat that advocacy of doctrine in the book is minimal. The studies of definition, of simplicity, of varieties of quality-predication, of order and measure, of tense and time, are in general neutral on broad philosophical issues. Despite my title, nothing in the book suggests that appearance has a unique structure; relativism runs throughout. The guiding principles are methodological: paucity of basis, maximization of system, discrimination of detail. Sometimes the results are unexpected: that extensional identity is too loose rather than too tight a criterion of definition; that extralogical postulates are eliminable wholesale; that several elements may not all be alike even though each two are; that a one-place predicate of classes may be more complex than any collection of one- or many-place predicates of individuals; that time is more static than space; and that certain strange paradoxes lie in wait for the unwary..

Many of the topics studied in this book, some for almost the first time in any detail, are actively discussed today; and some of my work has been absorbed, with or without credit, with or without confusion, into current philosophical literature. Other rather neglected parts of the book may perhaps eventually prove worth further attention. I am pleased that the book in its second quarter century has been welcomed into the *Boston Studies in the Philosophy of Science.*

Harvard University
April, 1977

NELSON GOODMAN

FOREWORD TO THE SECOND EDITION

Publication of a new edition will confound those fond of pronouncing obituaries over this book, while the number and nature of the changes made will give evidence of the ongoing obsolescence symptomatic of progress. The changes range from corrections of dozens of slips of pen, print, and mind, through appreciable improvements in some formulae and explanations, to the complete rewriting of most of one chapter and the addition to another of a new section. In Chapter III, old Sections 3 through 7 have been replaced by new Sections 3 through 10, incorporating results of the continuing investigation of structural simplicity; and the new Section 14 added to Chapter X outlines results of work on the problems of order by several mathematicians and the present writer. Some of these results are now first published.

For many corrections and improvements I am indebted to graduate students in my classes over the past thirteen years. Although I have tried to give explicit credit where due, I may sometimes have adopted a suggestion without having duly recorded its source.

The changes made have left the character and plan of the book, and indeed most of the text, substantially the same. I am not inclined to modify the basic approach or attitudes embodied in the first edition, or to withdraw any of its major tenets. On the whole, I have avoided controversy and have kept illustrative and historical passages to a minimum. Supplementary material will be found in some of my articles, which I hope to make available eventually in a volume entitled *Problems and Projects.*

Adding anything to counteract the misunderstandings warned against in the introduction to the first edition, yet still prevalent among those who have not read the book carefully, seems pointless; but the exposition of nominalism has been somewhat sharpened. I do find some justice in the complaint that the book begins so abruptly with a rather abstract and difficult chapter that the reader unfamiliar with my purpose and point of view may have trouble getting under way; and for him I have two suggestions derived from my experience in teaching the book. First, an informal discussion I have published under two different titles (as "The Revision of Philosophy", in *American Philosophers at Work*, New York: Criterion Books, 1956; and as "The Significance of *Der logische Aufbau der Welt*", in *The Philosophy of Rudolf Carnap*, La Salle, Ill.: Open Court, 1963) will provide general orientation. Second, Chapters IV and V may well be read before Chapter I so that abundant examples of constructional definition will be clearly in mind before a study of theoretical problems concerning it is begun.

The reader with inadequate time, patience, or technical equipment for the more exacting passages in the middle sections of Chapter III and some of the later sections of Chapter X may skip these without being unduly hampered in understanding what follows.

My present research assistant, Marsha Hanen, has helped greatly in the preparation of this new edition, as has David Meredith. They have also done most of the proofreading.

NELSON GOODMAN

FOREWORD TO THE FIRST EDITION

Some of the research for the present work was well under way by 1930, and a plan for the whole was drawn up not long afterward. The completed project in the form of a doctoral thesis entitled *A Study of Qualities* was deposited at Harvard University in November 1940. The war and other circumstances delayed publication; and in the meantime, continuing research led to the need for a number of revisions and additions. The work has therefore been entirely rewritten for publication under its present title.

It is particularly difficult, in view of the long period over which work on the book has been spread, to give proper credit to all who deserve it. While I have tried to indicate in footnotes the sources of any ideas that I have consciously borrowed and that have not already become current coin, I am well aware that some of my other results must also have been anticipated. I can only offer this blanket apology and the promise to make specific amends at the first opportunity concerning any matters of this kind that may be called to my attention. However, I feel no responsibility for crediting publications that were themselves anticipated by *A Study of Qualities.*

As for more personal acknowledgments, I owe lasting gratitude to the late Professor James Haughton Woods for the indispensable initial spark of encouragement. And I have profited much from the guidance and instruction, during my studies at Harvard, of Professor C. I. Lewis–although I am afraid he will find much in the book that is not to his liking. During the earliest years of research, I enjoyed the close collaboration of Henry S. Leonard; some of the first results of our joint investigations were reported in his doctoral thesis *Singular Terms* (Harvard, 1930). Intermittently since 1936 I have had the benefit of close association and collaboration with W. V. Quine; and the extent of his contribution to the finished book is not adequately represented by the footnotes referring to him. He and Rudolf Carnap read *A Study of Qualities* with great care and made innumerable valuable suggestions. I think it unlikely that those I have mentioned will be held responsible for all my ideas; and I trust that I shall not be held responsible for all of theirs.

To Professor Elizabeth F. Flower, I am indebted for generous and expert help in reading proof. So many other people have helped enormously by their encouragement, discussion, suggestions, and practical coöperation that

I have regretfully abandoned the attempt at any just listing; but I must mention Huntington Cairns, C. G. Hempel, Sidney Hook, Ernest Nagel, Glenn R. Morrow, C. L. Stevenson, and Morton G. White. And I make grateful acknowledgement to the following institutions: the John Simon Guggenheim Memorial Foundation, for a fellowship that enabled me to devote the academic year 1947–48 to preparation of part of the manuscript; the Bollingen Foundation, for a generous grant in aid of publication; the Department of Philosophy of Harvard University, for an additional subsidy; and the American Philosophical Society, for a grant for secretarial assistance. Material published as articles in the *Journal of Symbolic Logic* has been used with permission of the editors.

NELSON GOODMAN

GEOFFREY HELLMAN

INTRODUCTION

Along with some of the greatest classics of philosophy, this book is more widely known by description than by acquaintance. The descriptions generally make reference to Russell, Carnap, C. I. Lewis, and others whose work inspired it, thereby subsuming it under a time-worn umbrella covering a murky amalgam of constructivist-empiricist doctrine, which the purifying waters of ordinary discourse and common sense have, in the view of many, long since washed away. Those who do get close enough to acquire a glimpse of its pages, graced occasionally by some formulas of the quantifier calculus, sometimes discern that methodologically it is inspired by Russell and Whitehead's *Principia Mathematica* in its effort to bring logical systematization to bear on a variety of philosophical problems. But even actual readers of the book do not always realize how sharply, in fact, *Structure* breaks away from some of the main theses of its predecessors, and how strikingly it contrasts on a number of important substantive issues with longstanding views of the empiricist tradition while at the same time advancing several radical ideas of its own, currently of major interest in philosophy and other fields. The central purpose of these introductory pages will have been served if an overview can be provided affording a more accurate recognition of some of *Structure*'s original and lasting contributions.

An introduction-length introduction cannot possibly hope to do justice to the rich and varied content of this work. What it can accomplish is first, to provide a general setting of major themes running throughout Goodman's work that are usefully borne in mind in approaching *Structure*; second, to indicate in the broadest outlines its overall thrust, the main lines of its organization, and to provide some details on topics of special difficulty and current interest, potentially helpful to the reader; and third, to suggest some of the intimate connections between its content and a number of salient contemporary philosophical issues.

1. CONCEPTIONS

Goodman's corpus, from the perspective of major themes that emerge, constitutes a rather coherent–if scattered–whole. The most important for approaching *Structure* can be subsumed under four headings:

(1) the methodological outlook of constructionalism;

(2) an anti-foundationalist epistemology: rejection of the 'given', of any effort to sever perception from conceptualization (hence of all such approaches to an observation/theory dichotomy for science), and of the *a priori*, in favor of a modified coherence view of justification; emphasis on pragmatic considerations in choice of theory; and finally, the view that cognitive understanding is not the exclusive province of either science or linguistic systems but is an aim of the arts and is achieved through symbolic systems of great diversity;

(3) the emphasis on multiple systems and starting points adequate to their respective purposes along with renunciation of a single correct system embracing all knowledge or reality–methodological and ontological pluralism;

(4) the view that what are often taken as 'ultimate' metaphysical questions (concerning constituents or categories of 'reality') are pointless except when relativized to a system or 'way of construing' reference–a kind of metaphysical and ontological relativism.

For Goodman, progress in philosophy is often furthered by the careful formulation and development of constructional systems.[1] A constructional system is understood to be an interpreted formal system of definitions and theorems framed in the language of the (first-order) predicate calculus. (The restriction to first-order languages follows from Goodman's 'nominalism', which will be discussed below and need not be stressed here.) The definitions of a constructional system are to be thought of as 'real' definitions meeting some definite semantic criterion of accuracy in addition to the usual syntactic criteria (eliminability and non-creativity) imposed on purely formal or 'nominal' definitions. Thus, a constructional system is a formalization of some domain of (putative) knowledge which may be thought of as a set of sentences formulated in presystematic discourse (generally of a natural language), some of whose terms are to be appropriately defined in the system using logic plus a special set of terms adopted as primitive in the system (called the 'extralogical basis'). The primitives are to be thought of as already having an intended use or interpretation; if it is not obvious, it may be provided by an informal explanation, strictly not part of the system. Thus, for an uninteresting example, the presystematic domain might be (sentences concerning) human kinship relations and a constructional system might consist in accurate definitions of all the kinship predicates in terms of the primitives 'x is parent of y' and 'x is female' along with (recursively enumerable) specification of theorems in the system (via axioms and rules of logical inference), each being a translate (via the definitions) of one of the original presystematic sentences to be preserved.

It may well be asked, if we already have our presystematic 'knowledge' to which we must appeal in assessing both the *accuracy* of (the definitions of) a constructional system and its *adequacy* (i.e., whether its theorems form a sufficiently comprehensive set), then why bother developing such systems? All analytic philosophers must confront a 'paradox of analysis,' and the question simply raises a form of the paradox appropriate to Goodman's analytic approach. The answer is threefold. First, a successful constructional system tells us something we generally do not know in advance, namely, that certain primitives are an adequate basis for defining all the terms in question. In some cases, such knowledge is striking and far-reaching, to wit, the adequacy of set-membership (and set-theoretic axioms) for the complete battery of predicates (and theorems) of classical mathematics (as we, today, would sum up the import of Russell and Whitehead's *Principia*). In addition, a constructional system exhibits a host of relationships of logical and definitional dependence, many of which are not given in advance, but which may yield genuine insight as well as further interesting applications, not least among which will be the framing of new questions which could not even be anticipated prior to an attempt at systematization.

Second, in most cases of interest, the presystematic domain is not so clear as in our uninteresting example. Rather, perplexity and confusion at many points will motivate developing a constructional system in the first place. This is of course the case with respect to the systems with which SA is predominantly concerned, systems which we may provisionally call 'epistemological' to focus attention on their aim of representing some portion of our knowledge in relation to some relatively observational basis. Although, as will be explained, Goodman's epistemological purposes are radically different from those of his most influential constructionalist predecessors (especially Russell and Carnap), the point still holds that the presystematic domain is fraught with vagueness and obscurity which it is a major purpose of systematization to overcome. For example, a system which attempts to represent our knowledge of physical objects in terms of phenomena must confront at the outset the question what sort of phenomenal entities are to be countenanced (concrete particulars, such as momentary color spots in the visual field, abstract sensory qualities, time slices of the 'total stream of experience', and so forth), a question that no amount of appeal to ordinary usage or tradition can definitively settle. In fact, nothing short of the overall adequacy and fruitfulness of a system itself can settle such questions. (Of course, this kind of question may have no unique answer, as will emerge below.) This should make it clear that, while constructional systems may in the first instance be thought of as formalizations, they are not *just* formalizations: they are *theories*, and their development involves creative theory construction,[2] not simply formal mimicking of ordinary usage. Presystematic usage is at best a helpful guide at critical points; as Goodman stresses, it may be overruled by a system in the interests of coherence, simplicity, and other considerations, in much the manner that scientific 'data' is judiciously ignored or overruled by our best confirmed scientific theories.

Perhaps the most important consideration here is *theoretical tractability*. While preserving crucial features of presystematic discourse, constructional definitions permit a subtle kind of *replacement* of obscure and inexact by less obscure and more precise terms. If the aim were merely 'to capture' ordinary use, this would be inexcusable. From the point of view of theory construction, however, the procedure is indispensable: the more precise notions enter more readily into testable hypotheses and demonstrable results. Whole branches of mathematics have thus been erected. An example familiar to philosophers is recursion theory. In a sense, the intuitive notion, 'computable by algorithm', is replaced by a number of (demonstrably equivalent) technical notions ('general recursive', 'Turing computable', etc.), to whose superior tractability the richness of the field attests.

Indeed, the creative aspects of theory construction abound in SA, especially in Part I, Ch. 3, which develops an extensive formal account of logical simplicity (of classes of predicates) based on a single presystematic insight; and in Part III, which extends the realistic[3] phenomenalist system of Part II to the problem of ordering phenomenal qualities and paves the way toward defining predicates of shape and measure for this realm.

This leads to the third point in response to the alleged paradox of analysis. A number of philosophic questions of traditional importance can best be comprehended in terms of the adequacy of a certain type of basis. For example, the dispute between nominalism and opponents (nominalism in the traditional sense of 'no abstract entities') can be framed as the question whether a system which takes as basic only concrete individuals is capable of adequately representing all our knowledge claims worth representing. A great deal of debate may go on as to the scope of the latter phrase. However, even where agreement on this point exists, the matter can only be settled either by producing an adequate system or by somehow showing that it can't be done. Sometimes negative demonstrations can be given in advance. In philosophy, this has usually not been possible, partly because the pertinent issue of just what is admissible as a basic predicate is generally left woefully imprecise. (Consider traditional disputes such as phenomenalism vs. physicalism, behaviorism vs. its opponents, and so forth.) Constructional systems have the twofold advantage of necessitating decision on the question of admissible bases and of affording the best evidence that is generally obtainable, either a positive demonstration (the 'brute force' approach of *Principia*) or the negative evidence of failure (hopefully of the illuminating and instructive variety in which our Viennese colleagues have so sedulously specialized).

Before leaving the topic of constructionalism, it is worth pointing out that, although paradigm cases of constructional systems in the literature are often associated with single (herculean) individuals, developing such systems dealing with major philosophical problems is typically a collective enterprise. From this perspective, it will be evident that a great deal of work on current problems proceeds along constructionalist lines, for instance work on the semantics of natural language (formal semantics and pragmatics, especially treatments of the propositional attitudes, modalities, counterfactuals, and other idioms whose logical representations are problematic), work on fragments of theories of rationality, work in the foundations of physics and other natural sciences, some work on aspects of the mind-body problem and the unity of science, and in other areas that will occur to the reader. That completed systems for these domains do not exist should not obscure the fact that many contributions take the form of proposals for an appropriate basis for further constructions, tentative and partial constructions upon a proposed basis, challenges and defenses thereby generated, and metatheoretic inquiries concerning the standards of accuracy and adequacy of various types of systems.

Turning to epistemology, it must be stressed that, despite Goodman's indebtedness to Carnap and the positivists on constructionalism, *Structure* represents a sharp break away from the foundationalism that characterized the *Aufbau* and the work of other major predecessors (especially Russell and C. I. Lewis). First and foremost, the epistemological 'given' in experience is emphatically rejected along with any claim of epistemological priority of the extralogical basis of a constructional system.[4] Goodman has consistently been an original and leading opponent of the traditional empiricist dogma that all knowledge can be built up from some perceptual stratum free of conceptualization, for it is denied that such a stratum exists. In Lakatos' terminology, Goodman is an 'activist': the mind is active in perception at all levels; there is no such thing as unstructured, absolutely immediate sensory 'data' free from categorization. All perception is tainted by selection and classification, in turn formed through a complex of inheritance, habituation, preference, predisposition, and prejudice. Even phenomenal statements purporting to describe the rawest of raw feels are neither free from such formative influences nor incorrigible, in the sense of 'immune from revision for cause'. Even 'brown patch now' may reasonably be revised (without claim of 'linguistic mistake'!) in the interests of coherence with other judgments, some of which may describe particular experience, some of which may enunciate general principle.[5] If these considerations are combined with Goodman's position on 'meaning', according to which there are at best variable standards of relative likeness of meaning in natural languages, hence no significant analytic-synthetic dichotomy,[6] hence no substance to epistemological reductionist programs which seek to spell out 'the meaning' of all factual claims in terms of 'observational' entailments, then we have, in one fell-swoop, the full sweep of Goodman's foiling swipe at traditional foundationalism.

Thus, it should be clear that, although *Structure* is primarily occupied with phenomenalist systems–systems whose basic primitives are satisfied by phenomenal entities–*phenomenalism* as a foundationalist epistemological doctrine *is not espoused*. (Nor, as will be made clear, is phenomenalism as an ontological exhaustiveness claim espoused.) The pertinent question many are inclined to ask can be raised: What, then, is the epistemological relevance of such constructional systems? The answer stresses coherence: a system exhibits a network of interconnections among various parts of a conceptual apparatus. The foundationalist metaphor is replaced by Quine's 'web of belief'. As with hypothetico-deductive theories in natural science, definitions and theorems yield deductive relations among sentences which transmit rational support. And the simpler the basis, the tighter the systematic connections, the greater the overall coherence. This, of course, is no guarantee of ontological or other postulational economy (an outstanding problem which SA broaches (in Chapter 3, Section 12) but does not solve). Nevertheless, lest it be thought that coherent systems justify themselves floating freely in mid-air, it should be emphasized that a system is typically tested both against presystematic background knowledge that guides the definitions and theorems and against the achievements of other systems. Appeal to 'data' of these sorts pervades the system of SA developed in Parts II and III.[7]

While the epistemological relevance of coherence may be readily granted, it is also true that many traditional epistemological questions, motivated by the 'given' and the goal of securing everything upon incorrigible foundations, are simply being replaced by others of a different, frequently more specific kind. Thus, both the system of the *Aufbau*, critiqued in Chapter 5 of SA, and the SA system itself, are concerned with the relationship between the realms of abstract and concrete. They thus attempt to clarify a particularly murky set of distinctions that pervade our conceptual scheme. And the intriguing work in Part III on orderings of phenomenal qualities, with its definite bearings on cognitive psychology responds to problems only recently formulated. System-building need not be classical, grand-manner system-building.

In addition to coherence, however, there is a further major point on the epistemological relevance of constructional systems: such systems can provide rational support for what may be termed relative ontological adequacy claims, claims of the form, 'no entities beyond these (in the extensions of the extralogical primitives of the system in question) need be countenanced for these purposes (depending on the scope and aims of the system)'. Successful constructions upon a given limited basis show that ontological commitment beyond that of the basis is unnecessary for the purposes at hand. If a system were comprehensive enough to be adequate to all theoretical tasks (whatever that might mean), an ontological exhaustiveness claim would thereby be supported. As will be seen, Goodman takes a dim view of such claims, although they are traditionally associated with labels for types of systems distinguished in SA (such as 'phenomenalist', 'physicalist', and so forth). In general, absolute completeness (pretending for the moment we understand what that would be) is *not* to be expected of a constructional system; nor is it a demand that need be satisfied in order for a system to make an important contribution to knowledge. In particular, even if an ontological exhaustiveness claim (concerning entities of a given sort) is not supported, relative ontological claims stating just how far a given realm will take us are obviously significant and, possibly, the most we can hope for.[8]

Methodological pluralism is a natural corollary of Goodman's epistemological point of view. For if the aim of a comprehensive system linking everything to a unique 'given' is given up, multiple starting points are seen as plausible and valuable in their own right, even where the domain is one and the same and obviously inexhaustive. Thus Goodman develops a realist system covering roughly the same territory as the first stages of the *Aufbau*, but beginning with abstract phenomenal qualities ('qualia') rather than with Carnap's concrete *Elementarelebnissen* ('erlebs'); this is motivated not by *a priori* objections to the epistemological status of the latter, but rather because the relation between abstract and concrete is unsatisfactorily treated in Carnap's system, and the realist alternative promises a better solution to this problem. This does not mean the particularistic approach is rejected, for it may have advantages with respect to different problems.

There are a number of stronger 'pluralist' positions which Goodman endorses, positions which are metaphysical or ontological in character. These along with the fourth point, relativism, will be better appreciated after discussion of the metatheoretic content of SA, but we may note here the following stronger claims:

(i) Any subject matter may be systematized equally well in many ways which differ essentially in ontological commitment and which are on their face mutually incompatible (Multiplicity).

(ii) Because of multiple versions of the world in divergent symbol systems, it is futile to seek a complete description of reality (Essential Incompletability).

(iii) A fully realist attitude toward (the ontology of) any theory is (in view of (i)) arbitrary and unjustified (Anti-realism).

These lead to:

(iv) One can make sense of reference to 'the world' only if it is relativized to a system of description (or other mode of symbolization, such as depiction); similarly, ontological claims have truth value only relative to a 'construal of' or 'way of taking' objects, the world, reality, etc. (Ontological Relativism) (*We* do not see any way of completing the last sentence except by using these allegedly empty terms. But *we* are not Goodman. (This, we trust, holds in any system!))

The major sources for these views are 'The Way the World Is', 'Some Reflections on the Theory of Systems'[9] and 'Words, Works, Worlds'[10]. However, some of the kernel ideas and examples motivating the grander perspectives on metaphysics are contained in the opening sections of SA, to which we may now turn.

2. CONTENTS

Structure divides naturally into two tiers: there is the level of theory proper, the exposition, cirticism, comparison, and development of constructional systems treating the world of sensory phenomena; and there is the level of metatheory, concerned with constraints and desiderata for constructional systems generally. Part I is devoted to metatheory: the first three chapters deal, in order, with standards of definitional accuracy for constructional systems, the mathematical apparatus (with a sketch of the calculus of individuals as a nominalistic alternative to set theory), and the problem of choosing an extralogical basis (the bulk of this third chapter being devoted to a formal theory of logical simplicity of sets of predicates). Parts II and III are mostly on the level of theory proper, with Chapter 4, 'Approach to the Problems' (the first chapter of Part II), bridging the two levels by informally motivating the problems of abstaction and concretion to be treated in particular systems while containing important metatheoretic distinctions among types of systems (especially, physicalist vs. phenomenalist and realist vs. particularist), as well as an outline of the epistemological views already sketched. This ordering does not make for easy reading, for the opening three chapters on metatheory contain some of the most abstract and difficult material in the book. Thus, in his Introduction to the second edition, Goodman recommended (for those new to this material) beginning with Chapters 4 and 5 (the latter containing translation of portions of Carnap's *Aufbau* with Goodman's running commentary) in order to acquire a feel for the types of systems in question before turning to the analytically indispensable metatheory of Part I. He also recommended 'The Revision of Philosophy' (referred to above, n.1), for general orientation. In addition to these suggestions, we would add 'Some Reflections on the Theory of Systems'[11] as a good nontechnical introduction to the first chapter on definition, and 'A World of Individuals'[12] for further clarification and argumentation concerning Goodman's nominalism. Some remarks here may be of further help on these two important metatheoretic issues, to which space confines us.

We have already introduced the distinction between the notion of accuracy and that of adequacy of constructional systems. The question of adequacy is the question of completeness of a system–whether the set of definitions provides translates of all the presystematic sentences of interest, depending on the purposes of the system, and whether the set of theorems of the system is sufficiently comprehensive (relative to the same parameters–in many cases, of course, absolute completeness in the strict logical sense as well as categoricity are not obtainable in principle). The question of accuracy concerns rather the status of the real definitions of the system–the relationship between the defining terms ('definientia') and the defined terms ('definienda')–and the status of the theorems as true or otherwise acceptable. If the definitions are construed as axioms having the form of universally quantified biconditionals, any requirement of definitional accuracy is subsumed under a criterion of accuracy for theorems. In SA the crucial metatheoretic issue (of Ch. 1) is the standard or criterion of definitional accuracy, and this is treated independently. Rephrasing what is said here in the form of a requirement upon theorems is a matter of course.

What Goodman has to say on this topic constitutes one of the most controversial, interesting, and far-reaching contributions of *Structure*, and, as will be seen below, has direct bearing on the theses (i)–(iv) listed above and on a number of major current issues in contemporary philosophy. In effect, Goodman continues an important trend in the modern analytic tradition of relaxing the semantic criterion imposed on analyses. A major step in this direction was the move away from synonymy or analyticity (as was required implicitly by, for example, G. E. Moore and the early Russell) and toward a purely extensional criterion (as made explicit by Carnap in the *Aufbau*). Strikingly, Goodman argues that coextensiveness of definienda with their definientia is still too strong a criterion. Now in one respect this point was recognized by Carnap in his well-known view of the nature of *explication*:

presystematic usage is vague and inconsistent in many ways, and explication is not expected to–in fact, is expected not to–reflect these vagaries. Unclear cases cannot be decided according to any such demand as coextensiveness; moreover, occasional departure from presystematic clear cases is justified by desiderata of good theory construction. Goodman fully endorses these points, but goes on to raise a qualitatively different and deeper issue: There are many cases in mathematics and science of alternative construals of predicates, construals which are completely indistinguishable with regard to any criteria deemed relevant to the quality of the theories in which they are embedded; nevertheless, the alternatives are not merely non-coextensive – they are demonstrably *disjoint*.

The examples Goodman gives as paradigms are from geometry. Points may be taken as (i.e., defined as) suitable pairs of intersecting lines, or as triples of intersecting lines, or as certain classes of classes of volumes (as in Whitehead's construction), or in many (in fact, infinitely many) other ways. Presystematically, points are none of these things. Thus, if we were to adopt as our criterion of definition substitution in all non-intensional contexts *salve veritate*, all these constructions would be ruled out (since falsehoods (or indeterminates) such as 'Points are pairs of lines' are transformed into tautologies upon substitution). And if we were dogmatically to insist that points really are one of these sorts of things to the exclusion of all others, we face the double embarrassment of excluding all but one among many alternatives equally good in every respect that makes any difference whatever to our overall theory and finding ourselves utterly unable to say, except purely arbitrarily, which alternative is 'the right one'. The reader will be struck by the exact parallel between these examples and the much discussed alternative construals of natural numbers in set theory (either as members of Zermelo's ω-sequence, ϕ, $\{\phi\}$, $\{\{\phi\}\}$, ..., or as members of von Neumann's, ϕ, $\{\phi\}$, $\{\phi, \{\phi\}\}$, ..., or, indeed any other).[13]

Evidently, all that matters in such cases is that structural inter-relationships within the extensions of the definienda predicates (e.g., co-incidence, congruence, similarity, etc. in geometry, successor, addition, multiplication, etc., in the case of number theory) be exhibited within the extensions of the definientia. The absolute identity of the elements so related is of no importance. An appropriate criterion of definitional accuracy should thus abstract from absolute identity and focus on structure preservation. The mathematically inclined reader will by now have thought of 'isomorphism', of a one-one mapping from one domain onto another preserving[14] a set of relations expressed by the predicates of the theory being systematized. This is, in essence, what Goodman proposes, explains, and illustrates in Chapter 1, under the heading of 'extensional isomorphism' as the criterion of constructional definition. One point of detail requires attention here.

The kind of isomorphism just outlined may appropriately be called 'model-theoretic (MT) isomorphism', since it simply requires that the definientia be interpreted over a structure which is isomorphic in the model-theoretic sense to a structure over which the definienda are (presystematically) interpreted. Now one of the points of complexity of Chapter 1 is that the kind of isomorphism there specified (call it 'SA isomorphism') is not quite the same as MT isomorphism. The latter is symmetric (and transitive, hence an equivalence relation); SA isomorphism, however, is not symmetric. The reason, in brief, is as follows. SA isomorphism is designed for use in both platonistic and nominalistic systems. In the former, set theory (what Goodman refers to as 'the calculus of classes') is employed, making available reference to entities of great structural complexity (due to layerings of the set membership relation ($\in$)) as members of potential definientia extensions. From the constructional point of view, it is in general advantageous to utilize such complexity in building up a system of definitions. Section 3 of Chapter 1 illustrates how this works: by defining 'point' in a sample system as 'pair of intersecting lines' (the latter being set-theoretically more complex than the former), it becomes possible to define uniquely certain relations among the points (and even individuating predicates for each of the points themselves), due to relationships among the component elements (lines) of the pairs. This represents a significant systematization; in effect, certain definitions are derivable from earlier definitions and theorems and need not be separately adopted. Now consider what would happen if definitions of the opposite sort were admitted, that is, definitions such as that of 'pair of lines' in terms of 'point', in which the definiendum were more complex than the definiens. Overall systematization of the sort just illustrated could not be achieved by such definitions—at best, the degree of coherence among definienda taken presystematically could be duplicated by the systematic definitions; frequently it would be *decreased*. It is for this reason that SA isomorphism requires that the set-theoretic complexity of definientia elements be as great or greater than that of the definienda elements.[15] The result naturally is an asymmetric 'isomorphism' relation. For nominalistic systems, however, all elements satisfying any predicates are individuals, hence ultimate factors. Thus, restricted to these systems, extensional isomorphism reduces to MT isomorphism.[16]

We come now to a philosophically more interesting point. The very phrasing of all these isomorphism criteria presupposes that the predicates to be defined *have* extensions which can be mapped in a one-one fashion to extensions of the defining terms. Yet this is precisely what was not clear in the geometric and arithmetic examples of multiple admissible construals cited above, examples of the sort that motivate relaxing coextensiveness for more flexible criteria in the first place. To suppose, for example, that number words refer presystematically to items of one particular ω-sequence as opposed to any other would seem just as arbitrary and groundless as to require unique choice among systematic construals. We seem to be caught in the bizarre position of having to insist that number-theoretic predicates have definite extensions in order to apply a flexible criterion instituted because it seems obvious that they don't, *except relative to a system or construal*. One way out is as follows: read the isomorphism criterion as saying, "Let us suppose (pretend) that the extensions of the definienda are fixed over any domain whatever, in such a way as to reflect ordinary usage (satisfy intuitive axioms); then any structure isomorphic to such a 'comparison structure' will serve." Quine's way of handling a similar dilemma concerning ontological 'reduction' is parallel to this: proceeding as by *reductio ad absurdum*, we show that on the assumption that certain entities exist, there is no need to suppose that they do.[17]

Whether this move is satisfactory remains to be investigated. Something along these lines is presumably intended by Goodman, since, as noted above, on his view, the force of these mathematical examples is quite general: reference always has to be relativized to a system or a 'way of taking' some subject matter. This is thought to be necessitated by the multiplicity of equally good (in any ascertainable respect) theories incompatible in respect of ontology and otherwise, not just in the mathematical domain, but in any domain. The appropriate question to raise here is this: would it not be better to say that certain terms, especially from mathematics, do not have presystematic reference precisely because their usage is such that only certain structural interrelationships matter; and therefore, *of course*, alternative pairwise non-coextensive construals are equally legitimate, and, apart from any such, the question of reference simply does not arise (in other words, for these terms, absolute reference does not exist but relative reference does)? On this view, adopting a systematic construal amounts to adopting a *convention* which fixes the reference of the terms in question. The interesting thing about such cases is that usage *does* determine structural relationships, so that a systematic construal is only *partly conventional*, unlike cases of fixing the reference of newly introduced technical terms. But this is not the case with all terms. The reference of 'Julius Caesar' or of 'dog' *is* presystematically fixed as much as any structural relationship is. That all reference is fixed only within some language or symbol system is of course trivial, and is incorporated in all sophisticated versions of a correspondence theory of truth.

We shall return to these matters presently in order to draw links between these issues Goodman raises and current work in progress.

Save perhaps only his shunning of intensions, modalities, and counterfactuals, there has been in his writings no source of irritation more constant and provocative than Goodman's nominalism. As thoroughly explained in Chapter 2 (and in other writings on this subject cited above), the doctrine has to do with what Goodman conceives to be the mathematical or constructional apparatus of systems, specifically with what ontological commitments follow upon recognizing some individuals–whether just to wholes comprising those individuals as parts, or also to sets of them, sets of sets of them, and so on–, and not with the more general question (answered negatively by traditional nominalism), 'Are there any abstract entities?' As the reader will already have noted, Goodman–far from eschewing all abstract entities– takes a controversial category of such (phenomenal qualia) as basic in the construction of his own realistic system. (Chapter 6 on.) For many, especially the mathematically disposed reader sympathetic with the constructional approach, all this can become a source of unending frustration. For, not only does one have to learn a new use of the term 'nominalism' along with a number of intersecting dichotomies; once over that obstacle, one confronts the onerous task of trying to see the motivation for working with abstract entities as strange to ordinary discourse as phenomenal qualia, for which little if any scientific theory exists, and which seem quite dispensable for saying whatever many a scientific mind deems worth saying, while at the same time refusing to countenance entities apparently referred to by one means or another in many of the sentences of the average five-year old, a realm of objects apparently indispensable for mathematics and physics, indeed just those abstract objects for which we do possess a very highly refined, precise, and infinitely fruitful theory, and–what many regard as the last straw–just those entities that collectively constitute the sturdiest and most exquisitely honed instrument in the entire battery of constructionalist tools. (On this latter point, note the remarks above concerning the asymmetry of Goodman's isomorphism criterion on the domain of platonistic systems. Also, however, note Goodman's replies to many of these points in 'A World of Individuals', Sec. 3, *op. cit.*, n.12). Some will simply not make the effort. Others try and fail miserably. Others still become enchanted for a time with what one can accomplish boxing with naked fists only to return later in life, weary with scarred knuckles, to the Cantorian heaven beyond space and time, secretly hoping that in the end, all will be revealed.

It is clear that in SA, Goodman regarded the 'calculus of classes' less as a theory about a special domain than as a constructional tool common to many systems (as he also regarded the calculus of individuals with its only 'partially interpreted' part-whole relation, whose full interpretation always depends on application to a particular system). (Cf. *P & P*, p. 152.) In their Introduction to *Philosophy of Mathematics*, Benacerraf and Putnam suggested that this standpoint might motivate skepticism about sets (since 'building up' endless infinities of objects 'out of' a finite number would surely be 'compounding a felony'). Be this as it may, Goodman perceives and states clearly the issue of ontological commitment in Chapter 2.

Efforts at clarifying the distinction between nominalistic and platonistic systems are one thing; arguments for nominalism another. Dicta such as 'No distinction of entities without distinction in content', cannot serve in the latter capacity as they beg the question.[18] (For the platonist (set theorist), the distinction between an entity and its singleton is obviously a distinction in content–as basic as any there can be–whereas for the nominalist, it is illusory.) Despite all this, interest in nominalism and nominalistic systems continues.[19] In addition to some technical advantages of employing the calculus of individual (over set theory) in particular contexts (including, Goodman claims, solving the problem of concretion in the realist system of SA (Ch. 6, sec. 5)), there is always logical interest in how much can be accomplished on a slender basis (an interest which Section 3 of Chapter 2 arouses). Theoretically, platonists and nominalists alike share an interest in just how much mathematics can be done within a nominalistic framework: at just what point (if any), ontological commitment to sets is genuinely necessary.[20] Finally, even the most refined and agile mind[21] at moments boggles at the thought of a set-theoretic hierarchy with set-many objects that are, for us, literally unthinkable. Alternatives to set-theoretic realism that account adequately for mathematical truth and its physical applicability form an exciting topic of current inquiry. Recent papers by Putnam,[22] for example, explore a modalist interpretation of mathematics that is compatible with a nominalist ontology. (If, however, Putnam is right in regarding Platonism and modalism as 'equivalent descriptions', then one wonders what the debate over ontology is really all about.) Of course, modalism (which takes 'It is possible that' as primitive) is no more acceptable to Goodman than Platonism. The point is merely that mathematical realists are not all happy with the set theoretic picture.

It is worth bearing in mind that Goodman's approach to nominalism is also constructionalist. That is, his program seeks literally to translate mathematical discourse into a nominalist language whose only *mathematical* apparatus is one of the calculi of individuals, *salve* what we want to save. Presumably what we want to save is mathematical truth, at least insofar as it is established and especially if it is required for science. Now it is important to realize that this program is not committed to the clearly impossible task of translating set theory into the language of the calculus of individuals *simpliciter* (which mathematically, is just an elementary theory of partial orderings).[23] Rather the program is to provide a translation into this language plus additional predicates of individuals, and this is not ruled out by metamathematical facts concerning the obvious weakness of nominalistic calculi by themselves. Now the question can be raised: How does the isomorphism criterion of definition apply here? At first blush, the task would still seem hopeless since, by Cantor's Theorem, the domain of the definienda (for simplicity, say, '$\in$') outstrips the cardinality of any world of individuals. (Whatever the latter may be, on the Platonist picture, there is a set of that cardinality, and its power set exists and has greater cardinality.) But cardinality is one of the few properties that must be *absolutely* preserved on the isomorphism criterion. Does this doom constructive nominalism? Not quite, for there is the Löwenheim-Skolem theorem: any model can be mirrored precisely by a countable proper part thereof. (This strong form requires the Axiom of Choice. The nominalist will have somehow to justify this and the model theory used to prove the Löwenheim-Skolem theorem.) A countable infinity of physical objects or space-time regions could then serve as all the 'mathematical objects' it is ever really necessary to quantify over.[24] The pros and cons of such an approach, not to mention its details, have yet to be worked out. In particular, attention must be given to (1) the question of translation; (2) the problem of circularity, already noted, raised by the appeal to model-theoretic results. Even if the latter is surmountable, how one goes about specifying an actual translation is problematic. Is one, for example, free to employ the predicate (of individuals) '*x* is a member of *y* in the (a?) sense of 'is a member of' restricted to the (already specified) countable domain of individuals said to yield a model of set theory by the Löwenheim-Skolem theorem'? If not, and if actual translation proves impossible, does this necessarily defeat a nominalist 'interpretation' of mathematics. In short, does giving up constructionalist nominalism amount to giving up nominalism? Further work here seems in order.

3. CONTINUITIES

The ongoing research stimulated by *Structure* is extensive. With respect to a number of special topics, an overview can be gotten elsewhere.[25] In addition to further developments in the theory of simplicity, on topics relating to nominalism, constructional definition, and psychological orderings, much current work in diverse areas incorporates basic distinctions original with *Structure*. An important, frequently overlooked example is the distinction between indicator words and non-indicators (roughly, that between indexical and non-indexical expressions), introduced in the final chapter, 'Of Time and Eternity'.[26] Of special importance here is the sketch Goodman gives of how systematically to represent truth-conditions of tensed sentences in a tense-free language. Throughout, Goodman works on the level of particular utterances and inscriptions, taking 'is a replica of' and other predicates of such particulars as syntactic primitives and a relation of naming defined on them as semantic primitive. While one may be skeptical of the claim that language theory can in general dispense with universals or 'types', working with 'tokens' serves a definite clarificatory purpose when it comes to dispelling the perplexity of 'some very purple passages on The Past, The Present, and The Future' to which some metaphysicians have been prone. In particular, the analysis of temporal reference is used to dissolve notions of 'temporal flow' by means of systematic translation of sentences suggesting such 'flow' or 'passage' into sentences involving only reference to relative positions in temporal ordering.

A further example concerns the contrast between terms with the semantic property of dividing reference ('count' terms) and those that, while applying to the same sort of stuff, lack this property ('mass' terms). In Chapter 2 of *Structure*, a number of technical distinctions among types of predicates are framed, utilizing the calculus of individuals. Among monadic predicates of individuals, an important class are called 'collective': they satisfy the condition of being closed under nominalistic summation (that is, the sum of any two individuals satisfying such a predicate also satisfies it). This is indeed the semantic trait characteristic of mass terms, as has been noted,[27] and provides a nice illustration of how concepts of the calculus of individuals can be used to mark and analyze important distinctions quite apart from issues of ontological commitment.

Our aim is not, however, a comprehensive survey of such matters. Rather it is to draw attention to certain issues of major interest and importance that have definite links with *Structure*, not elsewhere given prominence, not in order to emphasize points of priority, but to highlight continuities that may help lead to further progress on the problems themselves. We focus on the following three issues, each arising in *Structure*, pressed further in later work by Goodman, and constituting major areas of current interest. (1) The critique of unrelativized similarity or resemblance; (2) consequences of epistemological activism and rejection of the given, especially for philosophy of science and philosophy of language; (3) the ontological pluralist and relativist views noted above, especially in relation to scientific realism, conventionalism, and theories of reference.

The first point occurs in *Structure* as an apparent matter of detail, and serves by the way as a fine illustration of what the constructional method can accomplish in point of uncovering major and far-reaching stumbling blocks all too easily glossed over in an informal setting. As explained in Chapter 5, Carnap attempted to account for qualities formally as certain classes of momentary phenomenal particulars, these classes being defined in terms of a primitive relation of resemblance or similarity among the latter. This was in line with the classical empiricist approach to abstraction going back at least to Locke. A number of formal difficulties with Carnap's construction, difficulties which Carnap treated as matters of detail to be handled by special, *ad hoc* assumptions, are argued by Goodman to be devastating to the whole approach. One of these, the 'problem of imperfect community', as Goodman calls it, encapsulates nicely the troubles encountered when one takes a relation of unrelativized similarity as primitive. In a nutshell, a collection of particulars may be such that each pair shares a common quality (in virtue of which the components stand in the similarity relation to one another), but there is no common quality shared by all. Thus, defining a quality as a maximal class of particulars such that each bears the similarity relation to every other member will not do. The problem is only complicated but not solved by taking a four-place comparative similarity relation as primitive ('*x* and *y* are more similar to one another than *u* and *v*') as opposed to the two-place relation. What one needs, rather, is similarity in some preferred respects; but to include a parameter for 'respects' is, of course, to beg the question, since these are just the qualities (or properties, or kinds, or natural kinds, etc.) that the program (in question) seeks to *construct*.

Still, in context after context, it is sought to base analyses of important, problematic notions on a relation of overall similarity or resemblance.[28] Some, explicitly cognizant of Goodman's point, still yearn for a treatment of natural kinds in terms of global (comparative) resemblance.[29] Others appeal to the vagaries of ordinary use of counterfactuals in defense of global similarity (on possible worlds) as the semantic key.[30] More significant, from Goodman's standpoint, are the widespread misconceptions about symbolic functioning in the arts, reflecting naiveté over 'resemblance'. These receive detailed attention in *Languages of Art*. Without repeating the arguments here, we note that Goodman is not denying obvious correlations between degrees of realism of representations and degrees of likeness in relevant respects between symbol and object. Rather, in addition to stressing the independent denotative core of representation, he is maintaining (i) that the relevant respects and their weights are culturally and otherwise relative, and (ii) that there is a kind of causal feedback from representational systems to standards of resemblance so that the latter cannot be taken as *independent* grounding of the former, even within a specified historical or cultural context.[31]

Ingrained views on resemblance as the independent criterion of representational realism have their parallel in empathy theories of expression. Similarity of psychological state of artist and audience is taken as the measure or aim of artistic achievement. In contrast, Goodman treats expression as a very broad symbolic relation between works and a subset of their metaphorically exemplified properties.[32] Accounts of metaphor in terms of similarity are found vacuous; instead, the need for a theory of transfer of symbols from old to new domains is emphasized. Moreover, expression is freed from narrow emotivist conceptions; predicates (or other labels) whose literal domains embrace the most diverse aspects of human experience and its environments are seen as expressible by art. Instead of mirroring external or internal reality, artworks are seen as performing a variety of important cognitive functions that place them in alliance with, rather than in opposition to, the works of science.[33]

Concerning epistemological developments, it is ironic that, perhaps because of its reputation associating it with Carnap's *Aufbau*, the close links between *Structure* and some of the major threads in recent philosophy of science and philosophy of language have been so little noticed. For, despite obvious indebtedness to logical empiricism on methodological questions, already emphasized, the epistemology of *Structure* is of a piece with the most significant modern ciriticisms of and departures from that tradition. Thus, the rejection of the given, of any theory and culture-neutral observational level constituting a foundation for knowledge, finds popular expression in that aspect of Kuhn's philosophy of science which stresses the theory-laden character of empirical testing and the changes in observational standards associated with paradigm shifts and scientific revolutions.[34] Comparison with some of the work of N. R. Hanson, P. K. Feyerabend, and others would also be germane here. In addition to the untenability of a sharp observation-theoretic dichotomy, the essential relativity of confirmation and inductive inference to the practices of a scientific community emerges in Goodman's later writing.[35] Nevertheless, I see no incommensurability thesis lurking in this quarter. Translatability across scientific revolutions is nowhere denied, although multiple and variable standards of translation are recognized.[36] (While uniqueness of translation is denied, existence is not.) Thus, the groups to which standards of rationality must inevitably be relativized may (as the last note suggests) be historically broad, and certainly they transcend a great many paradigm shifts. Moreover, it would square ill with Goodman's outspoken emphasis on ordinary human flexibility in mastering and adapting to new symbol systems[37] to find him denying that Aristotle and Galileo could forge some common means of talking about weights, strings, and periods of pendula.

In the area of language theory, it would be instructive to develop the links between the epistemological positions of *Structure* and the much debated thesis of Quine of the indeterminacy of translation.[38] Although Goodman does not address the problem of translational indeterminacy directly in *Structure*, he does raise the prospect that questions of epistemological priority may be ultimately indeterminate. (See e.g., Ch. 4, Sec. 4 and Ch. 8, Sec. 4.) Any system claiming that certain items of experience are the unanalyzable starting points of knowledge can be confronted with an alternative, adopting different primitives in whose terms the primitives of the first system are (possibly syncategorematically) construed. Further, these rivals will fit all determinable facts of experience equally well and will balance out in terms of relevant methodological criteria. Now this indeterminacy of epistemological priority can be used to argue that there is no uniquely identifiable starting point in language learning. What is the connection between this and the indeterminacy of translation? At least this much: rival translation manuals typically impute different orderings in the learning of semantic material (since we observe that certain sentences are with regularity mastered prior to others). For example, construing a word as a general term applying to concreta as against a singular term denoting a universal may, in the context of other assumptions, imply that certain abstract notions are learned later than certain concrete ones. Insofar as the latter sort of claim were determinate, there would be a basis for settling among the alternative translations. Indeterminacy of epistemological priority would block this route, in at least a significant class of cases.[39]

These themes of multiplicity of adequate descriptions and of the cognitive efficacy of non-linguistic symbol systems culminate in Goodman's ontological relativism and his challenge to realism (the third point raised above). As already explained, the geometric examples motivating the isomorphism criterion of definition are, for Goodman, paradigmatic. Any theory of any subject matter may be confronted with alternatives, equally good on all scores, but incompatible on the level of individual sentences (e.g., 'Points are classes of volumes', vs. 'Points are classes of lines'). This mutual incompatibility is crucial, since, otherwise there would be no obstacle simply to conjoining all equally good versions, at least where conjunction makes sense. Additional problems arise here due to the multiplicity of symbol systems, since conjunction across these is not even intelligible. This in turn leads Goodman to suspect that the very notion of a complete system of reality is unintelligible, since, presumably, such a system would have to be capable of translating or somehow representing the sound or 'right' contributions not just of scientific theories but of works in non-linguistic systems, such as works of art.

This matter of completability touches on some of the most fundamental questions currently being raised in metaphysics and the philosphy of science. Progress here depends crucially on distinguishing carefully a variety of notions of 'completeness'. On the one hand, there is 'expressive completeness', traditionally associated with reductionist versions of the unity of science. Secondly, there is 'ontological completeness', the matter of specifying an exhaustive ontology. Finally, there is a distinct notion of what may be called 'determinationist completeness', which has to do with whether a domain can be specified such that variation in any other domain whatsoever is lawfully dependent on concomitant variation in the original domain. It is of prime importance to recognize that the latter two types of completeness–apart from whether they are attainable or not–do not imply or otherwise require expressive completeness or reductionism in any sense that includes (even weak) definability.[40] This is important, because some of Goodman's arguments (in 'The Way the World Is' and 'Words, Works, Worlds', *op. cit.*) seem primarily directed against expressive completeness. Whether in fact they extend beyond remains to be investigated.

Even more serious for ontological claims than incompletability is the matter of relativity. Goodman's position here is instructively compared with that of W. V. Quine, who has promulgated an allied doctrine of 'ontological relativity'.[41] Whereas Quine bases his view on what he takes to be a fundamental indeterminacy affecting translation of the referential apparatus of a language, Goodman argues directly from the multiplicity of equally adequate theories of the (extra-linguistic) world, incompatible in point of ontological commitments. In fact, the positions are quite close, despite sharp differences on such matters affecting translation as behaviorism. Quine's multiple construals of reference may have been originally inspired by reflection on the alternative systems of definitions treated in Chapter 1 of *Structure*, and can be viewed really as a generalization of the latter. Not only are there multiple ways of taking what might be argued to be only 'partially (presystematically) interpreted' terms, such as 'point', 'natural number', etc.; there are multiple ways of construing the reference of any term whatever, such that there is no fact of the matter as to which construal is 'right'. Further, different construals corespond (in many cases) to different ontological commitments. (Construing 'gavagai' as a general term applying to rabbits involves attributing no commitment to sets; construing it as a singular term denoting the extension does.) Moreover, it is claimed, such multiple construals are always forthcoming with respect to our own language, so that any need to relativize statements of ontological commitment applies in our own case as well.

Nevertheless, there seems to be an important difference in the ways in which Quine, on the one hand, and Goodman, on the other, generalize Goodman's examples. Quine concludes that statements of reference (e.g., of the form ⌜*F* [a predicate] applies to *a*⌝) make no sense taken absolutely but that they must be relativized. Equally, *attributions* of ontology (ontological commitment) must be relativized. Now, since reference is the key to semantic versions of the correspondence theory of truth, the need (if such it be) to relativize reference raises serious questions about such a theory of truth. Nevertheless, if care is taken in the choice of the parameter of relativization, there is no obstacle to reconciling this Quinean multiplicity of referential schemes with a correspondence approach to truth.[42] Goodman, however, goes further: not only must ontological attributions be relativized, so must ontological claims themselves, and this seems irreconcilable with any meaningful talk of correspondence and spells trouble for any kind of metaphysical realism.[43]

This view, with its source in *Structure*, has been spelled out by Hilary Putnam.[44] Beginning with Goodman's geometrical examples of multiplicity, he has argued that our total theory confronts alternatives in the same fashion: irreconcilable on a sentence by sentence basis, yet so clearly equally adequate to every specifiable purpose except possibly truth itself that to regard one as 'the true one' would be otiose. Appealing to 'Goodman's principle',[45] that if you are committed to something about which you can say absolutely nothing interesting then your commitment is indefensible (not worth making?), he concluded that realism as a metaphysical position (with respect to, say, total science) is not coherent. (Presumably, this extends also to the correspondence theory of truth.) Now, it will be evident from what has already been said concerning Goodman's epistemology (the main lines of which Putnam endorses), that it is *not* being maintained that alternatives not decidable by observation are not really alternatives. This was the logical empiricist view, that empirical equivalents are semantic equivalents. Clearly, Goodman and Putnam are both urging some epistemological constraint on metaphysics, but just what it is has not yet been precisely formulated. Whether, for instance, it is being maintained that (absolute) ontological claims are meaningless, or merely idle, useless, etc., remains to be seen. What has emerged is the position that a full-fledged realism with respect to a given body of theory is as arbitrary as insisting that points are really individual components of lines as opposed to, say, sets of converging line segments. We also seem to have a neo-Kantian view that, because of this ontological relativism, there is some sense in which the objects of which we speak are conceptually constituted.

The reader may have the feeling that the art of philosophy as Bertrand Russell once described it[46] is here being all too skillfully practised. This is not, however, the place to critique a position only in the embryonic stage of development (and on which further work is in progress). In conclusion, it will perhaps be worthwhile instead to raise some questions suggestive both of the importance of this neo-Kantian outlook and of some avenues along which further research may profitably be directed.

The first group of questions concerns the nature and extent of the alternative theories or versions of the world that are alleged to undermine realism. That there are multiple, equally adequate versions, *apparently* conflicting in *some* respects, is a point everyone in his or her right mind concedes. Confining attention to linguistic versions, for which the notion of truth is applicable, one may pursue the usual line that the conflict *is* only apparent, being localized in sentences which reflect conventional choices in the use of language. This is the move sketched above (p. xxx) in connection with the geometric examples. While allowing for multiplicity, it meshes fully with a correspondence theory of truth à la Tarski, since the different versions come out, technically speaking, in different languages to which distinct yet fully compatible truth definitions are relativized. (Sticking to the paradigm, 'point' would be treated as an ambiguous expression, with a different extension in different systems.) This ties in with some of the deepest questions in the philosophy of language and the philosophy of science: what sort of separation of conventional from 'factual' components and of language from theory is it defensible to maintain, and how, without reviving a naive analytic-synthetic distinction, is it to be made? From Goodman's epistemological standpoint, interesting epistemic desiderata that could with justification be called upon to constrain metaphysical speculation (by showing it–or some of it–to be idle) concern properties of whole theories. (These would be the properties cited in spelling out 'adequate' as it occurs in the claim of Multiplicity.) Yet truth is a notion that applies to individual sentences. If convention is to be localized, it seems we need a theory of sentence meaning appropriate to this task, something we are far from having.[47] And it is noteworthy that significant efforts to demarcate the conventional within special scientific domains, such as spacetime physics, where the nature of admissible alternatives is relatively clear, have been a focus of much controversy.[48]

Further, it may be asked whether what is alleged and what needs to be alleged (to undermine realism) is that there exist alternatives merely with respect to certain, important questions traditionally thought of as 'corresponding to facts', or that for any putative factual claim there is an admissible alternative that denies it. Clearly, a number of logically quite distinct claims need to be separated here, and their respective merits assessed. Putnam, in the Address cited, focused on alternative interpretations of mathematics (as set theory vs. as a modal theory, cf. above p. xxxii) and alternative versions of micro-physics (framed in terms of particles or in terms of just fields). For such cases, it needs to be asked, first, whether the equal adequacy condition is really met; second, whether there is really incompatibility on questions of significance to realism; and, third, whether such examples really support the general thesis of Multiplicity upon which the thoroughgoing Anti-realist view would seem to depend.

Finally, on this first group of questions, there appears to be a serious danger that opponents in these debates will be talking past one another. To illustrate, take the two versions of mathematics cited by Putnam. Let us suppose that a modalist interpretation with a nominalist ontology really is adequate for all mathematical purposes. (It may seriously be questioned whether anything like this has yet been established.) Does *this* kind of 'equal adequacy' show there is no fact of the matter whether there are sets, or even that it makes no difference whether or not there are? Realists may balk here because the demonstration of 'equal adequacy for mathematics' (producing a mathematically satisfactory 'translation' between the two versions) abstracts from the ontological question, which may be conceded to be of no importance for mathematics, but which is the very point in the philosophical dispute. Is there any way of breaking out of this circle acceptable to both sides?

A second group of questions concerns the internal coherence of the anti-realist ontological relativist position. It should be stressed that, unlike some earlier pragmatists, Goodman does not renounce talk of truth, nor does he attempt to replace it with 'acceptability'. On the contrary, he has criticized a pragmatic formalization of his theory of projection on the grounds that "some such notion ... as 'determined to be true' [is] needed along with 'accept'."[49] Rather, Goodman has argued that truth is over-rated and naively exalted in the common-place about the scientist's quest.[50]

If it is not Goodman's purpose to quarrel with " 'There are dogs' is true (in English) if and only if there are dogs", what are we to make of the claim that ontological assertions themselves must be relativized? Clearly, if the right side must be relativized, so must all correct singular predications of 'dog' side must be relativized, so must all ordinary singular predications of 'dog' (since the latter logically imply the former, a matter Goodman has nowhere hinted is questionable). Evidently, all factual claims must be relativized. The question is, *to what* must they all be relativized? Not, apparently, to a language, since truth predications are already so relativized, and to insist on so relativizing ordinary predications would be either to confuse use and mention or to adopt a wholly redundant convention. Moreover, the point that all assertions make sense only within a language is a triviality. Rather, some kind of relativity to system or theory or 'way of taking' is intended. The idea is somehow to do justice to the fact that the recognition of, say, dogs in our ordinary ontology depends in part on our own selection and organization, that the world could be categorized in radically different ways, and so forth; to do justice to this without lapsing into Berkeleyanism, which Goodman regards as no more defensible than physicalism. Many questions arise here: Is it being claimed that a system, as acceptable as any we have, literally *denies* the existence of dogs (*ceteris paribus*, some would opt for such a system on these grounds alone!) as opposed to merely not affirming the existence (due to lack of the predicate or equivalent)? If not, why is it insufficient to grant the Kantian point concerning conceptualization, but hold to the view that *by means* of a certain, in many ways arbitrary, selection and organization process we achieve objective reference and predication *in a language*, and that no further relativization is called for; and that truly to describe the world, of necessity from within a certain framework, way of taking, etc., is as objective an enterprise as any realist requires? Moreover, if this does not suffice, how is relativization to be expressed? A further place in ordinary predications for variables taking systems as values raises the question, how it is that we can refer *simpliciter* to systems (or versions, or 'ways of taking'), but not to dogs. (If not '*simpliciter*', don't we have a vicious regress?) And if it is at least always meaningful to query the truth value of any sentence in a system, including the axioms, how is infinite regress to be avoided?

In some measure, some of these are old problems. But enough has been said to indicate that they are now being raised in a new light, on a more sophisticated plane than perhaps ever before, and in the context of developments in exact philosophy that hold out the promise of rich and illuminating future progress. *Structure* has already played a large role in fostering these developments, and undoubtedly it will continue in that capacity for a long time to come.

Indiana University
January 16, 1977

NOTES

[1] For a lucid exposition and defense of the constructionalist standpoint which also serves as a useful introduction to SA on this and related matters, see Goodman's 'The Revision of Philosophy", in *Problems and Projects* (Indianapolis: Bobbs-Merrill, 1972; new edition, Indianapolis: Hackett, 1975): 5–23 (hereinafter cited as *P & P*).
[2] Russell and Whitehead's *Principia* is again instructive here. Although the question, what sentences of classical mathematics need be preserved (i.e., be provable in the system modulo translation), *was*, in fact, quite precisely answerable, many fundamental questions concerning the primitives of the system were not. Rather they contributed to the flourishing of set theory as an independent mathematical discipline.
[3] Goodman distinguishes two traditionally conflated dichotomies among systems: nominalistic vs. platonistic, which turns on whether classes are *recognized*, and particularistic vs. realistic, which turns on whether concreta or repeatable abstracta (e.g., phenomenal qualities) are *taken as basic units*. See Ch. 4, Sec. 5.
[4] See especially, SA, Ch. 4, Sec. 4, and 'Sense and Certainty' in *P & P*, pp. 60–68; cf. also *Languages of Art* (Indianapolis: Bobbs-Merrill, 1968), Ch. 1 (hereinafter cited as 'LA').
[5] Cf. again 'Sense and Certainty', *op. cit.*, and SA, Ch. 4, Sec. 3.
[6] Cf. 'On Likeness of Meaning' in *P & P*, pp. 221–230.
[7] In addition, Goodman allows that individual statements may have initial credibility independently of relations to other statements, and that such initial credibility, rather than certainty or immunity from revision, should be seen as rendering other statements more or less probable. (See 'Sense and Certainty', *op. cit.*) The overall position is thus a modified rather than thoroughgoing coherentism (concerning justification). Similar qualifications must be made concerning 'holism' with respect to testing or confirmation.
[8] Note that support even for limited ontological claims derives not simply from the accuracy of the definitions of a system but from the system's overall adequacy. This is fortunate for, as we shall see, all that Goodman demands of accurate definitions is a kind of isomorphism of extensions (not necessarily coextensiveness), thereby weakening the traditional link between 'reduction' of language and theory on the one hand and 'reduction of ontology' on the other. For further discussion and illustration of the fundamental distinction between ontological exhaustion claims and reduction claims, see G. Hellman and F. W. Thompson, 'Physicalism: Ontology, Determination, and Reduction', *The Journal of Philosophy*, LXXII (1975): 551–564; cf. also, G. Hellman, 'Accuracy and Actuality', *Erkenntnis* (1977, forthcoming).
[9] Ch. 1, Secs. 2 and 3, in *P & P, op. cit.*
[10] *Erkenntnis*, 9 (1975), 57–73.
[11] *op. cit.*, n.9.
[12] In *P & P*, pp. 155–172; also reprinted in P. Benacerraf and H. Putnam, *Philosophy of Mathematics* (Englewood Cliffs, N.J.: Prentice-Hall, 1964).
[13] See, especially, P. Benacerraf, 'What Numbers Could Not Be', *The Philosophical Review*, 74 (1965).
[14] Normally, when the mathematician or logician speaks of a bijective mapping f from one domain D to another D' 'preserving' an n-ary relation R, what is meant is that for any ordered n-tuple $\langle d_1 \ldots d_n \rangle$, of objects d_i in D, it belongs to the relation R (construed set-theoretically as a set of n-tuples) if and only if the n-tuple, $\langle f(d_1) \ldots f(d_n) \rangle$, is also in R. Or, if we speak of relation-symbols (predicates) $\bar{R}$ of a language interpreted over D and D', then $\bar{R}$ is preserved by such a mapping f just in case $\bar{R}$ is satisfied by an n-tuple from D if and only if $\bar{R}$ is satisfied by the image of that n-tuple under f. Now, normally, in a constructional system, the structural relations that need to be preserved will be expressed by predicates for which we seek constructional definitions (e.g., 'x coincides

with y' or '$x + y = z$'). Yet such predicates will not in general even be interpreted over the domain of objects over which the definientia range. (E.g., presystematically, '+' is simply not interpreted over the domain of pure sets.) The constructional task is, in effect, to *assign* interpretations to such predicates. Thus, what is required of a set of definitions $D(\bar{R}_i)$ of predicates $\bar{R}_i$ is that the interpretations of the former over the definientia domain contain every n-tuple that is an image under a mapping f (as above) of an n-tuple belonging to the interpretation (extension) of the corresponding $\bar{R}_i$. Thus, isomorphism of a set of definitions requires that the relation-terms being defined be *relatively* preserved by an appropriate mapping (relative to the defining vocabulary), not that they be *absolutely* preserved (in the sense of the *same* relations holding of definienda and corresponding definientia items). For a more detailed exposition of these distinctions and their relevance for criteria of accuracy generally, see Geoffrey Hellman, 'Accuracy and Actuality', *op. cit.*, n.8.

[15] Goodman explains this in terms of retrieval of definientia extensions by a consistent plan of replacement of definienda ground level elements (individuals or the null set, urelements, or 'ultimate factors') by definientia elements (not necessarily ultimate factors). As he notes (below, p. 11, n.6), such talk of 'replacement' is figurative. For a precise account, framed in the language of set theory, see G. Hellman, 'Accuracy and Actuality', Sec. II, *op. cit.*, n.8.

[16] The reader will note that distinct set-theoretic construals of predicates (e.g., number-theoretic) are not SA isomorphic to one another (the only ultimate factor being the null-set). Thus, a purely set-theoretic concept, such as 'ordered pair of pure sets', could have only one accurate definition on the SA-criterion, contrary to the whole spirit of SA. In some such cases, MT isomorphism would have the desired flexibility. In others (such as the 'ordered pair' example), even this is too strong: any construal satisfying the single axiom of ordered pairs ('$\langle x, y\rangle = \langle u, v\rangle$ iff $x = u$ & $y = v$') will do. Evidently, no single accuracy criterion will work for all contexts.

[17] Cf. W. V. Quine, 'Ontological Relativity', in *Ontological Relativity and Other Essays* (New York: Columbia University Press, 1969): p. 58.

[18] The dictum occurs in SA, Ch. 2, sec. 2. The relevant notion of 'content', as Goodman sees it, is spelled out in 'A World of Individuals', *op. cit.* (It says, in brief, that content amounts to ground elements or atoms of a system.) When so spelled out, the dictum is clear enough. The platonist, however, is sure to argue that this cannot be the right notion of 'content' if the dictum is to be granted.

[19] Cf. *P & P* pp. 149–154.

[20] In this connection, see N. Goodman and W. V. Quine, 'Steps Toward a Constructive Nominalism', *Journal of Symbolic Logic*, XII (1947), 105–122, reprinted in *P & P*, pp. 173–198.

[21] Recall here a poem due, we believe, to W. V. Quine:

> The unrefined and awkward mind
> Of *Homo Javanensis*
> Could only treat of things concrete
> And present to the senses.

[22] See, e.g., Hilary Putnam, 'Mathematics Without Foundations' and 'What is Mathematical Truth', *Philosophical Papers*, Vol. I (Cambridge Univ. Press, 1975), Chs. 3, 4.

[23] Cf. R. Eberle, *Nominalistic Systems* (Dordrecht: Reidel, 1970).

[24] In Chapter 2 (below, p. 29), Goodman notes the possibility of construing natural numbers as individuals, but objects that this course does not provide a means of translating applied sentences (for example, 'there are three more F's than G's', where 'F' and 'G' stand for non-mathematical predicates). Now a concrete, denumerable interpretation of set theory solves this problem, since in the domain of such an interpretation are items that serve the role of functions, i.e., sentences of set-theory intuitively

referring to functions are simply reinterpreted (preserving truth-value) over the countable domain. In order to deal with applied sentences, we apply the Löwenheim-Skolem theorem to an applied set-theory, e.g., for the above sentence, ZF set theory, say, plus axioms asserting the existence of the set of all F's and the set of all G's. Now it is straightforward to express in ZF that there are three more F's than G's: "There is a 1-1 function from the set of F's to a subset of the set of G's and there is a 1-1 function from the difference between the set of G's and the range of the first function onto some set with exactly three members". Under the present strategy, this very sentence is interpreted over the countable domain of concreta: the functions and sets are simply some of those concreta, and '$\in$' (in terms of which everything in our quoted sentence can be rendered) simply designates a two-place relation among the concreta so that everything comes out right.

[25] See, especially, *P & P*, Forwards to the chapters.

[26] Goodman's notion of indicators corresponds roughly with Reichenbach's 'token-reflexives' and Russell's 'egocentric particulars', all having originated at roughly the same time. (Cf. *P & P*, p. 203) Current language theory, especially formal semantics and formal pragmatics, builds extensively on these distinctions and their analyses, which have proved indispensable for even the rudiments of a formal theory of natural language. (Cf., e.g. Donald Kalish's survey under 'Semantics', in *The Encyclopedia of Philosophy*, VII, 348, ff., esp. 355–357.)

[27] Cf. W. V. Quine, *Word and Object* (Cambridge: MIT, 1960), p. 91.

[28] An instructive task would be to sort out the interrelated formal obstacles to analyses based on unrelativized notions of similarity that have beset a number of fields in addition to those dealt with directly by Goodman. (See his 'Seven Structures on Similarity', *P & P*, 427–436.) For example, efforts in micro-economics to base consumer behavior on (unrelativized) *indifference* generally depend on *ad hoc* assumptions reminiscent of those called upon by the epistemologist taking unrelativized similarity as primitive. (For background on indifference-curve analysis, see, e.g., I. M. D. Little, *A Critique of Welfare Economics*, Second Edition (Oxford University Press, 1957), Ch. 2 and Appendices I, II. For a critical discussion that brings out the issue of relativization to context, see J. Sensat and G. Constantine, 'A Critique of the Foundations of Utility Theory', *Science and Society* (1975).)

[29] E.g., W. V. Quine, 'Natural Kinds', in *Ontological Relativity and Other Essays, op. cit.*, n.17.

[30] Though it now appears that David Lewis seeks to construct 'the right similarity relation' (i.e., relativized to the right respects, properly weighted) from our judgements as to truth-values of counterfactuals. See 'Counterfactual Dependence and Time's Arrow' (forthcoming).

[31] On both these points, Goodman relies heavily on many sources, for instance, E. H. Gombrich and others (cited in LA, Ch. 1). They are thus to be seen as points of departure for new developments in the symbol-theoretic approach to art. Nevertheless, it is worth pointing out that Goodman's formulation of the 'relativity of perception' thesis (along the lines of point (i) above, cf. LA, p. 39, n.31) does not lead to the paradoxes raised by the formulations of some of his predecessors. For example, some of Gombrich's remarks concerning cultural influences on 'the way the artist views the world' engender doubts as to why *we* should perceive a multiplicity of representational styles at all. For example, if Rubens painted subjects fatter than they were because he saw them as fatter (cf. E. H. Gombrich, *Art and Illusion* (New York: Pantheon, 1960), pp. 167–168 and *passim*), one wonders why he did not see the canvas images also as fatter so that everything would come out right (given, as we are by Gombrich, that Rubens was competent and aimed at faithful match). What bizarre psychological principle accounts for such systematic 'distortion' in perception of three-dimensional scenes but not of two-dimensional? On Goodman's formulation, such embarrassment is avoided:

perceptible differences (e.g., degree of chubbiness) count differently in different systems because of a difference in weighting of properties; a departure of picture from object, visible to any normal member of the species, may upset matching in one system while being accommodated–even aimed at for special purposes–within another. Note that the weighting (for matching purposes) will generally be inversely related to the importance for special purposes.

[32] LA, Ch. 2.

[33] LA, Ch. 6.

[34] See T. Kuhn, *The Structure of Scientific Revolutions* (Chicago: University of Chicago Press, 1970) 2d Edition. In this particular case, the irony is acute since Kuhn does in fact along the way refer to *Structure*, but the reference and quoted passage concern extensionalism as a criterion of definition and seem utterly beside the main points at issue. (See Kuhn, p. 127.)

[35] While this relativity is not made explicit in Goodman's *Fact, Fiction and Forecast* (Indianapolis: Bobbs-Merrill, 1965; new edition, Indianapolis: Hackett, 1977), it would seem to emerge from the essential role of past practice in the determination of projectibility of predicates, in conjunction with empirical facts concerning the historical variability of such practice. Note, however, that the scope of such communities depends on the theory in question and on the particular empirical facts concerning past projections. It would therefore appear that scientific communities for Goodman may range from quite narrow to very broad (the latter especially if we are allowed the resources of translation between languages), and do not in general coincide with the sociologically identified communities of Kuhn.

[36] See SA, Ch. 11, sec. 3; also 'On Likeness of Meaning', *op. cit.*, n.6.

[37] See, for example, his polemical remarks concerning N. Chomsky's views on language learnability in 'The Epistemological Argument' and 'The Emperor's New Ideas', in *P & P*, 69–79.

[38] See W. V. Quine, *Word and Object* (Cambridge: MIT Press, 1960), Ch. 2, and the vast subsequent literature.

[39] To what (if any) extent the claims of the Piagetian theory of stages are afflicted with such indeterminacies is an interesting topic for further investigation.

[40] These notions are explained and their independence demonstrated for the case of physicalism in G. Hellman and F. W. Thompson, 'Physicalism: Ontology, Determination, and Reduction', *op. cit.*, n.8.

[41] In 'Ontological Relativity', in *Ontological Relativity and Other Essays, op. cit.*, n.17, Ch. 2; cf. also W. V. Quine, *Word and Object*, *op. cit.*, n.27, Ch. 2.

[42] The main work here is that of Hartry Field, 'Quine and the Correspondence Theory', *The Philosophical Review,* **83** (1974): 200–228, where it is shown how problems for the correspondence theory, raised by Quine's relativization to a linguistic parameter (a translation manual) can be avoided if instead we relativize to admissible model-theoretic structures, corresponding to the different admissible ways of construing reference. (Field actually drops relativity talk for talk of 'partial denotation', 'partial signification', etc. As far as I can see, there are here two equivalent descriptions.)

[43] See 'Words, Works, Worlds', *op. cit.*, secs. 5 and 6.

[44] In his Presidential Address to the American Philosophical Association, Dec. 29, 1976, *A.P.A. Proceedings*, 1976 (forthcoming).

[45] Derived from 'Words, Works, Worlds', *op. cit*.

[46] As consisting in arguing from premisses so obvious that only a fool would question them to conclusions so outlandish that only a fool would accept them.

[47] A recent proposal of Quine's however, would short-circuit the problem: apparently contradictory sentences in rival systems, equally good in all discernible respects, are shown to be not genuinely incompatible in virtue of occurrence in such different systems. Put in traditional terms, the system in which we choose to operate becomes a

parameter affecting, not truth directly, but meaning. See the concluding remarks of his 'On Empirically Equivalent Systems of the World', *Erkenntnis,* **9** (1975) 313–328. Of course, Quine does not put it in traditional terms, thereby sidestepping demands for a theory of meaning that would deliver the desired result.

[48] A good set of bibliographical references for this field is contained in A. Grünbaum, 'Space, Time and Falsifiability', *Philosophy of Science,* Vol. 37, No. 4 (Dec., 1970), 469–588, which is a major locus of modern space-time conventionalism. For a critical review of Grünbaum's (pre-1970) work in this area, see A. Fine, 'Reflections on a Relational Theory of Space', *Synthese,* **22** (1971), 448–481.

[49] *P & P, op. cit.*, p. 405.

[50] Cf. LA, Ch. 6, sec. 7, and 'Words, Works, Worlds', p. 69.

[51] 'Words, Works, Worlds', p. 68.

ORIGINAL INTRODUCTION

This book attempts to use the techniques and example of modern logic in the investigation of problems of philosophy. Such a program does not make for speed, and some will regard my effort as intolerably narrow in scope, as excessively detailed, and as occupied with too many problems that do not properly belong to philosophy at all. But I look upon philosophy as having the function of clearing away perplexity and confusion on the most humble as well as on the most exalted levels of thought; and I hope the patient reader, much as he might like to take quick measure of the universe, may find that in philosophy as in science the microscopic method has its own fascination and rewards.

Nevertheless, one assumes a grave responsibility in setting out to apply symbolic logic to any subject matter; for even so promising a method can easily be discredited by a plague of overelaborate systems that do not repay the effort required to master them. I have accordingly sought to use symbolic notation only where it genuinely contributes to clarity and convenience, and have confined it to proportionately few pages, sections, and chapters. Since, furthermore, most of the logic used is rather simple and since full explanations in English are given throughout, the book should be accessible to anyone with a thorough grounding in elementary symbolic logic.

We shall find that systems of logical philosophy, which I call "constructional" to distinguish them from uninterpreted formal systems and from amorphous philosophical discourses, may be founded on many different bases and constructed in different ways. And several quite different programs may be equally correct and offer equal if different advantages. An impartial survey of alternative programs, however, will not by itself realize any of them. A passionate effort at construction is likewise needed and is wholly compatible with the dispassionate appraisal of results. To seek to develop a system is not to argue for its superiority to other systems. Perhaps the day will come when philosophy can be discussed in terms of investigation rather than controversy, and philosophers, like scientists, be known by the topics they study rather than by the views they hold.

The title of the book reflects the fact that I am concerned primarily with the analysis of phenomena and with the study of phenomenalistic systems. But the reader should not infer that I therefore advocate phenomenalism (cf. Chapter IV). Again, while it is true that I subscribe to nominalism in the particular version explained in Chapter II, it is not true that acceptance of the constructional system later offered depends upon acceptance of nominalism. As originally presented in *A Study of Qualities* (see p. ix) the system was not nominalistic. I feel that the recasting to meet nominalistic demands has resulted not only in a sparser ontology but also in a considerable gain in simplicity and clarity. Moreover, anyone who dislikes the change may be assured that the process of replatonizing the system–unlike the converse process–is obvious and automatic; and this in itself is an advantage of a nominalistic formulation. But the book is an argument for nominalism only incidentally, in that this system provides a sample of what can be done with the restricted means the nominalist allows himself.

Something of the detailed plan of the book can be gathered from the table of contents, but a more summary outline may be welcome here. Part One is devoted to a few salient problems concerning the nature and methods of logical philosophy. The three chapters deal, in order, with the requirements for an accurate logical construction or explicative definition, with practical and ontological aspects of what is ordinarily regarded as the logical apparatus of construction, and with factors governing the choice of a primitive extralogical vocabulary.

Part Two begins with a brief comparison of different types of system and then introduces the problem of the relation between qualities and concrete particulars. Attention is then turned to specific systems dealing with this and kindred questions. A system sketched by Carnap is discussed in detail. Then a quite different system is developed in outline, and the scope of the inquiry is somewhat enlarged.

In Part Three the latter system is extended to deal with the problem of the ordering of qualities among themselves and thus to lay the foundation for a calculus of shape and measure. The final chapter takes up special problems pertaining to time and the language of time.

Any effort in philosophy to make the obscure obvious is likely to be unappealing, for the penalty of failure is confusion while the reward of success is banality. An answer, once found, is dull; and the only remaining interest lies in a further effort to render equally dull what is still obscure enough to be intriguing. In this recognition that my book may be more stimulating for its failures than for its successes, I find some consolation for its shortcomings.

NELSON GOODMAN

PART ONE

ON THE THEORY OF SYSTEMS

CHAPTER I

CONSTRUCTIONAL DEFINITION

1. EXTENSIONAL IDENTITY

The definitions of an uninterpreted symbolic system serve as mere conventions of notational interchangeability, permitting the replacement of longer or less convenient definientia wherever they may occur by shorter or more convenient definienda. These conventions are theoretically unnecessary because the elimination of definitions and defined terms would affect the system only by making its sentences much longer and more cumbersome. Furthermore, notational conventions are of course arbitrary and cannot be disputed so long as certain formal requirements are satisfied, such as that no term be adopted as an abbreviation for two nonequivalent expressions of the system.

In a constructional system, however, most of the definitions are introduced for explanatory purposes. They may be arbitrary in the sense that they represent a choice among certain alternative definientia; but whatever the choice, the definiens is a complex of interpreted terms and the definiendum a familiar meaningful term, and the accuracy of the definition depends upon the relation between the two. In a formal system considered apart from its interpretation, any such definition has the formal status of a convention of notational interchangeability once it is adopted; but the terms employed are ordinarily selected according to their usage, and the correctness of the interpreted definition is legitimately testable by examination of that usage. This is true of many definitions in every system that is applied, hence in every system likely to be of much interest. In *Principia Mathematica* taken as an uninterpreted system, "1" is merely an abbreviation for "$\hat{\alpha}\{(\exists x) . \alpha = \iota' x\}$"; but the accuracy of the definition as part of the interpreted system depends upon whether "$\hat{\alpha}\{(\exists x) . \alpha = \iota' x\}$" bears an interpretation consonant with the usual arithmetical usage of "1".

A constructional definition is correct–apart from formal considerations–if the range of application of its definiens is the same as that of its definiendum. Nothing more is required than that the two expressions have identical extensions. Ordinarily, of course, extensional identity will be established not by inspection of everything to which the two expressions apply, but rather by bringing to bear all sorts of other knowledge. But if extensional identity is taken as the criterion of definitional accuracy, then our willingness to accept a proposed definition will be measured by our confidence that the definiendum and the definiens apply to exactly the same things, regardless of how that confidence is acquired or sustained. Considerations of possibility will in a sense enter into our choice of definitions. We shall have to consider whether it is possible that there are cases of either expression that are not cases of the other, and we shall adopt a definition without reservation only if we are certain there are not. But when we are sure that extensional identity obtains, no further question of possibility enters. We do not require that the definiendum and the definiens agree with respect to all cases that 'might have been' as well as to all cases that actually *are*. For example, if all and only those residents of Wilmington in 1947 that weigh between 175 and 180 pounds have red hair, then "red-haired 1947 resident of Wilmington" and "1947 resident of Wilmington weighing between 175 and 180 pounds" may be joined in a constructional definition (assuming, of course, that all terms in the expression taken as definiens have been previously introduced into the system). The question whether there 'might have been' someone to whom one but not the other of these predicates would apply has no bearing on the admissibility of this definition once we have determined that there actually is no such person.

Extensional identity may or may not be a sufficient guarantee of sameness of meaning;[1] but that point is not at issue here. I am not maintaining that extensional identity amounts to sameness of meaning, but merely that extensional identity is the most that is required for, or implied by, the definitions in the various systems we shall examine. It is fortunate that nothing more is in question; for the notion of 'possible' cases, of cases that do not exist but might have existed, is far from clear.

Often we cannot demand even that the extension of the definiens be exactly identical with that of the familiar term being defined; for ordinary usage is often ambiguous and scientifically inept, while a constructional language must be precise and scientifically efficient. The extension of the definiens will thus frequently be more precisely delimited and of greater scientific interest than the extension of the definiendum as ordinarily used. Such a divergence will occur not only where a word is commonly used in two or more different ways (as "cape", for example, is used for certain articles of clothing and for

[1] Cf. my articles "On Likeness of Meaning", *Analysis*, 10 (1949) pp. 1–7; and "On Some Differences about Meaning", same journal, 13 (1952), pp. 90–96; both have been reprinted in my *Problems and Projects* (hereinafter referred to as *P & P*) (Indianapolis: Bobbs-Merril, 1972) pp. 221–238.

certain bodies of land, but not for both at once), but also in cases of two other kinds.

In the first, common usage is indeterminate with respect to certain entities. The term "fern", for example, we unhesitatingly apply to certain plants and refuse to apply to others; but there are some plants about which we are undecided, not because we lack knowledge of these plants but because habits of using the term "fern" are not fully determinate. In such a case, we demand of a constructional definition only that it accord with common usage in so far as that usage is determinate. When any definition meeting this requirement is established, it will then serve as a means for classifying the hitherto borderline cases (e.g., as ferns or nonferns). Thus a definition accepted on the ground that it violates no manifest decision of ordinary usage becomes legislative for instances where usage does not decide.

In the second kind of case, the popular concept is inappropriate for scientific purposes. To most of us "fish" unquestionably applies to whales; if the biologist says whales are not fish, his use of "fish" differs from ours. To him a fish is a completely aquatic, water-breathing, cold-blooded craniate vertebrate, while to us a fish is any streamlined animal that lives in the water. Scientists and philosophers often thus trim and patch the use of ordinary terms to suit their special needs, deviating from popular usage even where it is quite unambiguous; and hence again we have to modify the demand for strict extensional identity in order to make possible the construction of an adequate systematic language.

The cases so far mentioned have called for no essential modification of the primary requirement of extensional identity, but rather for comparatively minor amendments. A much more difficult problem arises in the case of certain definitions where the extensional coincidence of definiendum and definiens is far from evident. Consider, for example, the definition of points as certain classes of volumes, along lines proposed by Whitehead.[2] It seems obvious that according to common usage a point is no class of volumes. Here we have a question not of the partial divergence of the extensions of the definiendum and the definiens but of their ostensible mutual exclusiveness. It may be held that systematic analysis will show geometrical points to be 'really nothing but' such classes of volumes. But in that case, instead of appealing to extensional identity as the justification for something in the system, we are seeking to establish extensional identity by appealing to the system. We are

[2] In, for example, *An Enquiry Concerning the Principles of Natural Knowledge* (Cambridge, England: Cambridge University Press, 1917), p. 76. Whitehead's actual definition involves further complications that in no way affect the point under discussion here.

then left with the original problem of finding the criterion by which the system is to be judged. In most constructional systems there are at least a few legitimate and indispensable definitions which, if they are taken to imply extensional identity, seem diametrically opposed to common sense. Our problem is to allow for them without so far relaxing our criterion as to admit actually wrong definitions.

We might propose to test extensional identity where it is not evident by determining whether the truth value of every sentence in which the definiendum occurs will remain unaltered when the proposed definiens is substituted for the definiendum. If, for instance, we substitute the description of the appropriate classes of volumes for the term "points" in the sentence

(i) "There are infinitely many points",

we secure a sentence that is likewise true. Substitutability in this one sentence proves only that points and the described classes–which we may call *p*-classes –of volumes are both infinite in number; to prove that they are extensionally identical we should have to show that the truth value would be preserved upon similar substitution in all sentences in which the term "points" occurs.[3] But a fatal difficulty with this criterion is revealed when we consider such sentences as

(ii) "Points and *p*-classes of volumes are the same".

The indicated substitution will give a true sentence, indeed a tautology; but is the truth value of sentence (ii) thus preserved? Obviously in order to apply the proposed test we must first know whether (ii) is true. Yet if we do know that at the outset, the whole test becomes useless; for its purpose was to determine whether points and *p*-classes of volumes are indeed the same. In other words, the fallacy of the proposed test is that universal substitutability with preservation of truth value requires such substitutability in the very sentence that is being tested, and in logically related sentences. Hence the truth value of the sentences being tested would have to be known before the test could be carried out; and so our proposed test entirely collapses. It would be futile to seek to repair the test by excluding only sentence (ii) from those in which substitutions are to be made; for there are obviously many other sentences such that knowledge of their truth value would directly involve knowledge of the extensional identity or nonidentity of definiendum and

[3] When I speak of a term's "occurring in a sentence" I mean "occurring otherwise than in quotation marks". Strictly, certain other reservations would have to be made.

definiens. For example, if "Points have no members" is true, then points cannot be *p*-classes of volumes.

Furthermore, we begin to perceive that the need for some method of testing definitions of the peculiar kind under consideration does not arise solely from the psychological difficulty of deciding whether the definiens and the definiendum do indeed have the same extension. Let us suppose it to be taken as true that points are *p*-classes of volumes. It is patently true that no *p*-class of volumes is a pair of intersecting lines; for however "line" and "volume" may be defined, a *p*-class of volumes will presumably have an infinite number of members while a pair of lines will have but two. Yet clearly any argument we might adduce to justify the interpretation of points as certain classes of volumes would overreach itself if it precluded the possibility of interpreting points as certain pairs of lines. Again, if a point may be defined as a certain pair of lines, it may equally well be defined as any other pair of lines which, according to ordinary usage, likewise intersect at that point–even though these several pairs of lines are acknowledged to be different. There are obviously many alternative ways of defining a term, all of them equally legitimate, and any criterion that bars any of them is thereby proved too narrow. We can allow for this sort of flexibility by ruling either that sentences like (ii) have no truth value apart from a given system, or that their truth value is not in itself a sufficient or necessary condition for the accuracy of a constructional definition. In either case, it is clear that extensional identity is *not* demanded of the terms conjoined as definiendum and definiens of a constructional definition; such a demand would be inconsistent with the fact that a single definiendum may be quite properly so conjoined with any of several extensionally nonidentical terms.

2. SUBSTITUTION CRITERIA

A natural suggestion would be, then, to adopt the substitutability criterion in a modified form as an independent and sufficient test of the accuracy of constructional definitions rather than as a test of denotative equivalence. But one of the necessary modifications would be the explicit exclusion of all troublesome sentences, such as (ii) above, from the sphere of sentences in which test substitutions are to be made. When a definition is formulated expressly for the purposes of a given science, then we need merely specify that the substitutability test is to apply solely to sentences within that science; for example, if Whitehead's definition of points is to be used solely for geometrical purposes, then test substitutions need to be made in the sentences of geometry alone. But where we are dealing with the broader problem of defining terms for the purposes of a general systematic reconstruction of experience, the difficulty of finding a satisfactory characterization of the relevant class of sentences seems insurmountable. Furthermore, even if we found a suitable general principle for excluding the troublesome sentences, the substitutability test would not be acceptable as stated; for in certain true sentences clearly within the sphere in which test substitutions would have to be made, the indicated substitution would result in sentences that are not true. For example, if we substitute "*p*-classes of volumes" for "points" in

(iii) "Many points lie within volume *A*"

we get the strange sentence

(iv) "Many *p*-classes of volumes lie within volume *A*".

If we remember that each of the *p*-classes of volumes in question has as members many volumes which contain *A* as a part, it will be evident that none of the classes can be said to lie within volume *A* according to any ordinary use of "lie within"; even under any normal figurative broadening of ordinary usage, we would say that volume *A* lies within (i.e., is a member of) these *p*-classes rather than vice versa. Hence while (iii) is true, (iv) is not.

This latter difficulty can be met by requiring that every nonsystematic term in a sentence must be replaced by its systematic counterpart before the truth value of the resultant sentence is to be considered.[4] Our criterion is then transformed from a substitutability test into a test in terms of the translation of sentences. According to this revised criterion, a constructional definition is admissible if and only if the truth value of every sentence in which the definiendum occurs remains unaltered when every nonsystematic term in the sentence is replaced by its systematic counterpart. But we are faced with two difficulties. In the first place, since the test requires systematic translation of every term in every sentence in which the given definiendum occurs and since the definiendum may be coupled in some sentence or other with any term whatsoever, it is obvious that no constructional definition could be tested until we had a *complete* definitional system– a system, that is, which contains definitions for absolutely all terms not taken as primitive. If any term that occurs anywhere were not formally introduced into the system as primitive or

[4] In this context I mean by a nonsystematic term any term containing anything but 'logical' signs, and the primitive signs of the system in question. For discussion of 'logical' signs, see Chapter II.

With respect to the test discussed in this paragraph compare Carnap, *Der logische Aufbau der Welt* (Berlin: Weltkreis Verlag, 1928; 2nd edition, Hamburg: Felix Meiner Verlag, 1961), Section 35.

defined, then *no* definition of the system could be tested. This difficulty might be at least partially obviated by requiring only that sentences in which the definiendum occurs and which are fully translatable into the system shall have the same truth value as their translations. But in the second place, the modification we have made in the fomulation of our test does not by any means succeed in evading the necessity for excluding from the sphere of the test such troublesome sentences as we first encountered. For example, if we translate

(v) "Every point is a *p*-class of volumes"

into a system for which volumes are basic units, and points are defined in Whitehead's way, we get the truism

(vi) "Every *p*-class of volumes is a *p*-class of volumes".

However, if we translate (v) into a system for which lines are basic units, and points are defined as certain *g*-classes of lines, and volumes as certain *r*-classes of points, we get

(vii) "Every *g*-class of lines is a *p*-class of *r*-classes of *g*-classes of lines",

which is meaningless according to the theory of types and false by intuition. Accordingly, no matter what we consider to be the truth value of (v), at least one of the two perfectly legitimate systematic definitions will fail to pass our test. Thus the test in terms of systematic translation, like that in terms of simple substitutability, proves quite unacceptable.

It should be remarked in passing that my argument here is not against translating nonsystematic into systematic sentences; such translation is integral to the very method and purpose of a constructional system. What I find unsatisfactory is the proposed method for testing the accuracy of the definitions that serve as the rules for systematic translation.

About the best we seem to be able to do toward a criterion along the lines so far considered is this: a definition must be such that every sentence we care about that can be translated into the system shall have the same truth value as its translation. But this is no criterion at all without some specification of what sentences we 'care about'. It is rather a criterion of criteria–a condition that must be met by definitions satisfying any acceptable criterion. And while the sentences we care about will vary somewhat with our purposes, what we want in a criterion is at least an approximately general characterization of at least a minimum class of sentences that we care about preserving in any constructional system.

3. EXTENSIONAL ISOMORPHISM

If we now look more closely at the very divergent definitions of a given concept that were equally legitimate, we find that they possess in common one feature that every illegitimate definition lacks; namely, that in each legitimate definition, the extension of the definiens is *isomorphic* to the extension of the definiendum. The necessary and sufficient condition for the accuracy of a constructional definition seems to be that the definiens be extensionally isomorphic to the definiendum. *More generally, the set of all the definientia of a system must be extensionally isomorphic to the set of all the definienda.*[5] I shall first explain and illustrate the kind of isomorphism I mean and then consider whether this criterion is satisfactory.

We may think of the extensions of the definienda and definientia in question as relations—that is, as classes of couples, classes of triples, and classes of longer sequences of any uniform length. While sequences may in turn be construed as classes, it is simpler to disregard this for our immediate purposes. A class of individuals or other one-place sequences may be considered as a monadic relation. By the *components* of a sequence I shall mean the elements that occupy entire places in the sequence. Thus the sequence

$$((a,b),c),(d,e)$$

is a couple; its components are $((a, b), c)$ and (d, e), not the couple (a, b) or any of the single individuals. On the other hand, if we progressively dissolve each component that is a sequence into its components, and every component that is a class into its members, and continue this until we reach elements that have no further members or components, we have what I call the *ultimate factors* of the sequence. Here they are a and b and c and d and e. The ultimate factors of a relation or other class are reached in similar fashion. For our purposes in the present chapter, a sequence is not considered to be identified, as by the Wiener-Kuratowski definition, with a class but symmetric relations are construed as classes of classes rather than of sequences. An ultimate factor is always either an individual or the null class.

A relation R is isomorphic to a relation S in the sense here intended if and

[5] Throughout this chapter, I assume that definientia are fully expanded, so that they contain no defined terms. Carnap, in the *Aufbau* (Sections 10–16, 153–155), discusses at some length a kind of isomorphism much stronger than that in question here. His criterion of constructional definition, however, is a translation criterion based upon extensional identity (Section 35).

only if *R* can be obtained by consistently replacing the ultimate factors in *S*.[6] Consistent replacement requires only that each not-null ultimate factor be replaced by one and only one not-null element; that different not-null ultimate factors be always replaced by different not-null elements; and that the null class be always replaced by itself. Since the replacing elements need not be ultimate factors (e.g., *h*, *k* might replace *t*), this sort of isomorphism is not symmetric; for if *R* is isomorphic to *S*, still there may be no way of replacing the ultimate factors in *R* so as to obtain *S*. Nevertheless, if *R* can be obtained by consistently replacing the ultimate factors in *S* by certain elements of *R*, it will also be true that *S* can be obtained by replacing those elements in *R* by the correlated ultimate factors of *S*. It is often more convenient to work in this direction in establishing that *R* is isomorphic to *S*, but it should be noted that this does not establish the isomorphism of *S* to *R*. Every relation is, of course, isomorphic to itself. Also any class having the same number of non-null members as a given class of individuals is isomorphic to it; but a class is not necessarily isomorphic to every class having the same number of members or of ultimate factors.

We can best make clear the operation of the proposed criterion and the nature of the isomorphism involved by using a very simple illustration. Suppose that our system deals with the diagram in Figure 1 and contains as undefined primitives one or more relations of the four lines in the diagram.

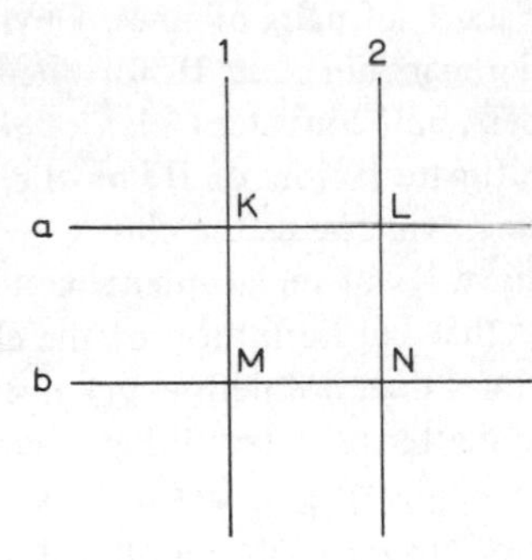

Figure 1. Universe for sample system.

Suppose further that some logical function of our primitives is discovered that will isolate the class of all those classes that have as their only members

[6] The notion of obtaining a relation by replacing elements in a given relation is, of course, purely figurative. For an indication of a more formal characterization, see Section 5.

two intersecting lines of the diagram. I shall for convenience speak of two-membered classes simply as "pairs", reserving the term "couples" (or where emphasis is desirable "ordered couples") for two-place sequences. For the couple consisting of x and y taken in that order I shall use the usual notation "x,y", but for the pair consisting of x and y I shall use "$x:y$", the colon indicating that the order is non-significant. Now the pairs of intersecting lines in our diagram are

(I) $a:1$
$a:2$
$b:1$
$b:2$.

Suppose that the function that has been found to isolate this class of pairs is adopted as our systematic definiens of "points" (which will here be understood to refer only to points of intersection shown in the diagram). The class of points

(II) K
L
M
N

is thus defined as the class, I, of pairs of lines. Obviously the definiens class, I, is isomorphic to the definiendum class, II; for since the two classes have the same number of members, any consistent plan of replacing the members of II (which are also the ultimate factors of II) by the members of I–and there are twenty-four such plans–will give us the class I.

The fact that so many ways of replacement would give the required class reminds us emphatically that our definition of the class of points as the class of pairs of intersecting lines does not define any given point as any given pair of intersecting lines; the matter of determining a particular correlation is left open. Though this is generally true of definitions of classes, not merely of the kind of definition we are now considering, it is often overlooked. For example, to define triangles as three-sided polygons is not to define a given triangle A as consisting of the three lines that actually constitute A. The definition of triangles sets up no unique correlation by itself; it would be equally consistent with the correlation of A with three other lines that actually constitute a different triangle, B. We may be thinking of, or intending, the normal correlation; but apart from other definitions such an intention is given no expression within the system and imposes no restriction upon its interpretation.

To proceed with our sample system, suppose that in order to treat the relation next-counterclockwise-to among the points in our original diagram, the relation we know to consist of the ordered couples

(III) *K,L*
L,N
N,M
M,K,

we define it by a logical function of our primitives having as its extension the class of couples

(IV) $(a:1), (a:2)$
$(a:2), (b:2)$
$(b:2), (b:1)$
$(b:1), (a:1)$.

Isomorphism is shown by the fact that replacement in III of *K* by $(a:1)$, *L* by $(a:2)$, *N* by $(b:2)$, and *M* by $(b:1)$ will give us relation IV. Another plan of replacement that would serve equally well is: *K* by $(b:1)$, *L* by $(a:1)$, *N* by $(a:2)$, and *M* by $(b:2)$; for this would give us the relation consisting of the couples

(V) $(b:1), (a:1)$
$(a:1), (a:2)$
$(a:2), (b:2)$
$(b:2), (b:1)$,

which is identical with IV since the order in which the couples are listed is not significant. Again the replacement of *K* by $(b:2)$, *L* by $(b:1)$, *N* by $(a:1)$, and *M* by $(a:2)$, or of *K* by $(a:2)$, *L* by $(b:2)$, *N* by $(b:1)$, and *M* by $(a:1)$ would give the same result.

Accordingly no unique correlation of couples of points with couples of pairs of lines, nor of points with pairs of lines, is established by our definition. Just as our definition of points did not say what pair of intersecting lines is to be correlated with *K*, so our present definition does not say what ordered couple of pairs of intersecting lines is to be correlated with the next-counterclockwise-to relationship of *K* to *L*. Yet it is to be observed that certain plans of replacement that would satisfy our first definition will not satisfy our second. For example, replacement in III of *K* by $(a:2)$, *L* by $(a:1)$, *N* by $(b:2)$, and *M* by $(b:1)$ would give the relation consisting of the ordered couples

(VI) $(a:2), (a:1)$
$(a:1), (b:2)$
$(b:2), (b:1)$
$(b:1), (a:2)$,

which is not identical with IV.[7] However, our criterion demands only that there be at least one plan of replacement that will give the definiens extensions when applied to the definiendum extensions throughout a system, and we have seen that there are four plans of replacement that will accomplish this for the two definitions so far considered in our sample system.

Suppose now that the relation of directly-above among the points in the diagram, which we know to consist of the ordered couples

(VII) K,M
L,N,

is defined in our system by an expression having as its extension the class of ordered couples

(VIII) $(a:1), (b:1)$
$(a:2), (b:2)$.

Isomorphism, of VIII to VII may be established either by replacing K by $(a:1)$, M by $(b:1)$, L by $(a:2)$, N by $(b:2)$; or by replacing K by $(a:2)$, M by $(b:2)$, L by $(a:1)$, and N by $(b:1)$. These are the only two plans that will serve. But we see that only the first of these plans is among those that will satisfy the two earlier definitions. Accordingly there is now but one plan of replacement that may be applied throughout all definiendum classes of the system to give the requisite definiens classes in every case. Hence a unique correlation between the points of intersection in the diagram and the pairs of intersecting lines in the diagram has been established for, or by, the system: namely, K is correlated with $(a:1)$, L with $(a:2)$, M with $(b:1)$, and N with $(b:2)$. No further definition of the system may be such as to require a conflicting plan of replacement. The relation of diagonal opposition in the diagram, for example, which we know to consist of the pairs

(IX) $K:N$
$L:M$,

---

[7] Even were VI isomorphic to IV (as erroneously stated in the first edition of this book), that would not be enough. Only if VI were *identical* with IV would correlation of VI with III be admissible in a system already having IV correlated with III.

will have to be defined in the given system as the class of pairs

(X) $(a:1):(b:2)$
$(a:2):(b:1)$,

and not as any other class.

When a unique correlation is thus established, we have the means for defining each of the several points and lines. For example, line *a* is distinguished by being a component of every first component of a pair in VIII; line 1 is distinguished by being paired with *a* to make up the first component of the only couple in IV that has *a* in both components; and *K* is the pair $(a:1)$. However, it must not be inferred that a system is valueless or hopelessly incomplete if it fails to establish a unique correlation of the sort set up by our sample system. A fixed one-to-one correlation may often be wanted, and our sample system illustrates how it may sometimes be achieved; but nothing in our general criterion requires that such a correlation must be sought or achieved.

Furthermore, the fact that the correlation established by our sample system is the most natural one–correlating each point with the pair of lines that pass through it–must not be taken to mean that this is the only one-to-one correlation that could have been established. Any other one-to-one correlation could have been set up by some other system. Suppose, for example, we had defined next-counterclockwise-to as the class of ordered couples

(XI) $(a:1), (b:2)$
$(b:2), (a:2)$
$(a:2), (b:1)$
$(b:1), (a:1)$.

Isomorphism with III could be established by replacing *K* by $(a:1)$, *L* by $(b:2)$, *N* by $(a:2)$, *M* by $(b:1)$. If we had then defined directly-above as the class of ordered couples

(XII) $(a:1), (b:1)$
$(b:2), (a:2)$,

the plan of replacement just mentioned would be the only one satisfying the system throughout, so that a correlation quite different from the previous one would be established. We might have defined next-counterclockwise-to or directly-above or both in some still different way, and so have established some still different correlation.

Yet observe that we could not, in a system in which next-counterclockwise-to is defined as XI, define directly-above as, for example, the class of ordered couples

(XIII) $(a:2), (a:1)$
$(b:2), (b:1)$;

for no plan of replacement that establishes the isomorphism of XI to III will establish also the isomorphism of XIII to VII; in other words, no plan of replacement would establish the isomorphism of the total set of definiens classes to the total set of definiendum classes. It must always be borne in mind that isomorphism of the whole is demanded by our criterion. Without this demand there would be no safeguard against such anomalies as the identical definition of several different but isomorphic classes. The demand for isomorphism of the whole enforces systematic respect for the differences among isomorphic relations that are to be defined; for these very differences must be paralleled in any set of definiens classes that is isomorphic with the set of definiendum classes.

In order not to complicate the explanation of isomorphism unduly, I did not specify a set of primitive relations and write out the defining functions in our sample system. This has obscured one important point. As has already been explained and illustrated, the ultimate factors in the definiendum relation may be replaced in the definiens relation by more complex elements. And so long as two different ultimate factors are in all cases replaced by different elements, it does not matter how the replacing elements may be interrelated by means of elements comprised within them. In our first sample system, for instance, the pairs of lines replacing *K* and *L* have a common component, while the pairs replacing *K* and *N* have not. Yet such facts are very useful in the building of a system. If we suppose intersection to be taken as primitive in the system in question, we can define diagonal opposition as the relation that every pair of intersecting lines bears to every other such pair that has no component in common with it. This will give us the appropriate class, X. Thus relationships below the level at which isomorphism is determined provide here the very means for defining a relation without using additional primitives, and so aid in the construction of a comprehensive system upon an economical foundation. The definitional substitution of increasingly complex systematic terms for relatively simple everyday terms goes hand in hand with simplicity of basis and coherence of structure.

4. CONSEQUENCES OF ISOMORPHISM AS A CRITERION

What now are the consequences of this criterion and to what extent is it satisfactory?

In the first place, since extensional identity of definiendum and definiens is no longer required, a given term may alternatively be defined by any of several others that are not extensionally identical with one another. Thus the flexibility wanted has been gained. Moreover, the kind of sentence that disrupted the operation of the substitution and the translation criteria no longer causes any trouble; for now we require neither that substitution of definiens for definiendum in every sentence leave the truth value unchanged nor that the translation of every sentence that can be translated into the system have the same truth value as the original sentence. For example, if we translate the sentence "L is identical with $(a:2)$" according to our first sample system, we get the true sentence "$(a:2) = (a:2)$"; while if we translate according to the second system suggested, we get the false sentence "$(b:2) = (a:2)$". Certainly, whatever the truth value of the original sentence is deemed to be, it is not preserved under both these equally good systems. But this no longer bothers us; for the test of a system is now in terms of isomorphism, and both systems satisfy it.

Yet it may be asked, "What is the good of a system if we cannot be sure that it gives us true translations of true sentences?" The answer is that a system is serviceable if its translations of such sentences as we care about are truth-value-preserving. The demand that its translations of *all* sentences be truth-value-preserving is incompatible with the very demand for flexibility that we have been seeking to meet in formulating a criterion of definition. That there are some statements we do not care about is immediately evident from the fact that in actual practice we accept alternative extensionally non-identical expressions as equally good definientia for the same term. Exactly what kind and degree of latitude we want and allow indicates which statements do not much concern us. Thus if the criterion of isomorphism succeeds in providing for just the appropopriate flexibility, it satisfies the 'criterion of criteria' suggested at the end of Section 2; for it then in effect formulates the general distinction between those sentences for which we do and those for which we do not insist that any systematic translation preserve their truth value.

If anyone holds that two terms are not extensionally identical, and if he wants to define both in his system, this nonidentity will constitute a structural feature of his set of definienda that will have to be reflected–according to the requirements of isomorphism–in his set of definientia. For example, suppose we hold that points and crosses in our diagram are not identical and want to define both in a system. Obviously we cannot then define both, within one system, as pairs of intersecting lines. A different way of defining one or the other will have to be found, if the system is to be adequate for its purpose. Naturally, isomorphism does not guarantee the *adequacy* of a system but only its *accuracy*. A system with but one definition may be accurate and provide a truth-value-preserving translation for every sentence we care about *for which it provides any translation at all*. An adequate system would have to provide such a translation for *every* sentence we care about.

Now it is clear that to adopt a criterion for constructional definitions is to make a decision concerning the significance of these definitions and indeed of any use in a system of a constructionally defined term. According to the criterion we have adopted, the symbol "$=_{df}$" serves notice that the extension of the definiens that follows the symbol is to be correlated in the system, in the manner above explained, with the extension of the definiendum that precedes the symbol. The definition is inaccurate if this correlation is precluded by the fact that the definiens is not extensionally isomorphic to the definiendum. Moreover, if the requisite isomorphism obtains for this definition taken in isolation, but the entire set of definientia of the system is not extensionally isomorphic to the entire set of definienda, then either this definition or some other in the system must be changed. This view of definition affects the significance of any systematic sentence containing a term so defined. For example, although the sentence:

$$(x)\,(\text{Point}\, x \supset (\exists y)\,(\text{Line}\, y \,.\, y \in x))$$

–occurring in a system in which "Point" is a defined term and "Line" a primitive one–may still be read in the usual fashion as:

Every point has some line as a member,

it is to be somewhat more explicitly rendered as:

Everything correlated (in this system) with a point has some line as a member.

And where "K" is a defined term and "$(a:2)$" consists of primitive terms, the sentence "$K = (a:2)$" says that what is correlated (in this system) with K is identical with $(a:2)$.

Once we understand this, we readily perceive the fallacy of certain objections against the flexibility of definition permitted by our criterion. For example, it might be argued that it is wrong to define L as $(b:2)$–in accordance with our second sample system–because this will give us "$b \in L$" as a theorem, whereas line b in the diagram is entirely separate from L. Now indeed "$b \in L$" does hold in this system, but means that b is a member of what is correlated in the system with L. That "$b \in L$" holds does not imply that b is a constituent part of or passes through L. In fact the relation of passing through, which presystematically[8] consists of the couples

(XIV)	a,K	b,N
	$1,K$	$2,N$
	a,L	b,M
	$2,L$	$1,M$,

would have to be defined in the system in question by the class of couples

(XV)	$a, (a{:}1)$	$b, (a{:}2)$
	$1, (a{:}1)$	$2, (a{:}2)$
	$a, (b{:}2)$	$b, (b{:}1)$
	$2, (b{:}2)$	$1, (b{:}1)$.

Accordingly the translation of "b passes through L" in this system would not be "$b \in L$" but rather "$b, (b{:}2) \in$ XV". It is simply not true that a line passes through a point if and only if the line is a member of the pair correlated with that point in this system.

The kind of care that, as this example shows, must be exercised in making systematic translations is especially needed when one word or phrase has both a systematic and a presystematic use. For example, "is a member of" is used in several different presystematic ways and also as the mere verbal reading of the systematic sign "$\in$". It would be as wrong to suppose that the words "is a member of" must always be translated into a system by the sign "$\in$" as to suppose that "point" has always to be translated by one and the same complex of symbols in every system. The sentences "John is a member of the Club" and "My hand is a member of my body" may well have different translations in the same system, and each may have many different translations in different systems.

As we have seen, the correlations established by systems admissible under our criterion may vary from the most natural, as in our original sample system, to the very unnatural, as in our alternative system. But the degree of naturalness, while it does not affect the legitimacy of any system that satisfies the criterion, may well affect our efforts to determine whether a system does

[8] That is, according to the understood or express informal explication of what is to be defined. This explication normally accords in general with ordinary usage, trimmed and patched in various ways and for various purposes as illustrated in Section 1 above.

satisfy the criterion. In our sample systems, we can quickly test for isomorphism by inspection of all cases of the definiendum and the definiens of each definition; but in the usual more comprehensive system any such complete examination is impossible. We therefore often have to rely upon all sorts of other knowledge–including whatever is regarded as knowledge of intensional connections–in order to decide whether isomorphism does in fact obtain. And the correlations we consider the most natural are in general just those that most readily engage our confidence. If we are to define the relation between every two intersection points that lie on some same ruled line on a large sheet of graph paper, we can rather easily satisfy ourselves that the requisite isomorphism obtains if our definition is based upon the natural and uniform kind of correlation exemplified in our first sample system; but we may well have considerable difficulty if our definition is based upon some such unnatural and irregular correlation as that exemplified in our alternative system.

Furthermore, the most natural and uniform correlation often serves technical as well as psychological convenience. For example, suppose that intersection is the primitive relation. If the correlations of our original sample system are set up, the relation of points to lines that pass through them can be defined simply as the relation of pairs of intersecting lines to their members. But as we have seen, if the correlations of our alternative system are set up instead, the relation of passing through will have to be defined in some much more complicated way that will isolate the class of couples listed above under "XV". More or stronger primitives may even be needed. Thus the more natural and uniform correlation is here technically advantageous as well.

Nevertheless, naturalness and technical efficiency alike are unprecise and entirely subsidiary considerations. Since efficiency is relative to what we want to define, we cannot be at all sure that the system which establishes the most natural or regular correlation will always be the most efficient. Moreover, what is most natural and most regular can hardly be determined objectively, but varies with different persons and even with different states of mind. And even when clear, these factors can at most only influence our choice among equally correct systems. The criterion of extensional isomorphism remains the necessary and sufficient condition for the accuracy of constructional definitions and systems. Any intuitive hesitancy we may have about accepting systems that set up irregular, or even bizarre, correlations arises as we have seen from misinterpreting the significance of constructional definitions.

To underline this point by an extreme example, suppose that there are four gorillas–Al, Bill, Cap, and Dan–in a certain zoo; that Al hates Bill, Bill hates Cap, Cap hates Dan, and Dan hates Al; that Al is the father of Cap, and

Bill is the father of Dan; and that no other hatreds or paternal relationships obtain. We might define these gorillas as pairs of intersecting lines in our diagram, and the relations of hatred and fatherhood among them as, respectively, the relations IV and X. From what I have said already it will be clear that this will not mean that any gorilla is a pair of lines in the diagram or has a line as a constituent part, or that hatred among the gorillas is the relation IV of these pairs of lines, or that fatherhood among them is the relation X. It means only that in this system the gorillas are correlated with pairs of lines, and the relations in question among gorillas are consequently correlated with certain pairs of these pairs. There is nothing here to offend intuition. While the correlation of such very diverse entities may often be impractical, for reasons already suggested, it involves no absurdity.

These extreme examples, however, must not obscure the fact that acceptance of a constructional definition or system on the ground that it satisfies the proposed criterion involves no commitment to the nonidentity, any more than to the identity, of the correlated entities. Since every relation and every system is self-isomorphic, nothing in any definition or system implies that the definiens relation or set of definiens relations is being correlated with anything but itself, or that the various complexes of primitive entities are being correlated with anything but themselves. In unusual cases, such as that where gorillas are correlated with pairs of intersecting lines, the nonidentity of the correlates is evident on extrasystematic grounds; but the system in itself is nevertheless noncommittal. In practice, when the system builder employs the sort of ostensibly inverted definition we have been especially considering, he frequently wants to avoid any commitment about the distinctness of the definiendum entities from the definiens entities. The phenomenalist, for example, may have observable regions among his basic units, and may define less-than-observable spaces as certain classes of those observable regions. We customarily say that in so far as his system is adequate, he is 'reducing' less-than-observable to observable regions. But we have seen that he is not proving that a less-than-observable region is identical with – is 'nothing but'– the correlated class of observable regions; for there are quite different, equally good, alternative definitions. What he is doing, rather, is proving that for the purposes at hand we need not assume that the less-than-observable region is anything other than the class of observable regions in question. This is quite compatible with similarly proving by means of alternative adequate systems that we could avoid the need for assuming that a less-than-observable region is anything but a complex of some other basic elements, or avoid the need for assuming that an observable region is anything but a certain class of less-than-observable regions. In other words, the reductive force of a constructional system consists not in showing that a given entity is identical with a complex of other entities but in showing that no commitment to the contrary is necessary.

5. ON SYSTEMS OF PREDICATES OF INDIVIDUALS

In discussing criteria of definition and explaining isomorphism, I have used the language of relations and other classes and have assumed that the systems under consideration are also framed in this language. This, indeed, is true of virtually all published systems. However, in stating the definitions of my system, and the extrasystematic rules and principles governing it, I shall confine myself as far as I can to language that speaks of no entities other than individuals. As we shall see in Chapter II,[9] such language must be free of bindable variables construed as taking classes or any other nonindividuals as values; it may contain only individual-variables, quantifiers binding these, individual-constants, the identity-sign, the truth-functional operators, the usual marks of punctuation, and predicates (of one or more places) of individuals. Thus all that can be said about a class of individuals or of sequences or even about a single couple is what can be said by speaking only of the individuals involved.

Our immediate concern is to see how the isomorphism criterion works for systems framed entirely in the language of individuals. The following discussion replaces the longer one contained in earlier printings of this book. Mr. Howard Burdick has shown that the more complicated proposal made there is unsatisfactory.[10]

Since all the extralogical terms of such systems will be one-place or many-place predicates of individuals, uniform replacement will require only that the individual replacing a given individual be the same in all occurrences and that the individuals replacing different individuals be different. Identities and differences among the arguments of all the definiendum-predicates of the

---

[9] Where I shall discuss more fully what is involved in the proposed restriction to the language of individuals, the reasons for such a restriction, and some of the typical problems raised by it. The ensuing description of the language is essentially due to Quine; see his *From a Logical Point of View* (Cambridge, Mass.; Harvard University Press, 1953), especially Chapter VI.

[10] "On a Nominalistic Criterion of Definition", *Journal of Philosophy*, vol. 66 (1969) pp. 382–383. In a second paper in the same journal, vol. 70 (1973), pp. 294–297, Burdick objects that operation of the criterion as now formulated might be hampered by a shortage of available inscriptions. Here he mistakenly assumes that a system in the language of individuals must be restricted to a finite domain. This is not the case (see below Chapter II, Section 4).

system will then be duplicated by identities and differences among the arguments of the definiens-predicates.

For such systems, isomorphism will be symmetric; and this may raise a question. The nearest analogue in such a system to a class of individuals is the whole that is the individual-sum of them. Where classes are admitted, we saw that while an individual *a* may be replaced by the class of two other individuals *b* and *c*, the converse replacement is prohibited. But *a* may replace as well as be replaced by the individual-sum of *b* and *c*; for that sum, like *a*, is an ultimate factor. Should we, then, disqualify such a sum as an ultimate factor in favor of the parts *b* and *c*, require uniform replacement for these, and thus ban replacement of the sum by *a*? Since *b* and *c* are not necessarily the only parts of the sum of the two, how are ultimate factors to be identified here? If we are to divide an individual into its least parts, the criterion cannot be applied where any individual is presystematically deemed to be infinitely divisible. If we are not to divide every divisible individual, where are we to stop?

Moreover, while identities and differences among classes of individuals are duplicated when their members are uniformly replaced, identities and differences among individuals are not in general duplicated when parts of them are uniformly replaced. Where *a, b, c*, and *d*, are uniformly replaced by *r, s, t*, and *u*, the sum of *a* and *b* may be different from the sum of *c* and *d* while the sum of *r* and *s* is identical with the sum of *t* and *u*; or *vice versa*. (For example, let the first four be vertical quarters of a square, the second four the vertical and horizontal halves of the same square, or *vice versa*.) This salient disanalogy overrides the analogy that suggested modifying the definition of ultimate factors. Thus the isomorphism criterion, for systems in either the language of individuals or the language of classes, calls for uniform replacement of ultimate factors as defined in Section 3 above; and the isomorphism defined is symmetric for systems in the language of individuals.

The criterion of definition applies only to what I shall call the special extralogical primitives of a system, and not to predicates belonging to what I shall call the general apparatus of a system. The interpretation of a primitive in the adopted general apparatus is assumed to be given for a system; and definitions solely in terms of such primitives are taken to be as set forth in the basic calculus of individuals or of classes.

CHAPTER II

THE GENERAL APPARATUS

1. GENERAL APPARATUS AND SPECIAL BASIS

Certain terms are used as primitives in many of the systems we shall consider. We can conveniently deal with these terms, and others definable from them, before we introduce the primitives that differentiate the several systems from one another. The present chapter is thus concerned with the study of terms belonging to the common apparatus of many systems.

Among these terms are some, like the stroke ("|") of truth-functional incompatibility, that we may think of as primitive in all the systems to be discussed. These and their derivatives–namely, the individual-variables, the quantifiers, the truth-functional connectives, and the marks of punctuation–may be called *logical* terms. This however, is purely a matter of convenience. I do not think that any terms can be distinguished as logical on the ground that they make up sentences that are decidable independently of experience, for I doubt whether there are any such sentences. The point need not be argued here, I want only to make it clear that my use of the term "logical" is intended to mark no epistemological distinction.

In addition to this basic logic, each of the systems under consideration will make us of either the calculus of classes (or functions) or the calculus of individuals or both. What I call the *general apparatus* of a system consists of its logic together with such of these two calculi as the system contains. No problem pertaining to the basic logic need concern us here, but some matters pertaining to the rest of the general apparatus of systems must be examined with care.

2. THE QUESTION OF CLASSES

It is often taken for granted that everything customarily called logic, including the calculus of classes, is purely neutral machinery that can be used without ontological implication in any constructional system. But this neutrality is preserved only so long as the machinery is uninterpreted. If we use variables that we construe as having entities of any given kind as values, we acknowledge that there are such entities.[1] In using variables that have individuals as

[1] This criterion of recognition or commitment is based upon Quine's (see his *From a Logical Point of View*, referred to in Chapter I, note 9, above). In my version (whatever may be the case in his), commitment occurs not with the use of certain characters called variables but only with express assignment of values to these variables.

values we are acknowledging that there are individuals–an acknowledgment we are not likely to be able or to want to avoid. If we also use variables that call for classes as values, we acknowledge that there are classes.

We may, indeed, use what are ostensibly class variables without any such commitment if we can show that our use of them amounts merely to a fictitious 'manner of speaking'. But this claim can be made good only in one of two ways. In the first place, it can be made good if we can explain how to eliminate all use of class variables–without, of course, turning to a completely different formulation of logic such as a combinatorial system that has no variables at all. Obviously, however, the calculus of classes is seldom used merely for the purpose of expressing more easily what one already knows how to express in the language of individuals. Translation is often very difficult and no one knows yet just how far it can be carried out. Accordingly, one who uses the calculus of classes is seldom in a position to show in this way he is not thereby conceding that there are classes. Or in the second place, the disclaimer of commitment can be made good if one consistently refuses to interpret the language of classes and provides a formulated syntax for manipulating that language like an abacus.[2] The syntax language itself must be free of class variables, must be framed within the language of individuals. But if we can solve the problem of framing such a syntax within the language of individuals, we can similarly solve many of our problems directly within this language; and the devious device of setting up and managing an additional and meaningless language recommends itself only where, as in the case of some parts of mathematics, direct translation is so difficult as to seem hopeless.

Thus when one uses and is unable to dispense with variables taking classes as values, one cannot disclaim the ontological commitment. Use of the calculus of classes, once we have admitted any individuals at all, opens the door to all classes, classes of classes, etc., of those individuals, and so may import, in addition to the individuals purposely admitted by our choice of the special primitives, an infinite multitude of other entities that are not individuals. Supposedly innocent machinery may in this way be responsible for more of the ontology than are the special frankly 'empirical' primitives.

Recognition of this fact will give pause to anyone who finds the notion of classes and other nonindividuals essentially incomprehensible. The nominalistically[3] minded philosopher like myself will not willingly use apparatus that

---

[2] As is done for mathematics in "Steps Toward a Constructive Nominalism", by Nelson Goodman and W. V. Quine, *Journal of Symbolic Logic*, 12 (1947), pp. 105–122; reprinted in *P & P*, pp. 173–198.

[3] In the course of this chapter, I shall be saying a good deal about what the nominalism here in question involves and does not involve, and providing technical means for a sharp formulation. I follow Quine in using "platonism", without claim of historical accuracy, for the opposite view.

peoples his world with a host of ethereal, platonic, pseudo entities. As a result, he will so far as he can avoid all use of the calculus of classes, and every other reference to nonindividuals, in constructing a system.

The nominalist's attitude stems in part, perhaps, from a conviction that entities differ only if their content at least partially differs. So far as individuals go, this is a truism; and any supposed exceptions, such as the case of two objects fashioned out of the same piece of clay at different times, clearly depend on the fallacy of ignoring the temporal or some other dimension. Further, it is clear that two classes, however defined, are indistinguishable if they have the same members; classes are in a sense distinguished only by what is comprised within them. But the nominalist goes still a step further. If no two distinct *entities whatever* have the same content, then a class (e.g., that of the counties of Utah) is different neither from the single individual (the whole state of Utah) that exactly contains its members nor from any other class (e.g., that of acres of Utah) whose members exactly exhaust this same whole. The platonist may distinguish these entities by venturing into a new dimension of Pure Form, but the nominalist recognizes no distinction of entities without a distinction of content.

The course of avoiding all dependence upon classes or other nonindividuals has the advantage that the resulting constructions should be as acceptable to the platonist as to the nominalist. For, assuming of course that the platonist recognizes individuals as well as nonindividuals, the nominalist is simply confining himself to one of the kinds of entities recognized by the platonist. Nor does the nominalist even show or claim to show that only individuals exist; he merely avoids committing himself as to whether anything else exists. The platonist may feel no need for avoiding commitment on this point, but he can hardly reject a system solely because it does not require him to grant all he is willing to grant. If he cannot be otherwise consoled, the offending nominalistic system can be readily turned into a platonistic one, while the reformulation of a given platonistic system to satisfy the nominalist is usually far from easy.

And those who are quite unconcerned with ontological commitments will hardly quarrel with the nominalist about his scruples in the choice of primitives so long as he succeeds in dealing with problems at hand.

3. NOMINALISM

Nominalism, then, consists of the refusal to countenance any entities other than individuals. Its opposite, platonism, recognizes at least some nonindividuals.[4] The nominalist's language contains no names, variable or constant, for entities other than individuals. It may contain individual-variables, quantifiers binding these, truth-functional connectives, marks of punctuation, and one-place and many-place predicates of individuals. It may also contain proper names of individuals, but for convenience I shall assume that these are supplanted by descriptions accomplishing the same purpose; "name of" is to be understood as "predicate applicable only to", and sentences containing "Hume" (e.g., "Hume is a philosopher") as abbreviations for sentences containing the predicate "is identical with Hume" (e.g., "The *x* such that *x* is identical with Hume is a philosopher"). But is there no further restriction upon the admissible predicates? May a nominalistic language contain even so platonistic-sounding a predicate of individuals as "belongs to some classes satisfying the function *F*"? If we use such a predicate and regard as true some sentences applying it, are we not acknowledging that there are classes? Strangely enough we are not–so long as we take this string of words as a single predicate of individuals. For then the words in the predicate are no more separable units of the language than are the letters in the words, and we cannot take the predicate apart and operate on a sentence containing it so as to derive such a consequence as "There are some classes satisfying the function *F*". Moreover, it would obviously be pointless to bar predicates containing such words as "classes"; for we could always circumvent this restriction by introducing some equivalent predicate that contains no such words–e.g., using, say, "is fective" instead of "belongs to some classes satisfying the function *F*". On the other hand, to bar every predicate that is equivalent to some predicate containing such words as "classes" would be to bar all predicates of individuals; for we can find such an equivalent for any given predicate–e.g., for "is wooden" we have "is wooden and belongs to some classes" or "belongs to all classes containing all wooden things". The distinction between nominalism and platonism thus depends not upon what predicates of individuals are employed but upon what values are admitted for the variables. The nominalist may use whatever predicates of individuals he likes. His choice will be governed not by the demands of nominalism but by such more general considerations as clarity and economy (see Chapter III).

[4] In "Steps Toward a Constructive Nominalism" (see note 2 above), nominalism was said with some reservations (see the third paragraph and the second footnote of that article) to renounce all abstract entities; but I prefer to characterize nominalism as renouncing all nonindividuals. Just what this formula means and my reasons for preferring it will become clearer as we proceed. A full explanation and defense is contained in my articles "A World of Individuals", in *The Problem of Universals* (Notre Dame: University of Notre Dame Press, 1956), pp. 15–30; and "On Relations that Generate", *Philosophical Studies*, 9 (1958), pp. 65–66; both are reprinted in *P & P*, pp. 155–172.

Furthermore, the decision to recognize nothing but individuals does not of itself specify what may be taken as an individual. Criteria other than the barest demand of nominalism operate in the selection of the individuals admitted under a system quite as much as in the selection of primitives; and even approximate general formulation of these criteria is difficult. Indeed, although nominalism is almost always intimately bound up with some views or other of what constitutes an individual, these views may differ widely while the central tenet remains the same. And this central principle of nominalism does not become inconsequential when it is brusquely isolated from these variant attendant views; for–as I have already pointed out–if we restrict ourselves to individual-variables, then we need not acknowledge any other entities than those we decide to admit as individuals, while if we use class-variables as well, we are also acknowledging classes of these individuals.

In other words, the nominalist countenances only individuals but may take anything as an individual. Whether a system is nominalistic depends not upon whether the entities admitted are in fact individuals (whatever that might mean) but upon whether they are construed in the system as individuals–that is, upon whether the system always identifies with one another entities that it generates out of exactly the same selection from among those admitted entities that it does not generate out of others.[5] Nominalism is defined not by independent standards of what constitutes an individual but by independent standards of what constitutes taking entities as individuals.

Thus while nominalism leaves us great freedom in our choice of predicates and individuals, it drastically curtails the means available to us for constructing a system. In foregoing all use of variables having nonindividuals as values, we give ourselves the task of retranslating a great many everyday statements that are customarily construed as pertaining to classes.

In some cases, there is little difficulty. Statements concerning such relations of classes as inclusion are readily interpreted by familiar means; for example, "All trees are plants" will be rendered not by "The class of trees is included in the class of plants" but by "Everything that is a tree is a plant", where "is a tree" and "is a plant" are taken simply as predicates of individuals. Similarly any statement to the effect that individuals belong to a certain class can be construed in terms of a predicate of those individuals; for example, "John and Bill and all their cousins belong to the Muskrat Society" is easily rendered (in abbreviation) as:

$$\mathrm{M}j \,.\, \mathrm{M}b \,.\, (x)(\mathrm{C}\,x,j \vee \mathrm{C}\,x,b \,.\supset \mathrm{M}x).$$

[5] For further explanation see the following sections of the present chapter, Section 11 of the next chapter, and the two articles cited at the end of note 4 above.

The problem of dealing with a sentence like "Every species of dog is exhibited" is a little more difficult, but in such cases we may often speak of wholes rather than classes. The species of dog may be regarded as certain discontinuous wholes composed of dogs. Then the sentence may be rendered: "For every x, if x is a species of dog then some y is a dog and is part of x and is exhibited". Sometimes, of course, use of this device is blocked by the fact that the several classes under consideration correspond to the same whole.[6] For example, suppose that several different planners have proposed dividing a plot of land in different ways and that it is said, "At least one group of lots into which it is proposed to divide this land violates city regulations". We cannot take "group of lots" here as referring to the whole that contains the lots in the group; for this whole is identical for all the groups in question. If we want to translate this sentence, we must do it in some other way.

Perhaps it is the problem of interpreting numerical statements, however, that best illustrates the problems confronting the nominalist. Obviously, he has deprived himself of the usual interpretation of numbers as classes of classes. To reconstruct in the language of individuals all of mathematics that is worth saving is a formidable task that need not concern us here. It will be enough to consider typical arithmetical statements used in ordinary discourse.

We might begin by identifying numbers with certain individuals–say with certain actual inscriptions of the various numerals–and thus bring them within the range of the values of our variables. A good deal can be done on this basis toward reconstructing a considerable portion of ordinary arithmetic, but such a development has little practical use. It leaves us without any means for speaking of numbers *of* things; we can relate 3 to 5, but not three things to five things. Some more applicable treatment is wanted.

Such statements as "There are three things" or "There are exactly two Poles" offer no difficulty. The latter, for example, is rendered by:

$$(\exists x)(\exists y)\{\mathrm{P}x\,.\,\mathrm{P}y\,.\,x\neq y\,.\,(z)(\mathrm{P}z \supset .\, z=x \vee z=y)\}.$$

Statements affirming that there are at least or at most or exactly so many dogs or desks or things in general can be handled similarly so long as time, space, and patience hold out.

It is a little more difficult to say, for example, that there are more cats than dogs without specifying the number of each. Since, however, we know how to say that there is at least one cat and not at least one dog, that there

[6] The general theory of wholes and parts will be discussed in the following section. A whole need not be continuous.

are at least two cats and not at least two dogs, and so on, we could say that there are more cats than dogs by means of a disjunction of enough such sentences. If we go far enough, one of the alternatives–and therefore the entire disjunction–will be true if and only if there are more cats than dogs. One trouble with this device is that we have to have some specific knowledge concerning the number of dogs before we can be sure that we have gone far enough to include the crucial alternative in our disjunction. And of course this method is intolerably cumbersome even if we omit as many of the earlier alternatives as we know surely to be false.

An easier course is simply to use the one-place predicate of individuals "has more cats than dogs as parts" along with the two-place predicate "is part of"; then we may write "Everything of which every cat and every dog is a part has more cats than dogs as parts" or, using obvious abbreviations:

$$(x)\{(y)(\mathrm{C}y \vee \mathrm{D}y \,.\supset \mathrm{P}y,x) \supset \mathrm{H}x\}.$$

There is nothing here repugnant to nominalism. The only difficulty is that a new predicate like "H" will be needed for every two kinds of things we want to compare numerically. This would result in a very uneconomical basis, with consequent difficulty in securing any generality for such principles as the transitivity of "more than".

Nevertheless, a problem of nominalism has now been transformed into one of economy, and there are promising directions in which simplification might be sought. For example, if we use the predicate "has more animals as parts than", along with the predicate "is part of", we can deal with sentences comparing any two classes of animals. The sentence "There are more lions than zebras" would be rendered: "Everything of which every lion and no other animal is part has more animals as parts than does anything of which every zebra and no other animal is part". The success of this device, however, depends upon the fact that animals are not parts of other animals. We might try to go a step further toward generality and translate these sentences about animals, as well as such sentences as "There are more human cells than humans", by using the predicate "has more organic units as parts than", where by "organic units" we mean all parts of any animal that consist of one or more cells. But this would not work for the sentence about human cells and humans, since everything that has all humans as parts also has all human cells as parts. Of course, we can deal very simply with a case like this, where every individual of the one kind is discrete from every other of the same kind and is made up of individuals of the other kind; for we need only say "Every human has at least some human cell as part, and some human has at least two human cells as parts". But then we are reverting to special devices for special cases.

A different procedure, which takes care of a wide variety of cases, makes use of the predicates "is bigger than"[7] (and "is of the same size", which is then easily defined) and "contains (as a part)". In order to translate such a sentence as "There are more townships than congressional districts" we first define "is a bit" so that it applies to those things that are of the same size as any smallest region that is either a township or a congressional district. Now if and only if there are more townships than congressional districts will it be the case that every individual that contains a bit of each township will be bigger than some individual that contains a bit of each congressional district. Thus our translation will be: "Every individual that contains a bit of each township is bigger than some individual that contains a bit of each congressional district". This method also works wherever, instead of "township" and "congressional district", we have any other two predicates such that the individuals fulfilling each are separate from one another. Thus it holds good for the sentence "There are more human cells than humans"; for the individuals satisfying either predicate need not be separate from those satisfying the other. Moreover, by a rather obvious change, the device can be made general enough to work wherever each individual fulfilling either of the two predicates has a part that it shares with no other individual fulfilling that predicate. But I do not know how to make it completely general.

In summary, then, we are never at a loss to translate a statement affirming a simple numerical comparison into the language of individuals so long as we are willing to introduce as many primitive predicates as needed, without worrying about economy. Moreover, we can describe economical procedures that enable us, with the use of but one or two unchanging auxiliary predicates, to deal with wide varieties of cases. But at present we have no uniform economical way of dealing with all cases.

Other problems of nominalism sometimes arise in a rather surprising way when we want to define a given predicate of individuals in terms of certain others. For example, the predicates "is a parent of" and "is an ancestor of" are both two-place predicates of individuals. Moreover, we can easily define "is a parent of" in terms of "is an ancestor of" if we know no parent of x is an ancestor of any ancestor of x. But if we have "is a parent of" as primitive and want to define "is an ancestor of", we find that the only familiar device available is Frege's definition of the ancestral of a relation; and this, since it

[7] I assume that this predicate is so interpreted that no indivisible individual is bigger than any other individual. Dr. John Myhill called my attention to this point.

uses bound class-variables, is definitely platonistic. Curiously, we here seem to depend upon platonistic means for defining one predicate of individuals in terms of others. However, a method of accomplishing the definition within the language of individuals is now known. It is like Frege's except for changes consequent upon using "is part of" in place of "is a member of". In translating "x is an ancestor of y" we must first stipulate that x and y are single whole organisms, not fragments of organisms or wholes containing something more than one organism; this can be done, without use of further predicates, by the requirement "x is a parent of some individual, and some individual is a parent of y". To make "is an ancestor of" irreflexive, as we want it to be for ordinary use, we also require "x is not identical with y". The rest of the translation, in direct analogy with Frege's definition, then runs: "Every individual z that has y as part and that has as parts all parents of parts of z, also has x as part". But here again, the method will not work for defining the ancestral of every two-place predicate of individuals; it fails at times where some individuals fulfilling the predicate are totally exhausted by the others.

Many other kinds of statement are difficult to translate into the language of individuals. I have sought only to illustrate some typical problems and procedures. The effort to carry out a constructive nominalism is still so young that no one can say exactly where the limits of translatability lie. We have seen above that some statements that look hopelessly platonistic yield to nominalistic translation, and the full resources available to the nominalist have not by any means been fully exploited as yet.

Except when explicitly concerned with platonistic systems (as in III, 10 and V), I shall so far as possible use only such language as I can translate into the language of individuals. I shall use platonistic language freely in extrasystematic contexts so long as a nominalistic translation is available. For example, "Some couple belonging to the relation R has the same individual as first component as some couple belonging to the relation S" is unobjectionable since it can be readily construed as "There is an x, a y, and a z, such that R x, y and S x, z", where "R" and "S" are two-place predicates of individuals. I may even make some extrasystematic use of platonistic language I cannot yet translate; but in actual systematic constructions, I shall use nominalistic language exclusively. Adherence to the language of individuals has been made easier by the fact that although I made use of the language of classes in framing definitions in *A Study of Qualities*, all the special primitives were predicates of individuals, and many predicates that might have been defined as predicates of classes were defined as predicates of wholes.

In the above translations of sample sentences into the language of individuals, repeated use was made of the predicate "is part of". This and kindred predicates tend to play in nominalistic systems a role as prominent as that of the analogous relations of classes in platonistic systems. But the utility of these predicates of individuals is not confined to nominalistic systems. In *A Study of Qualities* I found reason to introduce these predicates to serve important purposes, even though the calculus of classes was also available. The systematic treatment of these predicates–known as the 'calculus of individuals'–is common to virtually all nominalistic and some platonistic systems. As part of the general apparatus of systems it calls for treatment in this chapter.

4. THE CALCULUS OF INDIVIDUALS[8]

No one who uses the term "class" for technical purposes supposes that several people must take instruction together or be of similar social position or have some other special characteristic in common in order to constitute a class. The technical use of the term "individual" must likewise be freed of certain restrictive associations arising from popular usage. For example, although the most convenient illustrations of the application of such predicates of individuals as "is part of" and "overlaps" may be in terms of spatial regions, quite other applications will be explained later. Moreover, an individual may be divisible into any number of parts; for individuality does not depend upon indivisibility. Nor does it depend on homogeneity, continuity, compactness, or regularity. Indeed, nothing has intrinsic or absolute status as an individual or a class. Whether a system recognizes classes as well as individuals depends upon how it makes up entities out of others. The present section is concerned with formulating the kind of composition admissible in nominalistic systems.

One must, of course, indicate by description and example just what are the individuals to be countenanced by a proposed discourse or system–just what the universe of that discourse is considered to be. But since discourses or systems may differ a good deal in what individuals they recognize, we can best leave this until we come to consider special systems. We do not need to know just what individuals are admitted before we set up certain general

---

[8] A calculus of individuals was published by Lesniewski in 1916, and again–revised and expanded–in 1927–31. Both versions were in Polish and remained untranslated. The calculus to be outlined here was developed by Henry S. Leonard and Nelson Goodman. It was presented first in Leonard's doctoral thesis *Singular Terms* (typescript, Widener Library, Harvard University, 1930); and, in a later version, read before the Association for Symbolic Logic in 1936 and published in "The Calculus of Individuals and its Uses", *Journal of Symbolic Logic*, 5 (1940), pp. 45–55. In 1937, Tarski published a simplified version of Lesniewski's calculus as an appendix to J. H. Woodger's *Axiomatic Method in Biology* (Cambridge, England: Cambridge University Press, 1937).

apparatus to deal with them, any more than we need to know just what business transactions are done before we set up a system of double-entry bookkeeping for them.

The variables of the calculus of individuals are lower-case italic letters taking individuals as their only values. The sole primitive needed is the two-place predicate "overlaps",[9] which I abbreviate in symbolic contexts by "**o**". Two individuals overlap if they have some common content, whether or not either is wholly contained in the other. The predicate "**o**" is symmetric and reflexive but not transitive. Moreover, it is ubiquitous among individuals in that all and only those things that overlap something are individuals. Thus "individual" may be easily defined in terms of overlapping; but in a nominalistic system, "(is an) individual" is a universal predicate and so of little use.

I shall not give here anything like a complete account of the calculus of individuals, but merely present and explain the definitions briefly, and note a very few propositions of the calculus without regard to their deductive order or their status as postulates or theorems. The plan of numbering, to be followed also in later chapters, is this:–The number of a definition is preceded by a "D", which together with the main "=" performs the usual function of "$=_{df}$". The number itself indicates the chapter and section in which the definition appears, as well as the serial position of the definition in that section; for example, "D9.053" indicates the third definition in Section 5 of Chapter IX. The labeling of postulates and theorems follows the same plan except that the initial "D" and the zero following the point are omitted.

If and only if two individuals x and y overlap is there some individual z (viz., any individual wholly contained within x and within y), such that whatever overlaps z also overlaps both x and y; that is:

2.41 $\quad x \mathbf{o} y . \equiv (\exists z)(w)(w \mathbf{o} z \supset . w \mathbf{o} x . w \mathbf{o} y).$

Two individuals that do not overlap are said to be discrete from each other. The predicate "is discrete from" is defined as follows:[10]

D2.041 $\quad x \wr y = \sim x \mathbf{o} y.$

Any two individuals that have no common content are discrete; for example, if two spatial regions are completely separate, they are discrete whether they

[9] The reason for choosing a symmetric predicate rather than the more familiar "is part of" will be made clear in Chapter III. The reason for choosing "overlaps" rather than "is discrete from"–aside from the fact that the former applies to all individuals–is a resulting expository convenience in the following section of the present chapter.

[10] Each symbolic definition may be thought of as accompanied by another that joins the symbol and its verbal reading; e.g., along with D2.041 goes: "$x \wr y =_{df} x$ is discrete from y".

are near together or far apart. The predicate "$\imath$" is symmetric but irreflexive and nontransitive. While every individual overlaps some individual, every individual except the whole universe is discrete from some individual.

In terms of the primitive, we can define the predicate "is part of" as follows:

D2.042 $x < y = (z)(z \circ x \supset z \circ y)$.

That is, one thing is part of another if and only if whatever overlaps the former also overlaps the latter. Obviously, the predicate is transitive but not symmetric. The definition is purposely so framed as to make every individual a part of itself. Parts less than the whole are said to be *proper* parts. "Is a proper part of" is defined thus:

D2.043 $x \ll y = x < y \,.\, {\sim} y < x$.

The predicate is asymmetric, irreflexive, and transitive.

If the calculus of individuals is used along with the calculus of classes in a platonistic system, the identity of individuals may be defined in the usual Leibnizian way. But since this usual definition says, in effect, that *a* and *b* are identical if and only if they belong to exactly the same classes, it is not open to us if we are to restrict ourselves to the language of individuals. However, a definition of "is identical with" can readily be provided within the calculus of individuals: *a* and *b* are identical if and only if they overlap exactly the same individuals:

D2.044 $x = y = (z)(z \circ x \equiv z \circ y)$.

The definition is to be accompanied, of course, by the usual rule permitting substitution of either side of a true identity statement for the other side in any context.

Beside the predicates so far defined, we shall often need terms for the sums, products, and negates of individuals. These might be treated as predicates –letting "P x,y,z", for example, mean "x is the product of y and z"–but descriptive functions are more convenient to use. The product of two individuals is that individual which exactly contains all that is common to the two:

D2.045 $xy = (\imath z)\{(w)(w < z \equiv . \, w < x \,.\, w < y)\}$.

Two individuals have a product if and only if they overlap; i.e.,

2.42 $(\exists z)(z = xy) \equiv x \circ y$,

as may be seen from 2.41.

The negate of an individual contains exactly all that is discrete from that individual:

D2.046 $\quad -x = (\imath z)\{(y)(y \wr x \equiv y < z)\}$.

Every individual that is discrete from some individual–in other words, every individual except the universe itself–has a negate:

2.43 $\quad (\exists y)(y = -x) \equiv (\exists z)(z \wr x)$.

No special definition needs to be introduced for a difference of two individuals since according to preceding definitions $x-y$ is the product of x and $-y$. The difference $x-y$ exists unless x is part of y:

2.44 $\quad (\exists z)(z = x-y) \equiv \sim x < y$.

If neither x nor y is part of the other, then x and y have two differences: $x-y$ and $y-x$.

The sum of two individuals is that individual which exactly and completely exhausts both. Yet we cannot define the sum of *b* and *c* as that individual which has the parts of *b* and the parts of *c* as its only parts, for the sum may have other parts as well. The sum of the regions of Massachusetts and Vermont has as parts not only regions that are parts of each but also regions containing parts of both. The sum of two individuals may, however, be defined as that individual which overlaps just those individuals which overlap at least one of the two:

D2.047 $\quad x+y = (\imath z)\{(w)(w \circ z \equiv . w \circ x \vee w \circ y)\}$.

Although not every individual has a negate and not every two individuals have a product, every two individuals *do* have a sum. Bearing in mind that only individuals are values of our variables, we can affirm the unconditional statement:

2.45 $\quad (\exists z)(z = x+y)$

as a postulate or theorem of our calculus. This statement sometimes arouses surprise and opposition. The usual objection is to name some two very different and widely separated individuals and ask if it is reasonable to suppose that they have a sum that is an individual. Such an objection misses the point. If the Arctic Sea and a speck of dust in the Sahara are individuals, then their sum is an individual. As I have already explained, an individual need not be organized or uniform, need not be continuous or have regular boundaries. The supposition that bizarre instances demonstrate that two individuals can fail to have a sum betrays a misunderstanding of the range of our variables.

A different objection sometimes raised is that the truth of 2.45 depends upon stretching the notion of an individual so far as to make it include all classes. It is argued that unless individuality implies some minimum of coherence or homogeneity, an individual having two utterly disparate and disconnected parts will be indistinguishable from the class of the two. The answer is that even assuming that classes exist, we cannot identify an individual with a certain class of its parts, for there will be many such classes. The class having as members Caesar's nose and the state of Utah is different from the class having as members Caesar's nose and the counties of Utah, yet the individual that exactly exhausts the members of the first class is the same as the individual that exactly exhausts the members of the second. If the classes are different, they clearly cannot be identified with the same individual. In short, if we recognize classes at all, an individual made up of scattered parts cannot be the same as the class of those parts; for the relation of class to individual-exhausting-the-members-of-the-class is not one-one but many-one (see Section 5).

We often want to speak of the sum of all the individuals satisfying a certain predicate. The sum will be the individual that exactly exhausts all such individuals. The schema[11] for the definiens of such an individual for any predicate is as follows, the predicate to be written in for the string of dots:

$$(\imath x)\,\{(y)(x \circ y \equiv (\exists z)(\ldots z \,.\, z \circ y))\}.$$

For example, the sum of all Dalmatians is that individual which overlaps all and only those individuals which overlap some Dalmatian. If any individual satisfies a given predicate, then there is an individual that is the sum of all individuals satisfying the predicate.

Similarly, the product of all individuals satisfying a given predicate could be defined by use of the following schema:

$$(\imath x)\,\{(y)(y < x \equiv (z)(\ldots z \supset y < z))\}.$$

The product, which exists only if some individual is a common part of all the individuals in question, is the individual that has as parts all and only those individuals that are parts of all the individuals in question.

No more of the actual calculus of individuals need be presented here except for two theorems that permit convenient abbreviations:

---

[11] To give a definition rather than a schema, without violating the tenets of nominalism, would require the introduction of syntactical predicate-variables and consequent recourse to a metalanguage.

2.46 $x < yz \equiv . x < y . x < z,$
2.47 $x + y < z \equiv . x < z . y < z.$

But note especially that we do *not* have as a theorem either the statement "$xy < z \supset . x < z . y < z$" or the statement "$x < y + z \supset . x < y . x < z$".

Although no specific postulates have been chosen here, we may assume that postulates are adopted that will yield a calculus approximately equal in strength to the various published calculi. The chief difference will be that since our calculus is framed in nominalistic language, we shall not have any postulate stating that every class of individuals has one and only one sum but must either adopt an indefinite number of postulates satisfying a corresponding schema or accomplish a similar result in some other way (see, however, III,12). Minor differences in the set of postulates required will result from our choice of primitive and from the fact that identity is defined within the calculus. It should be especially noted that the calculus is to contain no postulates implying that the number of individuals is either finite or infinite; thus, for example, it will contain no statement affirming that every individual has a proper part.

The calculus of individuals calls attention to certain rather important but little-noticed differences among predicates. The extrasystematic terms introduced in the following paragraphs need by no means be memorized, but are explained here for later reference.

A one-place predicate is said to be *dissective* if it is satisfied by every part of every individual that satisfies it. Since every part of everything that is smaller than Utah is also smaller than Utah, the predicate "is smaller than Utah" is dissective. Fewer predicates than might be supposed are unreservedly dissective; for example, if no further restriction is placed upon the kind or size of the parts to be considered, we cannot even say that if a block of metal is pure silver, then every part of it is pure silver, since the component electrons are hardly pure silver. In practice, we are usually concerned only with dissectiveness under some special or systematic limitations–for example, with the fact that if a block of metal is pure silver then every *metallic* part of it is pure silver. A similar comment applies in the case of the other terms explained below.

A one-place predicate is *expansive* if it is satisfied by everything[12] that has a part satisfying it–or in other words, if it is satisfied by every whole consisting of anything satisfying it added to anything else. The predicates "is large" and "is populated" are expansive.

[12] Although "thing" is ordinarily used only for an entity of a certain kind, "everything" and "anything" may be used for "every individual" and "any individual".

That a one-place predicate is *collective* means that it is satisfied by the sum of every two individuals (distinct or not) that satisfy it severally; examples are "is pure gold", "is in Utah", "is owned by Shakespeare". Such a predicate will also, it turns out, always be satisfied by any sum of more than two individuals that satisfy it severally. For example, if a predicate is collective and is satisfied by x and y and z, it must then be satisfied by $x+y$ and $x+z$ and $y+z$; and since it is satisfied by all these it must also, because it is collective, be satisfied by $x+y+z$.

A one-place predicate is *nucleative* if it is satisfied by the product of every two overlapping individuals that satisfy it severally.

In view of the obvious theorems "$x < x$", "$x = x + x$", and "$x = xx$", no predicate that is satisfied by at least one individual can be undissective, unexpansive, uncollective, or unnucleative. A predicate could be uncollective for example, if and only if it were not satisfied by any sum of individuals that satisfy it; but if x satisfies a predicate, then $x+x$ satisfies it. Similar reasoning applies in the other cases. Predicates may, of course, be *non*dissective, *non*-expansive, etc.

As the number of places increases, so does the number of such differences among predicates as we have been considering; but there is no need for burdening ourselves with terms for all the possible variations. The following three, however, will prove important later. I am concerned with them only in application to symmetric predicates and hence will ignore the variations resulting from nonsymmetry.

A two-place predicate is *pervasive* if whenever it applies between[13] x and y, it also applies between every two parts of $x+y$. An example is the predicate "lies within the same state as". Some predicates are pervasive only in the limited sense that whenever they apply between x and y, they also apply between every two discrete parts of $x+y$.

A two-place predicate is *cumulative* if whenever it applies between x and y and between x and z it also applies between x and $y+z$. The predicate "is made entirely of the same metal as" is cumulative. By an argument like that outlined for collective one-place predicates, it can be shown that if a predicate is cumulative and applies between x and each of three or more others, it also applies between x and the sum of those others. But what is especially important is that many predicates–for example, "has the same shape as" and "weighs either more or less than"–are *not* cumulative. Thus we cannot always infer that a predicate applies between an individual x and the *sum* of two or more others from the fact that the predicate applies between x and *each* of

[13] The predicate "P" applies between x and y if and only if P x,y.

the others. Obvious as this seems, fallacies resulting from ignoring it are not unknown.

A two-place predicate is *agglomerative* if whenever it applies between x and y and between x and z and between y and z, it also applies between x and $y+z$. All cumulative predicates are agglomerative; one noncumulative but agglomerative predicate is "has as a proper part or is a proper part of". But here again, the important point is rather that many predicates are not agglomerative; that we cannot in general infer that a predicate applies between x and the *sum* of two others *even* when it applies between x and each of the others *and also* between these others. The predicate "is of the same size as", for example, is noncumulative and nonagglomerative. Another illustration, more like some that we shall meet in later chapters, is the predicate "T", where "T" applies between every two lots of land that can be together completely enclosed by a single circle with a radius of one mile. There are certainly three lots a, b, and c such that although T a,b, and T a,c, and T b,c, still it is not the case that T $a,b+c$.

Finally, two traditional but not always intelligible terms can be explained in a very straightforward and literal way. Two entities are sometimes said to be internally related if neither could be completely destroyed without affecting the other, but no criteria are given by which to determine where such interdependence obtains. In the absence of any other clarification of the notion, it seems to me that to say that two individuals stand in such a relation is just to say that they have a common part. In other words, an *internal* predicate is one that applies between overlapping individuals only. An *external* predicate applies between discrete individuals only. We shall see later that the predicate "is similar to" is sometimes used as an internal predicate (i.e., when similarity is interpreted as actual part-identity), and sometimes as an external one.

5. THE CALCULUS IN SYSTEMS

If states, counties, and congressional districts are thought of as geographical individuals, then a state is at once the sum of all its counties and the sum of all its congressional districts. An individual may be at once the sum of all the individuals satisfying each of several predicates, even though no individual satisfies any two of these predicates. Or in the language of classes, an individual may be the sum of the members of each of several classes–as well as the sum of the sums of the members of each of several classes of classes, and so on–even if the classes have no common members. This gives rise to a certain ambiguity in almost every term. The predicate "is a state", taken as a geographical term, may be interpreted as applying to certain region individuals, or alternatively–if we recognize classes–to certain classes of counties, or to certain classes of congressional districts, and so on. Thus more than a look at the map is involved in deciding whether counties are proper parts of states. County individuals are indeed proper *parts of* state individuals. But if states are construed as classes of counties, then counties are *members of* states; while if states and counties alike are construed as classes of smaller regions, then counties are *included in* states. Whether the sentence

$$\text{“}(x)(\text{County } x \,.\supset (\exists y)(\text{State } y \,.\, x \ll y))\text{”}$$

holds in a given system thus depends not only upon whether every county individual is a proper part of some state individual, but also upon whether "County" and "State" (i.e., "is a county" and "is a state") are construed in the system as predicates of individuals. In few cases of this sort does common usage decide; the choice ordinarily has to be made for each system on the ground of practical and technical convenience. That the statement "Windham County $\ll$ Vermont" is true in one system and not true in another does not mean that one of the systems must be wrong but merely that a certain ambiguity in the predicates "(is identical with) Windham County" and "(is identical with) Vermont" has been differently resolved in the two systems.

Not all apparently similar cases of divergence among systems can be explained as the result of such a choice. Some are due to the different universes of discourse of the different systems. In the systems we shall consider, the range of the individual-variables is to be described not as embracing every individual that satisfies the predicate "overlaps" but rather as embracing every individual that satisfies, or that is a sum of individuals that satisfy, one or more of the special primitives. It is the special primitive predicates that constitute the peculiar vocabulary the system builder proposes to use in explaining his subject matter; and the individuals he is to deal with must usually be constructed out of individuals satisfying these predicates. (See further III,11 on basic units.) He may or may not have doubts about whether in fact there are any other individuals; restriction of the values of his variables to the specified individuals frees him of the need for making any commitment concerning other individuals, just as restriction to individuals in general frees the nominalist of commitment concerning classes.

A result of restricting the realm of individuals for a system is that some individuals that overlap presystematically may not overlap for the system. Systematic overlapping may be a proper subrelation of general overlapping; in other words, even though the predicate "overlaps" as used in a given system applies only between individuals between which it applies presystematically, still there may be individuals between which the predicate applies presystematically but not as used in the system. Suppose, for example, that in a given system dealing with our diagram of Chapter I (Figure 1), the only admitted individuals are sums of one or more of the four lines. For such a system, lines *a* and 1 do not overlap, for the system recognizes no individual such as *K* that is the product of the two. On the other hand, restriction of the realm of individuals will not conversely result in the systematic overlapping of individuals that fail to overlap presystematically. Obviously just the opposite situation obtains in the case of "is discrete from" (which as used in a system is to be regarded as defined in terms of the systematic predicate "**o**" rather than its presystematic counterpart); for example, "*a* is discrete from 1" is true in this system even though presystematically false. Such differences between a system and the presystematic, or between two different systems, are readily traced back to their source in the restricted interpretation of the primitive "**o**" resulting from the special systematic restriction imposed on the realm of individuals.

Yet other apparent discrepancies are not directly accounted for either by systematic restrictions upon individuals or by the kind of ambiguity we noted in such terms as "state". For example, if, in a system dealing with our diagram, points of intersection are defined as crosses (i.e., as sums of two intersecting lines), then some such points overlap others, although presystematically no point of intersection overlaps any other. Surely this is not because points are construed as classes in one case and as individuals in the other, for points are here taken as individuals in both cases. Nor can the difference result merely from a systematic restriction of "**o**"; for here we seem to have systematic cases of overlapping that are not presystematic cases. The explanation will be clear, however, if we remember that to assert in a system the sentence:

(i) There is a point that overlaps another point

is in effect to assert that what is correlated in the system with some point overlaps what is correlated in the system with some other; or rather that when "is a point" is replaced by its definiens in the system the result will be a true sentence. Obviously the resulting sentence here:

(ii) There is a sum of two intersecting lines that overlaps another sum of two intersecting lines,

is true. This does not conflict with the presystematic discreteness of the points in question. And the system may be quite compatible with another in which, as the result of a different definition of the points in question, the apparent contradictory of (i), namely:

(iii) No point overlaps another point,

is true. As we saw in Chapter I, such apparent discrepancies are often the natural outcome of differences between equally acceptable definitions.

In summary, then, a good many differences among systems may be traced back to legitimate differences in definition or in the interpretation of the primitives. Considerable latitude in definition is permitted by the criterion of definition that we have found to be applicable. Differences in interpretation of primitives may arise from simple ambiguity in ordinary usage, or from the general ambiguity that leaves most predicates open to interpretation as predicates of either individuals or classes, or from differences in the realm of individuals acknowledged.

Some uses of the calculus of individuals were illustrated in Section 3. But the calculus also has its uses in platonistic systems, along with the calculus of classes. In such systems, of course, the sum (or whole or 'fusion') of a class can be directly defined, since class-variables are available. The definition follows the schema given in Section 4 for defining the sum of the individuals satisfying a certain predicate:

$$S^{i}{}^{\text{‘}}(\alpha) =_{\text{df}} (\iota x)\{(y)(y \circ x \equiv (\exists z)\ (z \in \alpha \,.\, y \circ z))\}.$$

The decision whether to define a given term in a platonistic system as a predicate of individuals or of classes may be influenced by two factors. On the one hand, since there are in general many classes to every whole, class predicates are often more specific than predicates applying to the corresponding wholes; for example, "is a squad" might be defined as a predicate of classes because we are normally concerned with the members of a squad to the exclusion of the other parts (such as the cells of the men, or one man plus part of another) into which the squad as an individual may be divided. On the other hand, predicates of individuals allow us to remain within a single logical type; for example, if townships, counties, and states are all construed as individuals, then townships are parts of counties, and counties of states, and townships of states–no type differences are involved.

But such considerations would hardly justify the introduction of an added primitive, "overlaps", into a system in which a calculus of classes was already available. I have already mentioned that although platonistic apparatus was used in *A Study of Qualities*, the calculus of individuals was nevertheless introduced for an important constructional purpose, which will later be amply explained. An inclination toward nominalism indeed favored this course, since the alternative method considered for accomplishing the same result required adopting predicates of classes as primitives. But quite apart from that, and more to the point in a platonistic system, it can be shown (VI,7) that adoption of the primitives required by the alternative methods would be a far less economical course than adoption of the primitive needed for the calculus of individuals. And economy, as we shall see in the next chapter, is not entirely a vague nor by any means an unimportant consideration.

CHAPTER III

EXTRALOGICAL BASES

1. THE NATURE OF PRIMITIVE TERMS

The extralogical basis of a system consists of all its primitives that are not in our list of basic 'logical terms'. It thus may include, in addition to primitives peculiar to the system, primitives like "overlaps" which are common to many systems.

To adopt a term as primitive is to introduce it into a system without defining it. In so far as its interpretation is not clear from ordinary usage, an explanation–which is not part of the formal system–must be provided. Familiar terms in familiar contexts–as, for example, "triangle" in a system of plane geometry–may need little explanation. The interpretation of newly invented words or symbols, on the other hand, depends entirely upon the unofficial explanation in terms of words whose usage *is* familiar. Often, the interpretation of a primitive is given partly by ordinary usage, partly by an explanation designed to resolve any ambiguities in that usage.

Insistence upon extensionality has led some[1] to consider the adoption of a primitive predicate as in effect the adoption of a list of everything that satisfies the predicate. To take "is a man" as primitive is thus said to amount to taking the list labeled "man" consisting of "Thales", "Hume", etc. This notion is acknowledged to be a fiction since we seldom in fact have anything like a complete list, but the aim is to stress the point that at least we have nothing more than the list–that the list, which sets forth the extension of the predicate, is the most that is assumed. But useful as the fiction may be for this purpose, it is somewhat misleading even when recognized as a fiction. For in adopting a predicate as primitive not only do we ordinarily have no complete list of its applications but we are not adopting as primitive even one item on such a list. To do that would amount to making that item an additional primitive. Thus, in taking "is a man" as primitive we are not taking as primitive either the list of men or any name on that list. It is true, however, that all that is demanded of the interpretation of a primitive predicate is a determination of its range of application; and in explaining a primitive, we may use enumeration or listing wherever convenient.

---

[1] For example, Carnap in the *Aufbau*, Section 102.

But what, now, are the implications of choosing one term rather than another as primitive? It is not because a term is indefinable that it is chosen as primitive; rather, it is because a term has been chosen as primitive for a system that it is indefinable in that system. No term is absolutely indefinable. Indefinability sometimes means incomprehensibility; but incomprehensible terms have no place at all in a system. In general, the terms adopted as primitives of a given system are readily definable in some other system. There is no absolute primitive, no one correct selection of primitives. Attention is therefore directed to the factors that affect the choice of primitives for systems.

2. THE CHOICE OF BASIS

Some of those factors are so obvious as to call for the briefest mention only. For example, plainly inapplicable predicates like "circular and noncircular" (taking the component words in their ordinary usage) will hardly be chosen as primitive; nor will predicates be chosen that are regarded as quite obscure, as "predisposed to telepathic communication with supernatural aid" might be. Moreover, such a predicate as "denotes", even construed as applying between inscriptions and other individuals, will be chosen only with the greatest hesitation, if at all, in view of the paradoxes that have been found to result from its incautious use.

However, the choice is not merely a matter of eliminating such unlikely candidates; the considerations involved are often much more complex. Clarity is commonly thought to be the cardinal factor; but actually one frequently takes as primitive an unfamiliar or even a coined term, where the ratio of decided to undecided presystematic applications is very low. Any term chosen as primitive must indeed be *made* clear for the system by means of an explanation; but a term need not be antecedently very clear to be eligible as primitive. Of course if we decide that a given term (say "ectoplasmic"), because of its presystematic usage, is not amenable to clarification, we do not take it as primitive–but then neither do we define it in the system. The antecedent clarity or clarification of this sort demanded of primitives, then, is only that demanded of any term that is to be admitted into a system at all, as either defined or undefined. If clarity is gauged not by ratio of decided to undecided cases but by ease of explanation and understanding, then obviously in the *process* of building a system we may like to begin with terms that are clear. This makes for ease not only in the construction but also in the comprehension of a system. And it diminishes the risk of error; for faulty use of a primitive may infect all the definitions in the expansion of which the primitive appears, while misconstruction of a defined term ordinarily affects a smaller part of the system. But the psychological clarity in question here is a rather elusive consideration, for ease of understanding varies with persons, situations, and even moods.

Limitations imposed by the very problem or set of problems a system is designed to meet constitute a more definite factor. Occasionally, the primitives to be employed are uniquely indicated in the statement of the problem, as in "Define points in terms of the intersection of lines". More often, the problem merely circumscribes a sphere of eligible primitives. The problem of defining certain predicates of geometry in terms of predicates of perceptible volumes, the problem of defining sensations in terms of objective entities, the problem of defining qualities in terms of things, all carry with them express limitations upon the kind of primitives that may be used. Once the problem is accepted for investigation, the question of the merits of such primitives as compared with others is no longer to the point, for to take other primitives would be to deal with a different problem. Sometimes a problem or set of problems may impose no explicit conditions upon the kinds of primitives that are relevant, but ordinarily at least tacit conditions are in force. A demand for definitions of certain predicates is usually a demand for definitions in terms of primitives of a certain kind, as becomes evident when otherwise satisfactory definitions are rejected as trivial or irrelevant.

Since, however, the problems to be dealt with seldom determine uniquely the primitives to be used, other factors must almost always enter into the choice of primitives from among all eligible predicates. The primitives chosen must, of course, form an *adequate* basis for all the definitions required; but adequacy, in so far as it is attainable at all, could readily be insured by adopting as primitive all predicates not excluded by the conditions of the problems at hand. Not merely an adequate basis but the minimum or *simplest* adequate basis is wanted. The simplicity of bases is both more difficult to measure and more important than might be supposed.

3. SIMPLICITY[2]

Most generally speaking, the purpose of constructing a system is to interrelate

---

[2] In the first edition of this book, the discussion of simplicity was based mainly upon three of my articles: "On the Simplicity of Ideas", *Journal of Symbolic Logic*, 8 (1943), pp. 107–121; "The Logical Simplicity of Predicates", same journal, 14 (1949), pp. 32–41; and "An Improvement in the Theory of Simplicity", same journal, 14 (1949), pp. 228–229. In the second edition, the treatment was very considerably revised and expanded (see note 3 below). See also *P & P*, pp. 275–321.

its predicates. The same purpose is served by reducing to a minimum the basis required. Every definition at once both increases the coherence of the system and diminishes the number of predicates that need be taken as primitive. Thus the motive for seeking economy is not mere concern for superficial neatness. To economize and to systematize are the same. Some economies are relatively unimportant and some apparent economies are spurious, but the inevitable result of regarding all economy as trivial would be a willingness to accept as primitives all the predicates that are clear enough to be admitted into the system at all. Not even the limitations imposed by a specific problem of construction would remain; for such a problem is in effect a specific problem of economy.

We might, of course, construct a system by using, instead of definitions, the corresponding biconditionals; but this will not circumvent the question of economy. Whether our basic equations of predicates are set forth as definitions or as postulates, they are in effect licenses for the use of alternative terms in describing certain facts. In either case, there are one or more least sets of predicates that will, if we take the fullest advantage of such licenses, do service for the whole set of predicates contained in the system. And the articulation of the system is improved as the minimum vocabulary needed is reduced–as economy, measured by the paucity of any such least set of predicates, is increased. In the systems we shall consider, definitions rather than the corresponding biconditionals will be used, so that the set of primitives for each system will be uniquely determined.

Just how to measure the simplicity of a set of primitives is not very obvious, however. The mere counting of primitives is plainly unsatisfactory; for if the number of primitives were the only concern, maximum economy could quickly be achieved in most systems simply by compounding all the predicates into one having many places. The natural proposal that primitive *predicate places* be counted will also prove inadequate if we find (as we shall) that there are purely mechanical and therefore trivial means of replacing certain kinds of primitive predicates by others having fewer places. We unhesitatingly reject an apparent economy as not genuine in some such cases; but we must still formulate rules that will distinguish in general between true and false economy.

Fortunately we need concern ourselves only with what might be called the *formal* simplicity of bases–simplicity, that is, only in so far as it is affected by those differences among predicates that are expressible by using only the basic logical terms and the identity-sign (in addition to the predicates themselves). Thus, how many places a predicate has and whether or not it is symmetric are relevant considerations, but the number of words in the predicate and the complexity of what it expresses are not. For example, "P" and "Q" may be equally simple for our present purposes even if "P" means "is crimson" while "Q" means "is crimson and metallic and exhibits fluorescence in a degree equal to the square of its electric charge".

Sometimes the relevant kind of simplicity is thought to be the direct inverse of defining power. It seems reasonable enough to consider basis *A* as simpler or at least more economical than basis *B* if *A* is definable solely in terms of *B* (and the basic logical terms) while *B* is not similarly definable in terms of *A*. But this formulation provides at most a very incomplete criterion, for it does not enable us to compare two bases if neither is definable from the other or if they are interdefinable. And we certainly cannot take *equal* defining power as a criterion of equal economy. That would make any full system of predicates as economical as any adequate basis for it. For a system is, of course, definable from any adequate basis, and the basis is contained in and therefore definable from the system; hence a system and any adequate basis for it are interdefinable and thus equal in defining power. Accordingly, by the proposed criterion, it would be as economical to let all the predicates of a system be primitive as to use any narrower basis. This reveals the basic absurdity of identifying weakness with economy. The most economical basis, like the most economical engine, is the one that accomplishes most with least. Simplicity–or low fuel consumption–is a different factor from power and has to be taken equally into consideration. And power, far from being inversely proportionate to economy, is directly proportionate to it where simplicity is constant; the stronger of two equally simple ideas is the more economical. Moreover, where we are concerned with comparing interdefinable and thus equally powerful alternative bases for a system, as is often the case, simplicity is the sole determinant of economy. But what is the measure of simplicity?

Replacement of one basis for a system by another basis effects no genuine economy if the replacement is purely routine. For example, economy is never achieved by adding a primitive but is often achieved by eliminating one; for we can always add but not always subtract a primitive without destroying the adequacy of our basis. Similarly, where we do not merely drop or add a primitive but change our basis in some other way, the simplicity of the basis is increased only if the change is not a purely mechanical one, of a sort that can always be made. The reason that merely replacing a primitive predicate by another having more places effects no saving is just that a predicate can always be traded for one having more places.[3]

[3] From this point on through Section 10, the text was rewritten for the second

A rough general rule, then, is that if every basis of a given kind can be replaced by some basis of a second kind, then no basis of the first kind is more complex than every basis of the second kind. But obviously a good deal of clarification is needed to make this into an explicit and useful principle. Under what circumstances are two bases considered, for present purposes, to be of the same kind? And what constitutes replacement?

Although only structural characteristics are admissible in determining the appropriate kinds, not all structural characteristics can be admitted. For then, since every replacement depends upon structural characteristics, routine would not be distinguished from non-routine replacement. Clearly, in spelling out the notion of routine replacement, the relevant specifications are those most commonly used in classifying predicates and bases. But ordinary practice here as elsewhere is neither very uniform nor very attentive to the requirements of a calculus; and our final selection, as is so often the case in setting up a system of measurement, must be guided by a judicious mixture of faithfulness to practice, concern for systematic coherence, and arbitrary decision. For a start, we may say that the relevant specifications are those pertaining to the number of predicates in a basis, the number of places in each predicate in a basis, and the reflexivity, symmetry, and transitivity of each predicate. Thus "consists of one 2-place predicate" is a relevant specifica-

---

edition, replacing the text through Section 7 of the first edition. Although many of the revisions and expansions have been explained in articles I have published in the meantime, many others postdate even the most recent of those articles. Before the first edition actually appeared, I had discovered that the complexity-values assigned were appropriate only for irreflexive predicates (and predicates equivalent to them) and that some other predicates are much more complex. The required change was explained in "New Notes on Simplicity", *Journal of Symbolic Logic,* 17 (1952), pp. 189–191. The scale of assigned values has remained substantially the same since then. But a drastic reorganization of the treatment of complexity came with its axiomatization in "Axiomatic Measurement of Simplicity" (hereinafter referred to as *AMS*), *Journal of Philosophy*, 52 (1955), pp. 709–722. This was improved, and various matters pertaining to it and to criticisms and alternative proposals discussed, in "Recent Developments in the Theory of Simplicity" (hereinafter referred to as *RD*), *Philosophy and Phenomenological Research*, 19 (1959), pp. 429–446, (reprinted in *P & P*, pp. 295–318) which contains a bibliography of all the preceding articles. The present version incorporates many further changes and much new material, and though it can hardly be considered complete and final, I think it goes a good deal further toward an articulate and finished system. Some of the earlier articles, especially *AMS* and *RD*, contain useful supplementary and expository material not included here.

Whether or not set forth in the text, proofs are available for all numbered theorems and some numbered theorem-schemata. Where a full proof has not yet been worked out for a numbered theorem-schema, at least a general procedure can be outlined for proving any instance of the schema.

In preparing these sections for the new edition I have been much helped by David Meredith and Howard Burdick.

tion; and so also is "consists of one 2-place symmetric predicate". Indeed, every conjunction or disjunction of relevant specifications, and every negation of a relevant specification, is itself a relevant specification. If K and L are relevant specifications, then KL is a (relevant) subspecification of each, as each is of *K-or-L*. However, we shall find presently that definition of the appropriate varieties of reflexivity and symmetry is a delicate and intricate matter, especially as the number of predicate-places increases; and that ordinary transitivity must be supplanted for our purposes by something stronger.

Now replacement is routine when, given only elementary logic including identity together with the information that a basis b answers relevant specification K, we can define a basis b', answering relevant specification L, and can redefine b from b'. Where finding a b' or recovering b requires anything more, replacement is not routine. In what follows, the replacement under discussion is always routine replacement unless otherwise indicated.

I shall use "$\mathbf{v}b$" as short for "the complexity-value of the basis b". Again, "$\mathbf{v}K$" likewise reads "the complexity-value of the relevant specification K", but since complexity-values are initially assigned to bases rather than to specifications, a further ellipsis is involved here: $\mathbf{v}K$, the number associated with the relevant specification K, is the lowest number such that under the postulates of our calculus, K guarantees that no basis answering K has a higher value.

With these preliminaries, our initial clue can now be formulated as the fundamental postulate of our calculus:

P3.31 *If every basis answering a relevant specification K can be replaced by some basis answering a relevant specification L, then* $\mathbf{v}K \leqq \mathbf{v}L$.

A few elementary theorems that follow obviously from the first postulate or from the definitions of "$\mathbf{v}b$" and "$\mathbf{v}K$" are:

3.311 If every basis answering relevant specification K can be replaced by some basis answering relevant specification L, and every basis answering L can be replaced by some basis answering K, then $\mathbf{v}K = \mathbf{v}L$.

3.312 If a basis b answers relevant specification K, then $\mathbf{v}b \leqq \mathbf{v}K$.[4]

3.313 $\mathbf{v}K \leqq \mathbf{v}(K\text{-}or\text{-}L)$

3.314 $\mathbf{v}(KL) \leqq \mathbf{v}K$

---

[4] Hereafter, "basis", "relevant", and even "specification" will often be omitted as understood. Furthermore, until Section 10 all predicates being discussed are understood to be predicates of individuals; and until Section 8, the complexity under consideration is what we shall there distinguish as *primary* complexity.

3.315 $\mathbf{v}K < \mathbf{v}L \supset \mathbf{v}L = \mathbf{v}(L\text{-}not\text{-}K)$ (called *Subtraction*)

3.316 $\mathbf{v}K \leqq \mathbf{v}L \supset \mathbf{v}(K\text{-}or\text{-}L) = \mathbf{v}L$ (called *Summation*)

3.317 $\mathbf{v}K < \mathbf{v}(K\text{-}or\text{-}L) \supset \mathbf{v}(K\text{-}or\text{-}L) = \mathbf{v}L$.

Also since every specification of 1-place extralogical predicates that is to be admitted as relevant is satisfied by the same predicates as any other, 3.311 will also yield:

3.318 All extralogical 1-place predicates have the same complexity-value.

The objection that the complexity of a 1-place predicate should depend upon its range of application is untenable; for the complementary of a primitive predicate is always immediately available. Any extralogical 1-place predicate effects one cut in the universe of discourse; and where that cut occurs does not matter.

Our fundamental postulate needs supplementation by an auxiliary stipulating that the complexity-values to be assigned are integral and additive:

P3.32 *The complexity-value of every extralogical predicate is a positive integer; and the complexity-value of a basis is the sum of the values of the extralogical predicates in it.*

A predicate is extralogical unless definable solely in terms of elementary logic including identity. Only extralogical predicates are counted as contributing to the complexity-value of a basis; for the logical apparatus may be regarded as constant for all the systems in question. Thus a basis consisting solely of logical predicates has the value 0. Justification for the limitation to integral values is evident enough. The number of predicates in a basis, the number of places in a predicate, the number of different sequential patterns exhibited by the arguments of a predicate, and the number of places or place-sequences with respect to which a predicate is reflexive, symmetric, etc. are all integral. No intermediate numbers are needed; and we leave no gaps—that is, where K specifies only number of predicates and of places in each, and $\mathbf{v}K = n$, there will be for every i less than n some subspecification L (obtained by adding further restrictions) such that $\mathbf{v}L = i$.

The second clause of P3.32 in effect excludes from relevant specifications all information concerning relationships—even of coextensiveness—between predicates in a basis. Ultimately of course (see Section 8 below), we shall have to take into account certain relationships among predicates as well as the analogous relationships within a predicate.

A useful immediate consequence of P3.32 is:

3.321 If the extralogical predicates in a basis b are some but not all of the extralogical predicates in a basis b', then $\mathbf{v}b < \mathbf{v}b'$.

In what follows, use of square brackets will result in a name for the relevant specification abbreviated within them. Thus the full reading of "[2-pl. irref.; two 1-pl.]" is "the relevant specification 'basis consisting[5] of one 2-place irreflexive predicate and two 1-place predicates' "; and the full reading of "**v**[2-pl. sym.]" is "the complexity-value associated with the relevant specification 'basis consisting of one 2-place symmetric predicate' ". But I shall normally give much less cumbersome elliptical readings such as, for the first case, "basis consisting of one 2-place irreflexive and two 1-place predicates"; and, for the second case, "the complexity-value of a basis consisting of one 2-place symmetric predicate"–or even just "the complexity-value of 2-place symmetric predicates". Where "*m*", "*n*", etc. occur within the square brackets, we have in effect a specification-schema that becomes a specification when these letters are supplanted by integers. By way of illustration, consider such elementary theorem schemata as:

3.322 If $n \leqq m$, then $\mathbf{v}\,[n\text{-pl.}] \leqq \mathbf{v}\,[m\text{-pl.}]$

3.323 If $n \leqq m$, then $\mathbf{v}\,[n \text{ preds.}] \leqq \mathbf{v}\,[m \text{ preds.}]$.

The consequent of the former reads: "the complexity-value associated with the relevant specification-schema 'basis consisting of an *n*-place predicate' is no higher than the complexity-value associated with the relevant schema 'basis consisting of an *m*-place predicate' "; or elliptically, "the value of *n*-place predicates is no higher than the value of *m*-place predicates". Both 3.322 and 3.323 follow directly from P3.31 since we can always replace a predicate by one containing more places, and a basis by one containing more predicates. Nothing here denies that some predicates may be more complex than others having more places, or that some bases may be more complex than others containing more predicates.

We have seen (3.318) that all extralogical 1-place predicates have equal value. Pretty obviously they are at least as simple as any other extralogical predicates, and so are entitled to be assigned the lowest value permitted by P3.32. Since logical 1-place predicates of course have in effect the value 0, our third postulate,

---

[5] In addition to the primitives, including identity, of logic; the latter need not be mentioned since they belong to all bases. However, bases may differ in the logical predicates they contain; for a predicate other than a primitive of logic may, even though definable in terms of such primitives, nevertheless be taken as primitive in some systems. See further Section 4 below.

P3.33 $\mathbf{v}$ [1-pl.] = 1,

sets the value of extralogical 1-place predicates at 1. However, the substantive force of this postulate lies in putting the value of 1-place predicates at the bottom of the scale of complexity-values. Assignment of the integer 1 is a dispensable abbreviational convenience; we could have assigned an arbitrary constant, which would then have to be carried through most of our formulae. What matters are such consequences of P 3.33 as:

3.331 If a basis *b* contains any extralogical predicate, then $\mathbf{v}$ [1-pl.] $\leqq \mathbf{v}b$.
3.332 Unless *K* guarantees that all predicates in every basis answering it are logical, then $\mathbf{v}$ [1-pl.] $\leqq \mathbf{v}K$.

4. REFLEXIVITY AND COMPLEXITY

Reflexivity has to do with the occurrence of repetitions within the sequences (belonging to the extension) of a predicate. The familiar properties here are total reflexivity, ordinary reflexivity, and irreflexivity; but this is a haphazard lot. Our purposes demand a much more systematic treatment.

Every pair of elements is either an identity-pair or a diversity-pair. The important properties of reflexivity depend upon the relationship between the given predicate and the logical predicates "=" and "≠", and between the identity-pairs and diversity-pairs of the predicate itself. If all pairs of a predicate are diversity-pairs, the predicate is *irreflexive*; if all are identity-pairs, it may be called *redundant*. A predicate having at least one identity-pair and one diversity-pair is *composite*. If all identity-pairs are pairs of a predicate, it is *totally reflexive*; if all diversity-pairs are pairs of a predicate, it may be called *totally diversive*.

A 2-place predicate that is both redundant and totally reflexive is coextensive with "="; and a 2-place predicate that is both irreflexive and totally diversive is coextensive with "≠". Thus the complexity value of such a predicate is 0 and we have the theorems:

3.411 $\mathbf{v}$ [2-pl. t.r., red.] = 0.

3.412 $\mathbf{v}$ [2-pl. t.d., irref.] = 0.

Again, only inapplicable 2-place predicates–those with null extensions–are both redundant and irreflexive; and only universal 2-place predicates are both totally reflexive and totally diversive. Specification of either combination of properties thus results in logical definability, and hence:

3.413 **v** [2-pl. red., irref.] = 0

3.414 **v** [2-pl. t.r., t.d.] = 0.

A 2-place predicate having no or all diversity-pairs in its extension can always be replaced by a 1-place predicate; and conversely, a 1-place predicate can always be replaced by such a 2-place predicate. Hence:

3.415 **v** [2-pl. red.] = 1

3.416 **v** [2-pl. t.d.] = 1.

And, since a 2-place predicate having no identity-pairs is always interreplaceable with a 2-place predicate having all the identity-pairs,

3.417 **v** [2-pl. irref.] = **v** [2-pl. t.r.].

The principle of Subtraction (3.315), together with 3.411 and 3.415, yields the theorem:

3.418 **v** [2-pl. red., ~t.r.] = 1.

Likewise, the principle of Summation (3.316), together with 3.411 and 3.416, yields:

3.419 **v** [2-pl.-t.r.-red.-or-2-pl.-t.d.] = 1.

Many similar examples are readily constructed.

Every non-composite 2-place predicate is either redundant or irreflexive. Accordingly, **v** [2-pl. non comp.] will be the same as **v** [2-pl. red.] or as **v** [2-pl. irref.], whichever is the higher. Since [2-pl. irref.] is a specification not restricted to logical predicates, its value must (by 3.332) be at least 1, while (by 3.415) **v** [2-pl. red.] = 1. Hence:

3.421 **v** [2-pl. non-comp.] = **v** [2-pl. irref.].

A 2-place predicate not covered by the above theorems may have a value higher than the value of a 2-place irreflexive predicate, but higher by 1 only; for any 2-place predicate can be replaced by a 2-place irreflexive and a 1-place predicate, and vice versa. Thus:

3.422 **v** [2-pl.] = **v** [2-pl. irref.; 1-pl.] = **v** [2-pl. irref.] + 1.

By Subtraction again, these two latest theorems yield:

3.423 **v** [2-pl. comp.] = **v** [2-pl. irref.] + 1.

In summary, the value of a 2-place predicate is at most that of a 2-place irreflexive and a 1-place predicate. The 2-place irreflexive predicate takes care of the diversity-pairs; the 1-place predicate takes care of the identity-pairs. Where the original predicate has no or all diversity-pairs among its sequences, the 2-place irreflexive predicate drops out; where the original predicate has no or all identity-pairs among its sequences, the 1-place predicate drops out. But the 1-place predicate may also drop out when the identity-pairs can be defined in certain ways from the diversity-pairs of the original predicate. For example, suppose that the identity-pairs contain all and only those elements that occur in the diversity-pairs. We may call such a predicate *join-reflexive*; and, admitting this as a relevant specification, we have the theorem:

3.424 **v** [2-pl. j.r.] = **v** [2-pl. irref.] .

Likewise, we may admit as relevantly specifiable: *meet-reflexivity*, where an element occurs in an identity-sequence of the predicate if and only if that element occurs as left component of at least one of the diversity-sequences and as right component of another; *left-reflexivity*, where an element occurs in one of the identity-sequences if and only if it occurs as left component of at least one of the diversity-sequences; and *right-reflexivity*, analogously defined. Then:

3.425 **v** [2-pl. m.r.] = **v** [2-pl. irref.]

3.426 **v** [2-pl. l.r.] = **v** [2-pl. irref.]

3.427 **v** [2-pl. r.r.] = **v** [2-pl. irref.] .

Where any of the above relevant specifications for a basis consisting of a 2-place predicate has a value less than **v** [2-pl. irref.; 1-pl.] , we call such a basis *regular*. Regularity thus enters as relevant by definition as the disjunction of total reflexivity, irreflexivity, redundancy, total diversification, meet-reflexivity, join-reflexivity, left-reflexivity, and right-reflexivity. And by Summation:

3.428 **v** [2-pl. reg.] = **v** [2-pl. irref.] .

Obviously in many other cases the identity-pairs of a predicate can be defined from the diversity-pairs, or vice versa, and we might go on adding further specifications as relevant until we had covered all such cases. But this would hardly be in keeping with our program of spelling out routine replacement. The guiding policy is to refine and supplement familiar classifications enough to meet systematic requirements. Beginning with the standard notions of total reflexivity (inclusion of Identity) and irreflexivity (inclusion in Diversity), we added for balance the parallels: redundancy (inclusion in Identity) and total diversification (inclusion of Diversity). In combination these give us specifications for the several varieties of logical 2-place predicates. We need further, for reasons that will soon appear, a relevant specification for each variety of reflexivity that may arise from the simple composition of two 1-place predicates–that is, from defining "R x,y" as "Px . Qy". Every such "R" will be either irreflexive, redundant, meet-reflexive, left-reflexive, or right-reflexive. Sometimes "R" may have more than one of these properties, and it may also answer other already admitted specifications; but never will it fail to answer one of these five.

Join-reflexivity is not strictly needed, but is admitted as the parallel of meet-reflexivity and as the nearest approximation to ordinary reflexivity that is systematically significant. Join-reflexive predicates are reflexive in the ordinary sense, but so is a predicate having as its pairs:

$$\begin{array}{llll} a,b & a,a & c,c & e,e \\ c,d & b,b & d,d & \end{array}$$

where the identity-pairs cannot be defined from the diversity-pairs or vice versa. Thus ordinary reflexivity has no effect on complexity; admission of it as relevantly specifiable would yield the vapid theorem "**v** [2-pl.] = **v** [2-pl. ord. ref.] = **v** [2-pl. not ord. ref.]".

However, so long as the few required specifications are admitted, decisions about others, while altering evaluation of certain bases, will little affect the general operation of our calculus. For example, if join-reflexivity is not admitted as relevant, then a join-reflexive predicate will have the value of a 2-place irreflexive predicate plus 1; but the value of a 2-place regular predicate, under a narrowed interpretation of "regular", will nevertheless remain (as in 3.428) the same as the value of a 2-place irreflexive predicate.

Compositeness is systematically significant, since all 2-place non-composite predicates are regular, and **v** [2-pl. comp.] is greater by 1 than **v** [2-pl. non-comp.]. But the positive reason for admitting compositeness becomes evident only when the treatment of reflexivity is extended to 1-place predicates. All 1-place predicates are degenerately irreflexive, redundant, join-reflexive, meet-reflexive, left-reflexive, and right-reflexive. A totally reflexive 2-place predicate that is also totally diversive is logical, having universal extension. But whereas 2-place predicates with null extension can be distinguished as those that are both redundant and irreflexive, all applicable and inapplicable 1-place predicates alike are both redundant and irreflexive. However a 1-place predicate is composite if and only if it is applicable. An identity-sequence, in general, embraces no differences; a diversity-sequence, no repetitions; thus monads are degenerately both identity- and diversity-sequences. Hence any applicable 1-place predicate has at least one identity-monad and one diversity-monad, and so is composite. Here as elsewhere, otherwise inconsequential changes in definitions would give different results for the degenerate cases; but my choice is a calculated one. We can now amplify P3.33 by the theorems:

3.431 **v** [1-pl. non-comp.] = 0

3.432 **v** [1-pl. comp.] = 1 (by Subtraction).

With predicates of more than two places, the kinds of reflexivity multiply rapidly. A triad of a 3-place predicate, for example, may be of any of five different patterns: identity (one component), fully variegated[6] (three components), or 2-variegated with the two components distributed in any of three ways:

$$x,x,y \quad x,y,x \quad y,x,x.$$

A 3-place predicate is thus always interreplaceable with a set of five irreflexive predicates; and we have

3.433 **v** [3-pl.] = **v** [3-pl. irref.; three 2-pl. irref.] + 1.

Where a 3-place predicate is regular with respect to one or more of these patterns–that is, where all or no sequences of the patterns in question are among the triads of the predicate, or where the sequences of these patterns are definable from others in certain ways–some of the predicates may be dropped from the equivalent set of irreflexive predicates. For example, where only identity-sequences occur, so that the predicate is redundant, only the 1-place predicate is needed, and:

3.434 **v** [3-pl. red.] = 1.

And if all triads present are of a single 2-variegated pattern, the value is that of a single 2-place irreflexive predicate.

Specifications covering the patterns exhibited by the triads of a predicate have here been tacitly admitted as relevant. And other specifications admitted earlier may be broadened in obvious ways to cover 3-place predicates. A 3-place predicate is composite if it has triads of two or more patterns. Redundancy, total reflexivity, irreflexivity, and total diversification need no further explanation for predicates of any number of places. A 3-place predicate is left-reflexive if its identity-triads contain all and only elements

[6] I use "variegated" in application to sequences and sequential patterns, "diversified" in application to predicates. All fully variegated sequences are among the sequences of a totally diversified predicate.

occurring in the first place of fully-variegated triads of the predicate; and right-reflexivity, meet-reflexivity, and join-reflexivity are defined analogously. We must, of course, add middle-reflexivity. And then among the many other ways that the triads of one pattern of a predicate may be definable in terms of the triads of one or more other patterns, which ways are to be admitted as relevantly specifiable? The number of candidates is bewildering and mounts steeply with each increase in number of places. I shall not attempt to review or decide upon all these cases. Rather I shall suppose them to have been decided in accord with the policy of adding only those needed for covering in sufficient detail the varieties of reflexivity that can arise from simple composition of a predicate out of shorter ones; and shall take regularity with respect to a pattern to be definability, in one of these admitted ways, of the sequences of that pattern from others, which may be called the *prime patterns* of the predicate. This will little affect the general operation of our calculus, and while it leaves decisions to be made in evaluating some bases, most of the many-place primitives encountered in serious systems exhibit a rather high degree of regularity of obvious kinds.

A redundant predicate, regardless of its number of places, is interreplaceable with a 1-place predicate; and 3.415 and 3.434 are instances of the theorem-schema:

3.435 **v** [n-pl. red.] = 1.

A 4-place predicate may have sequences of fifteen different patterns. The analogue of 3.433 is:

3.436 **v** [4-pl.] = **v** [4-pl. irref.; six 3-pl. irref.; seven 2-pl. irref.] + 1.

The general schema is:

3.437 **v** [n-pl.] = **v** [n-pl. irref.; h_1 $n-1$-pl. irref.; ... h_{n-2} 2-pl. irref.] + 1,

where for each j, the number h_j of $n-j$-pl. irreflexive predicates is

$$\sum_{r=0}^{n-j} (-1)^r \frac{(n-j-r)^n}{(n-j-r)!\,r!}.$$ [7]

For any n-place predicate that is regular in one way or another, some predicates can be dropped from the equivalent set of irreflexive predicates. But the value of any relevant specification can be equated with the value of some specification of bases consisting entirely of irreflexive predicates. Hence if

[7] This formula was derived for me from Fine's formula (see 3.743 below) by Patricia Savage.

values are determined for irreflexive predicates of all kinds, the value of any basis may be computed. Let us therefore confine our attention for a time to irreflexive predicates.

5. TRANSITIVITY, SELF-COMPLETENESS, AND COMPLEXITY

A 2-place predicate R is transitive in the usual sense if and only if it satisfies the condition:

$$(x)\,(y)\,(z)\,(\mathrm{R}\,x,y\,.\,\mathrm{R}\,y,z\,.\supset \mathrm{R}\,x,z).$$

This property does not affect complexity, and so is not relevantly specifiable. But a kindred property is highly relevant. We remove the requirement that the first two pairs be linked, and add the requirement that all three be diversity-pairs. Predicates answering the resulting specification:

$$(x)\,(y)\,(z)\,(w)\,(\mathrm{R}\,x,y\,.\,\mathrm{R}\,w,z\,.\,x \neq y\,.\,w \neq z\,.\,x \neq z\,.\supset \mathrm{R}\,x,z),$$

may be called *self-complete*. Were the "≠" clauses omitted, only meet-reflexive predicates would be self-complete. With them inserted, we can ignore identity-pairs in determining self-completeness, which is thus rendered largely independent of reflexivity-properties. In particular, an irreflexive predicate may be self-complete even though the first components of some of its pairs are the second components of others.[8]

The relevance of self-completeness is immediately evident from the fact that 2-place irreflexive self-complete predicates–unlike 2-place predicates in general, irreflexive or not–can always be replaced by two 1-place predicates. We can now show, furthermore, that **v** [2-pl.irref., s.c.] = 2. Consideration of the route of proof leading to this central theorem will explain the reasons for some of our earlier decisions. We begin by showing:

3.511 **v** [2-pl. irref., s.c.] $\leqq 2$.

Given that "R" is 2-place, irreflexive, and self-complete, we can define:

(*a*) $\mathrm{P}x =_{\mathrm{df}} (\exists y)\,(\mathrm{R}\,x,y)$

(*b*) $\mathrm{Q}x =_{\mathrm{df}} (\exists y)\,(\mathrm{R}\,y,x)$

and recover "R" from these by:

(*c*) $\mathrm{R}\,x,y =_{\mathrm{df}} \mathrm{P}x\,.\,\mathrm{Q}y\,.\,x \neq y.$

---

[8] Totally diversified predicates are of course self-complete; so also, vacuously, are redundant predicates.

Hence by P3.31, **v**R is not greater than **v** [two 1-pl.]; and by P3.32 and P3.33, **v** [two 1-pl.] = 2.

Proof of the needed converse of 3.511 is much less easy; for we cannot always replace two 1-place predicates by a 2-place irreflexive self-complete predicate. In the first place, if of the 1-place predicates "P" but not "Q" is non-composite (i.e. inapplicable), and if "R" is defined by using (*c*) above, then "R" will be inapplicable and the attempt to recover "Q" by using (*b*) will fail, yielding an inapplicable predicate. In the second place, even if the 1-place predicates are specified at the start to be composite (applicable), we cannot always replace them by a 2-place irreflexive self-complete predicate. For if "P" applies to one element only, and "Q" also applies to that element and "R" is defined by (*c*), then (*a*) will not recover "P". Hence proof of the converse of 3.511 must be more devious.

We can readily show:

3.512 $\quad 2 \leqq$ **v** [2-pl. reg., s.c.];

for given that "P" and "Q" are 1-place composite predicates we can define:

$$\mathrm{R}\,x,y =_{\mathrm{df}} \mathrm{P}x \,.\, \mathrm{Q}y,$$

and recover "P" and "Q" by (*a*) and (*b*); and from what has been said about regularity, we know that "R" will be regular.

Next we need to establish:

3.513 $\quad$ **v** [2-pl. irref., s.c.] = **v** [2-pl. reg., s.c.];

but this takes some trouble; for given that a 2-place predicate "R" is regular and self-complete, we cannot automatically replace it by one that is irreflexive and self-complete. We have to know what sort of regularity "R" has. Thus our proof takes a number of steps paralleling those leading to 3.428. Each of the first five steps follows by simple interreplaceability (i.e., by two applications of P3.31).

(1)	**v** [2-pl. irref., s.c.]	=	**v** [2-pl. l.r., s.c.]
(2)		=	**v** [2-pl. r.r., s.c.]
(3)		=	**v** [2-pl. m.r., s.c.]
(4)		=	**v** [2-pl. j.r., s.c.]
(5)		=	**v** [2-pl. t.r., s.c.]

The next two steps follow from 3.415 and 3.416 and the fact that our definition of self-completeness makes all redundant and all totally diversive predicates self-complete.

(6) $1 = \mathbf{v}$ [2-pl. red., s.c.]

(7) $1 = \mathbf{v}$ [2-pl. t.d., s.c.]

The disjunction of all seven right-hand specifications is equivalent to

[2-pl. reg., s.c.].

The value of each of the seven has here been shown to be either 1 or **v** [2-pl. irref., s.c.]. Thus by Summation, this disjunction must have whichever of either is the higher of these values. We do not yet know the numerical value of "2-pl. irref., s.c." but we do know, by 3.332, that it cannot be less than 1. Hence we have completed proof of 3.513.

From 3.512 and 3.513 together, we get immediately the converse of 3.511:

3.514 $2 \leqq \mathbf{v}$ [2-pl. irref., s.c.];

and from 3.511 and 3.514 together we arrive at the wanted equation:

3.515 **v** [2-pl. irref., s.c.] $= 2$.

In general, an irreflexive *n*-place predicate is self-complete if it has among its sequences every *n*-variegated sequence of components taken in order from its other sequences. Theorem 3.515 is an instance of the theorem-schema:

3.516 **v** [*n*-pl. irref., s.c.] $= n$.

Bothersome complications are avoided by letting the self-completeness of a non-irreflexive many-place predicate rest entirely upon the self-completeness of the irreflexive predicates in its equivalent set. One-place predicates are degenerately self-complete.

An irreflexive predicate that is not self-complete with respect to all its places may still be self-complete with respect to mutually separate and jointly exhaustive combinations of its places. For example, suppose that an irreflexive 3-place predicate is such that for every three different elements x and y and z, if some triad of the predicate has x and z as end-components, and some triad has y as middle component then x,y,z is a triad of the predicate. This predicate is self-complete with respect to its middle and the combination of its end places–or briefly, (1,3) (2) self-complete. Its triads might be:

$a,b,c \quad a,s,c$

$r,s,t \quad r,b,t.$

In this case, the predicate will not be self-complete with respect to its three places taken severally; for such triads as *a, b, t* are missing.

The combination of places with respect to which a predicate is self-complete may be called its *partitions*. Any combination of partitions–and thus trivially the whole predicate–is also a partition.[9] A *minimal* partition contains no other partition. A predicate whose minimal partitions are all single places is also self-complete with respect to any grosser partitioning; but, as shown by the 3-place predicate just described, the converse does not hold.

To generalize further upon 3.516, an *n*-place irreflexive predicate has the same value as a set of irreflexive predicates corresponding to the minimal partitions of the predicate–a set, that is, containing for each minimal partition an irreflexive predicate of the same number of places as that partition:

3.517 **v** [*n*-pl. irref. with minimal partitions of *k*, *k*′, etc. places] = **v** [*k*-pl. irref.; *k*′-pl. irref.; etc.].

If we think of a partition as having the complexity-value of its corresponding predicate in such a set, the value of the *n*-place predicate here is the sum of the values of the minimal partitions. But as we shall see, where other relevant properties are specified the value of the whole may be less than this sum.

The *degree of self-completeness*, **sc**, of a predicate is simply the number of divisions between minimal partitions–or the number of minimal partitions minus 1. All one-partition predicates, hence all 1-place predicates, have **sc** = 0. A 3-place predicate has **sc** = 1 if it has a 1-place and a 2-place minimal partition, **sc** = 2 if it has three 1-place partitions. The maximum degree of self-completeness for an *n*-place predicate is of course $n-1$.

6. SYMMETRY AND COMPLEXITY

Reflexivity and self-completeness enable us in effect to take apart and put together again certain predicates and so equate them in complexity-value with sets of other predicates. For any reflexive predicate we find an equivalent set of irreflexive predicates. For any irreflexive self-complete predicate we find an equivalent set of shorter irreflexive predicates. Symmetry of an

[9] Thus any predicate is technically and trivially self-complete with respect to itself as a whole, so that attribution of self-completeness is significant relative to a given partitioning only. In what follows "not self-complete" or "~s.c." will mean "not self-complete except with respect to itself as a whole", while "self-complete" or "s.c." unmodified will normally mean "self-complete with respect to all places taken severally"–i.e., having single places as minimal partitions. Where other varieties of self-completeness are in question, they will be expressly indicated.

irreflexive self-complete predicate has the effect of making readily calculable eliminations from the latter set.

A 2-place predicate is symmetric if the reverse of each of its pairs is also among its pairs. We have seen (3.515) that 2-place irreflexive, self-complete predicates have the value 2; but 2-place irreflexive, self-complete, symmetric predicates have the value 1.

3.611 **v** [2-pl. irref., s.c., sym.] = 1.

Proof is easy. Since such a predicate is always replaceable by a 1-place predicate by means of the definitions:

$Px =_{df} (\exists y)(R\, x,y)$
$R\, x,y =_{df} Px \,.\, Py \,.\, x \neq y,$

v [2-pl. irref., s.c. sym.] is less than or equal to 1; and since the specification does not guarantee that all predicates answering it are logical, the value (by 3.332) cannot be less than 1.

In general, an *n*-place predicate is symmetric if every permutation of each of its sequences is also among its sequences. Quite obviously any such predicate that is also irreflexive and self-complete can always be replaced by a 1-place predicate; and a proof like that just given establishes:

3.612 **v** [*n*-pl. irref., s.c., sym.] = 1.

An *n*-place predicate may, however, be symmetric with respect to some rather than all its places. For example, a 3-place predicate "Q" such that

$(x)\,(y)\,(z)\,(Q\,x,y,z \supset Q\,x,z,y)$

is symmetric with respect to its second and third places–or, briefly, (2) (3) symmetric–but not necessarily with respect to all. An irreflexive predicate of this kind can always be replaced by a basis consisting of a 2-place irreflexive self-complete symmetric predicate (value 1) and a 1-place predicate, and can be shown–by a roundabout proof like that for 3.515–to have the value 2.

The *degree of symmetry*, **sy**, of an *n*-place predicate symmetric with respect to k of its places is $k-1$; thus the maximum degree of symmetry is $n-1$. For 2-place symmetric predicates, **sy** = 1; for 3-place symmetric predicates, **sy** = 2; and for partially symmetric 3-place predicates (e.g., "Q" just above), **sy** = 1. A 3-place predicate symmetric with respect to each of two pairs of places is symmetric with respect to all three places,[10] and has

[10] On the other hand, a 3-place predicate that meets the condition "$Rx,y,z \supset Rz,x,y$" is

sy $= 2$. A 1-place predicate is degenerately symmetric with respect to its only place, thus having **sy** $= 0$.

A predicate of more than three places may be symmetric with respect to each of two or more disjoint sets of its places. The degree of symmetry for each such symmetric subset of k places is $k-1$; and the degree of symmetry for the whole predicate is the sum of the degrees of its maximal symmetric subsets–i.e., those contained in no others. For example, a 4-place predicate may be (1) (2) and (3) (4) symmetric without being symmetric with respect to all or to any three or to any other two of its places; here **sy** $= 1 + 1 = 2$. An asymmetric n-place predicate is in effect symmetric only with respect to each of its places taken singly; and has **sy** $= n(1-1) = 0$.

If an n-place predicate is symmetric with respect to all in a certain consecutive or nonconsecutive subset of its places, but is not symmetric with respect to these places together with any others, then all places in such a subset must have exactly the same occupants. In other words, if some sequence of the predicate has x in place of p of the subset, then for each other place p' of the subset, some sequence of the predicate has x at p'. Obviously such coextensivity of places may occur otherwise than as the result of symmetry, but what matters in the present context is how far such coextensivity is guaranteed by a specification of symmetry. The maximal subsets of places guaranteed coextensive by the symmetry of a predicate may be called *segments* of a predicate. The degree of symmetry of an n-place predicate with k segments is then $n-k$.[11]

The need for this method of determining **sy** becomes evident when we encounter higher-level symmetries. Consider, for example, a 4-place predicate "R" that is (1,2) (3,4) symmetric, meeting the condition:

$$(x)(y)(z)(w)(\mathrm{R}\, x,y,z,w \supset \mathrm{R}\, z,w,x,y).$$

not therefore symmetric with respect to any two of its places. For this reason, such a cyclic shift, although often considered a variety of symmetry, is not admitted as a relevant property for our purposes.

[11] The definition of **sy** here given differs somewhat from that in *AMS* and *RD*. Not only does the earlier version become extremely complicated in some cases where higher-level symmetries are involved, but also **sy** as so defined varies for higher-level symmetries with the degree of selfcompleteness of the place-subsequences with respect to which the symmetry obtains. Under the present definition, **sy** is readily calculable even where several intricately related symmetries at different levels are involved; and **sy** (which now never exceeds $n-1$ for an n-place predicate) remains constant irrespective of variations in self-completeness of the place-subsequences in question–the effect of combining symmetry and self-completeness now being provided for (see the final paragraph of the present section) by a combination-factor. The notion of 'segment' used above was used in a somewhat different way in the first edition of this book but dropped in my most recent articles on simplicity.

This predicate may happen not to be symmetric with respect to any two of its places; for though it clearly ties places 1 and 3 together, and also ties 2 and 4 together, still if its only sequences are *a,b,c,d* and *c,d,a,b*, "R" is neither (1) (3) symmetric (since it lacks, for example, *a,d, c,b*) nor (2) (4) symmetric (since it lacks, for example, *c,b,a,d*). What, then, is the degree of symmetry of "R"? The symmetry specified guarantees coextensivity of places 1 and 3, and also of places 2 and 4. Thus "R" has two segments; and **sy**R = $n-2=2$. Although the pattern of symmetry is quite different from that of a predicate that is (1) (2) and (3) (4) symmetric, the degree of symmetry is the same, and so is the effect of the symmetry upon the complexity-value of an irreflexive self-complete predicate: reduction by exactly the degree of symmetry. For while the complexity-value of 4-place irreflexive self-complete predicates is 4 (by 3.516), the complexity-value of 4-place irreflexive self-complete predicates that are (1,2) (3,4) symmetric is 2. Proof depends on the fact that such a predicate is always replaceable by a 2-place irreflexive self-complete predicate, and that a 2-place irreflexive self-complete predicate is always replaceable by a 4-place self-complete (1,2) (3,4) symmetric predicate that is regular with respect to all its not-fully-variegated patterns.

Some predicates embrace lower-level symmetries within higher-level ones. The degree of symmetry of a 4-place predicate that is (1) (2) and (3) (4) and (1,2) (3,4) symmetric is not the sum $(1+1+2)$ of these three symmetries. Rather, since the specification of this layered symmetry guarantees all four places coextensive, the predicate has but one segment, and **sy** $= 4-1=3$. Here again, the reduction in complexity if the predicate is irreflexive and self-complete is exactly the degree of symmetry, since such a predicate will also be symmetric with respect to all its places and replaceable by a 1-place predicate. The layered symmetry described does not, however, always thus imply full symmetry; unless the predicate is self-complete, some permutations of its sequences may be missing. But what counts for our purposes here is not presence of all permutations but symmetry-guaranteed coextensivity of all places. In this respect, hence in number of segments and degree of symmetry, and also in effect upon the complexity-value of irreflexive self-complete predicates, this layered symmetry is equal to full symmetry. Nevertheless, the difference between saying that an n-place predicate is (fully) symmetric and saying that it has the maximal degree of symmetry (or has **sy** $= n-1$) will need to be observed in some later contexts.

In all cases dealt with so far, the value of an irreflexive self-complete predicate is reduced by exactly its degree of symmetry, and these are but instances of the general theorem-schema:

3.613 **v** [n-pl. irref., s.c.] $-$ **v** [n-pl. irref., s.c., **sy**$=k$] $=k$;

and its corollary;

3.614 **v** [n-pl. irref., s.c., **sy**$=h$] − **v** [n-pl. irref., s.c., **sy**$=k$] $= k-h$.

The proofs are laborious in detail, and I shall omit them; but the methods employed are already familiar.

These theorems cover only fully self-complete predicates (with **sc** $= n-1$); and for such predicates, numerical complexity-values are determined by the postulates and definitions already given. The relative if not numerical effect upon complexity-value of symmetry in combination with only partial self-completeness can also be determined for all predicates symmetric with respect to partitions only. The only cases of this sort calling for consideration are those where **sy** is *between* 0 and $n-1$; for predicates with **sy** $= 0$ have already been taken care of, and predicates with **sy** $= n-1$ that are symmetric with respect to partitions only are fully self-complete. If a predicate is symmetric with respect to two or more disjoint partitions, these partitions will all be exactly alike and have equal complexity-value; and the predicate can be replaced by and can be shown to have the same value as a single one of them:

3.615 **v**[n-pl. irref., sym. with respect to k disjoint h-valued partitions] $= h$.

Layered symmetries can be accommodated by iterated application of this theorem; for example, if an 8-place predicate is (1,2) (3,4) (5,6) (7,8) self-complete and is (1,2) (3,4) and (5,6) (7,8) and (1,2,3,4) (5,6,7,8) symmetric, then it has the value of the partition consisting of (e.g.) its first four places, and this in turn has the value of the partition consisting of (e.g.) its first two places.

Where symmetry is not with respect to partitions only but is or contains symmetry with respect to the places within partitions, then whatever (not as yet determined) effect the latter symmetry may have on the value of the partitions will be reflected in the h of 3.615.

Now compare, say, 4-place irreflexive (1,2) (3,4) self-complete predicates that are not symmetric with those that are also (1,2) (3,4) symmetric. The difference in **sy** is 2. But since (by 3.517) the complexity of predicates of the former sort is 2**v** [2-pl. irref.] while (by 3.615) the complexity of predicates of the latter sort is **v** [2-pl. irref.], the difference in *values* is **v** [2-pl. irref.] – and this we have not yet determined numerically. Thus where partial symmetry is combined with partial self-completeness, and the self-completeness remains fixed while the symmetry varies, the difference in values may not always equal the difference in **sy**. This discrepancy may be accounted a *combination-factor*. This combination-factor for a given irreflexive predicate

"P" can be determined by comparing "P" with an irreflexive predicate "Q", identical with "P" in number of places and in self-completeness but without any symmetry; then **cf**P = (**v**Q−**v**P)−**sy**P. In the example just given, the combination-factor for irreflexive predicates that are both (1,2) (3,4) self-complete and (1,2) (3,4) symmetric is whatever may be the difference between **v** [2-pl. irref.] and 2. Where **sy** = 0 or **sc** = $n-1$, the combination-factor is obviously 0.

7. FINAL FORMULAE FOR PRIMARY COMPLEXITY

We now have a way of finding for any predicate a minimal equivalent set consisting solely of irreflexive predicates, and for any irreflexive predicate a minimal equivalent set consisting solely of predicates without any self-completeness. What remains is to find a way of evaluating such one-partition predicates, and hence also the minimal partitions of other predicates.

The first step is to determine the effect of symmetry on such predicates and partitions. When a predicate is fully self-complete we have seen how to measure directly and numerically the effect of symmetry upon complexity-value. But where symmetry is within minimal partitions we cannot determine its effect by means already at hand. We proceed more deviously by applying to these inaccessible cases our findings in the accessible cases; that is, we postulate that the reduction in value resulting from symmetry where self-completeness is 0 is the same as where self-completeness is maximal. Thus although we here postulate rather than prove, what we have already proven points the way to what we postulate.[12] Theorems 3.613 and 3.614 established the difference in values between n-place irreflexive self-complete predicates to be the negative of the difference in degrees of symmetry. Our new postulate and a corollary provide the parallel for predicates without self-completeness:

P3.71 **v** [n-pl. irref., **sc** = 0] − **v** [n-pl. irref., **sc** = 0, **sy** = k] = k

3.711 **v** [n-pl. irref. **sc** = 0, **sy** = h] − **v** [n-pl. irref., **sc** = 0, **sy** = k] = $k-h$.

The postulate derives support from its consequences as well as from its antecedents. For one thing, the combination-factor should obviously turn out to be 0 whenever either of the two participating factors is 0. That **cf** is 0 where **sy** is 0 follows from the definition of **cf**; our new postulate sets **cf** at 0

---

[12] Here, and again below, adoption of a postulate has been postponed until what has been derived from other postulates points a direction and provides some rationale. Such a dynamic development, with successive additions to the foundations being guided by the results of prior assumptions, may not be characteristic of axiomatizations of mathematical theories but seems peculiarly well suited to our undertaking here.

where **sc** is 0. We saw also that **cf** is 0 for every fully self-complete irreflexive predicate. The new postulate enables us to show that **cf** is 0 for every fully symmetric irreflexive predicate,[13] since such a predicate will always either be fully self-complete or have **sc** = 0. The combination-factor thus can exceed 0 only where partial but not full self-completeness and partial but not full symmetry are combined, though it will not exceed 0 in all such cases.

Since a 2-place predicate has either maximal or zero self-completeness, and if symmetric has **sy** = 1, we have:

3.712 **v** [2-pl. irref.] − **v** [2-pl. irref., sym.] = 1.

And, since a fully symmetric n-place predicate has **sy** = $n-1$:

3.713 **v** [n-pl. irref., **sc** = 0] − **v** [n-pl. irref., **sc** = 0, sym.] = $n-1$.

Though we can now measure the *effect* of symmetry upon the complexity of irreflexive one-partition predicates, we cannot yet measure the complexity of any one-partition predicate having more than one place. The next step is to determine the effect of self-completeness on complexity. We know that an n-place irreflexive self-complete predicate has the value n, but how much does n differ from the value of an n-place irreflexive predicate with **sc** = 0? Once this difference is numerically determined, then of course the value of an n-place irreflexive one-partition predicate will also be numerically determined as n plus this difference.

As before, we look first to see how much can be proved. Unfortunately, we cannot show that when **sy** = 0, the difference in value is the negative of the difference in **sc**; but we can prove an equally pertinent, though weaker, theorem relating the ratio of differences in complexity-values to the ratio of differences in self-completeness. Let the as yet undetermined value of n-place irreflexive predicates (or partitions) without symmetry or self-completeness be represented by "x_n". Then, by virtue of 3.517, the value of any n-place irreflexive predicate with **sy** = 0 is readily expressed in the form illustrated by "$3 + x_2 + 2x_3$". If all the unknowns cancel out of a given function of given values so expressed, let us call that function of these values *directly numerical.* Now consider four different 6-place irreflexive predicates with **sy** = 0 but with self-completeness as follows:

"P" is (1,2) (3,4) (5,6) self-complete; hence **v**P = $3x_2$.
"Q" is (1) (2) (3) (4) (5) (6) self-complete; hence **v**Q = 6.

[13] But not for every irreflexive predicate that has maximal **sy**; e.g., for a 4-place irreflexive predicate that is (1) (2) and (3) (4) and (1,2) (3,4) symmetric and is (1,2) (3,4) self-complete, the combination-factor will turn out to be 2.

"R" is (1,2) (3,4) (5) (6) self-complete; hence $\mathbf{v}\mathrm{R} = 2x_2 + 2$.

"S" is (1,2) (3) (4) (5) (6) self-complete; hence $\mathbf{v}\mathrm{S} = x_2 + 4$.

The ratio of the difference in values between "P" and "Q" to the difference in values between "R" and "S" is directly numerical; for

$$\frac{3x_2-6}{2x_2+2-(x_2+4)} = \frac{3x_2-6}{x_2-2} = \frac{3(x_2-2)}{x_2-2} = 3.$$

Furthermore, this is exactly the ratio of the difference in **sc** between "Q" and "P" to the difference in **sc** between "S" and "R"; for

$$\frac{\mathbf{sc}\mathrm{Q} - \mathbf{sc}\mathrm{P}}{\mathbf{sc}\mathrm{S} - \mathbf{sc}\mathrm{R}} = \frac{5-2}{4-3} = 3.$$

We have here an instance of the significant general theorem:[14]

3.714 If K, L, M, and N are relevant specifications of bases consisting of single n-place irreflexive predicates with $\mathbf{sy} = 0$, but $\mathbf{sc}K \neq \mathbf{sc}L$, and $\mathbf{sc}M \neq \mathbf{sc}N$, then if $\frac{\mathbf{v}K - \mathbf{v}L}{\mathbf{v}M - \mathbf{v}N}$ is directly numerical, $\frac{\mathbf{v}K - \mathbf{v}L}{\mathbf{v}M - \mathbf{v}N} = \frac{\mathbf{sc}L - \mathbf{sc}K}{\mathbf{sc}N - \mathbf{sc}M}$.

That is, in all cases meeting the stated requirements, the two ratios are equal wherever the first is numerically determinable by means already at hand. But this ratio is not thus numerically determinable in all cases meeting these requirements.

The next step is to project the result for the accessible cases onto the inaccessible cases. Just as P3.71 parallels and complements 3.613, so P3.72 parallels and complements 3.714.

P3.72 *If K, L, M and N are relevant specifications of bases consisting of single n-place irreflexive predicates with* $\mathbf{sy} = 0$, *but* $\mathbf{sc}K \neq \mathbf{sc}L$, *and* $\mathbf{sc}M \neq \mathbf{sc}N$, *then if* $\frac{\mathbf{v}K - \mathbf{v}L}{\mathbf{v}M - \mathbf{v}N}$ *is not directly numerical,* $\frac{\mathbf{v}K - \mathbf{v}L}{\mathbf{v}M - \mathbf{v}N} = \frac{\mathbf{sc}L - \mathbf{sc}K}{\mathbf{sc}N - \mathbf{sc}M}$.

Hence the two ratios are now established as equal for all cases meeting the requirements in question.

To illustrate the force of P3.72, let K, L, M, and N specify bases consisting

[14] A simple and elegant proof of this theorem has been provided by David Meredith (see footnote 3). His proof, indeed, establishes a much stronger theorem, without limitation to equality in number of places among the predicates involved. But the weaker consequence stated above is of greater interest here; for projection of the stronger theorem to all cases where the left-hand ratio is directly numerical would be manifestly absurd while projection (see P3.72 below) of the weaker theorem is both reasonable and sufficient for our purposes.

of single 3-place irreflexive predicates with **sy** = 0, and let K specify (1,2) (3) self-completeness, L specify (1) (2) (3) self-completeness, M specify zero self-completeness, and N be the same as L. Then

$$\frac{\mathbf{v}K-\mathbf{v}L}{\mathbf{v}M-\mathbf{v}N}=\frac{(x_2+1)-3}{x_3-3}=\frac{x_2-2}{x_3-3},$$

while $\frac{\mathbf{sc}L-\mathbf{sc}K}{\mathbf{sc}N-\mathbf{sc}M}=\frac{2-1}{2-0}=\tfrac{1}{2}$.

By P3.72, the two ratios are equal:

$$\frac{x_2-2}{x_3-3}=\tfrac{1}{2};\text{ hence } x_3-3=2x_2-4;\text{ and } x_3=2x_2-1.$$

Thus we have an evaluation of x_3 in terms of x_2. In like manner, we can show that $x_4=2x_3-x_2$, and hence that $x_4=4x_2-2-x_2=3x_2-2$. And in general, $x_n=(n-1)\,x_2-(n-2)$. Further, since by P3.71 **v** [2-pl. irref., ~s.c., sym.] is less than **v** [2-pl. irref., ~ s.c.], and so by Subtraction x_2 or **v** [2-pl. irref., ~s.c., ~sym.] = **v** [2-pl. irref., ~s.c.], the foregoing abbreviated formulae become the theorems:

3.721 **v** [3-pl. irref., **sy** = 0, **sc** = 0] = 2**v** [2-pl. irref., ~s.c.] – 1

3.722 **v** [4-pl. irref., **sy** = 0, **sc** = 0] = 3**v** [2-pl. irref., ~s.c.] – 2

3.723 **v** [n-pl. irref., **sy** = 0, **sc** = 0] = (n – 1) **v** [2-pl. irref., ~s.c.] – (n–2).

Now the value of any predicate is determinable relative to at most a single unknown: **v** [2-pl. irref., ~s.c.].[15] We have seen earlier how to find for every predicate a minimal equivalent set of irreflexive predicates, and for each irreflexive predicate a minimal equivalent set of irreflexive predicates without self-completeness. For any in the latter set that are without symmetry, the values are given by 3.723, and for any with symmetry these values need only be reduced by the degree of symmetry. Thus all that is needed for a fully quantitative calculus is to fix numerically (i.e., relative to **v** [1-pl.]) the value of 2-pl. irreflexive not-self-complete predicates.

Since **v** [2-pl. irref., ~s.c., sym.] is at least 1, and is 1 less than **v** [2-pl. irref., ~s.c.], the latter (i.e., x_2) must be at least 2. Assignment of any integer not less than 2 is compatible with all that has so far been proved. The question which integer to assign amounts to the question what effect self-completeness has on the value of a 2-place irreflexive predicate. For since we

[15] Or alternatively, of course, relative to any of many other unknowns; e.g., **v** [2-pl. irref., ~s.c., sym.], **v** [3-pl. irref., ~s.c.], etc.

know that **v** [2-pl. irref., s.c.] $= 2$, then self-completeness here (where **sc** $=$ 1) has the effect of reducing the value by $x_2 - 2$.

To set x_2 at 2 would thus be to set the reducing effect of self-completeness at 0. A non-self-complete 2-place (or *n*-place) irreflexive predicate would thus have the same value as two (or *n*) 1-place predicates; and a non-self-complete *n*-place irreflexive symmetric predicate would have the same value as a single 1-place predicate. These results are clearly inacceptable. A non-self-complete 2-place predicate "R" is more complex than a self-complete one–hence than two 1-place predicates–in that "R" joins certain among the occupants of its first place with certain among the occupants of its second. In other words, if "P" and "Q" are 1-place predicates applying respectively to all and only the occupants of the first and to all and only the occupants of the second place of "R", then "P" and "Q" select certain elements from the universe of discourse, while "R" further selects some among all pairs x, y such that $Px . Qy$. Now my argument here is not at all that "R" is more complex because not always replaceable by two 1-place predicates; for this argument quickly leads to contradiction. Such a 2-place predicate cannot, indeed, always be replaced by two or by any number of 1-place predicates,[16] but neither can two or any number of 1-place predicates always be replaced by such a non-self-complete 2-place predicate. Hence if mere lack of replaceability of every basis answering one specification by some basis answering another were sufficient grounds for assigning a higher value to the first specification, we should be forced to conclude that a 2-place non-self-complete irreflexive predicate is both more and less complex than any number of 1-place predicates. Far from using any such argument, I am merely suggesting that the additional selectivity of the 2-place predicate here over that of two 1-place predicates calls for assigning x_2 a value greater than 2.

But how much greater? All that has been said so far is compatible with

---

[16] This follows from an important theorem due to Lars Svenonius. In "Definability and Simplicity", *Journal of Symbolic Logic*, 20 (1955), pp. 235–250, he has in effect proved that a basis containing at least one *n*-place predicate of individuals can*not* always be replaced by some basis containing any finite number of fewer-than-*n*-place predicates of individuals. This once and for all refutes objections to my calculus of complexity on the ground that every basis can always be replaced by a basis consisting solely of 2-place predicates or even of a single 2-place predicate. Any such replacement–achieved, say, by Quine's device in "Reduction to a Dyadic Predicate", *Journal of Symbolic Logic*, 19 (1954), pp. 180–182; or Tarski's in "A General Theorem Concerning the Reduction of Primitive Notions" and "On the Reduction of Generators in Relation-rings" (abstracts), same journal, same volume, pp. 158–159–either yields predicates of higher type or depends upon existential or other assumptions not permitted in establishing always-replaceability in the sense I have defined. For a brief discussion of this theorem and two other pertinent results obtained by Svenonius, see *RD*, cited in note 3 above.

setting $x_2 - 2$ at any positive value. A general policy of assigning no higher value than is required recommends itself as less arbitrary than any other policy that yields a unique decision. And setting $x_2 - 2$ at 1 has a good deal else in its favor, for it gives **sc** the same effect as **sy** upon complexity-value; that is, the difference in value between *n*-place irreflexive predicates that are without symmetry, or that are fully symmetric, becomes the exact negative of the difference in **sc**. Furthermore, as is surely desirable, the combination-factor in a partially self-complete, partially symmetric, *n*-place irreflexive predicate "P" can be computed either (as stipulated in the definition of **cf**) by comparing "P" with an *n*-place irreflexive predicate having the same self-completeness and no symmetry, or alternatively by comparing "P" with an *n*-place irreflexive predicate having the same symmetry and no self-completeness. Finally, setting $x_2 - 2$ at 1 leaves no gaps: as self-completeness ranges from $n-1$ to 0, the value of extralogical *n*-place irreflexive predicates with full symmetry ranges over all integers from 1 to *n*, and for such predicates without symmetry, over all integers from *n* to $2n-1$. These consequences, some of them to be embodied in theorems, are considerations leading to the final postulate for primary complexity. Among many equivalent versions, I choose the following:

P3.73 **v** [2-pl. irref.] – **v** [2-pl. irref., s.c.] = 1.

This obviously implies that $x_2 = 3$, and therefore that $x_2 - 2 = 1$. Nothing further is needed for the fully quantitative calculus of primary complexity now to be summarized.

Proofs of the theorems

3.731 **v** [2-pl. irref.] = 3

3.732 **v** [2-pl. irref., sym.] = 2

are obvious; and we have seen in 3.515 and 3.611 that

v [2-pl. irref., s.c.] = 2

v [2-pl. irref., s.c., sym.] = 1.

Furthermore, from 3.731 and 3.515 we obtain by Subtraction:

3.733 **v** [2-pl. irref., ~s.c.] = 3.

Using this to put "3" for "**v** [2-pl. irref., ~s.c.]" in 3.723 gives:

3.734 **v** [*n*-pl. irref., **sc** = 0, **sy** = 0] = $2n-1$;

and applying 3.711 to this yields:

3.735 $\mathbf{v}$ [n-pl. irref., $\mathbf{sc} = 0$, $\mathbf{sy} = k$] $= (2n-1) - k$.

This is a key theorem. Since we already had means of finding for every predicate, and hence for every basis, an equally valued set of one-partition predicates; and since 3.735 provides a numerical complexity-value for each one-partition predicate, *we can now determine numerically the complexity-value of any basis whatever.*

Some further useful theorems follow. Elementary mathematical considerations enable us to show that the complexity-value of an n-place irreflexive predicate without symmetry is reduced by the degree of self-completeness; thus, as a parallel of 3.735 we have:

3.736 $\mathbf{v}$ [n-pl. irref., $\mathbf{sc} = h$, $\mathbf{sy} = 0$] $= (2n-1) - h$.

Also 3.516 and 3.613 together yield

3.737 $\mathbf{v}$ [n-pl. irref., $\mathbf{sy} = k$, s.c.] $= n - k$;

while the parallel:

3.738 $\mathbf{v}$ [n-pl. irref., $\mathbf{sc} = h$, sym.] $= n - h$

follows from 3.516 and 3.612 together with the fact that since a fully symmetric predicate has either zero or full self-completeness, h here will be either 0 or $n-1$.

Where an n-place irreflexive predicate "P" is partially but not fully self-complete and partially but not fully symmetric, the reduction in value may not be equal to $\mathbf{sc}$ plus $\mathbf{sy}$; for as we have seen, a combination-factor may enter in such cases. This factor, and hence the total reduction, is not a constant function of $\mathbf{sc}$ and $\mathbf{sy}$, but varies with the particular combination of symmetry and self-completeness.[17] Formulae can be given for computing the combination-factor from the latter information; but the easier course is to compute the value of the predicate first. Then **cfP** is

$\mathbf{v}$ [n-pl. irref.] $-$ [**vP** + **scP** + **syP**].

The two short general formulae:

3.741 $\mathbf{v}$ [n-pl. irref.] $= 2n-1$

3.742 $\mathbf{v}$ [n-pl. irref., sym.] $= n$

are easily proved; and we have already seen in theorems 3.516 and 3.612 that:

[17] For example, consider two 4-place irreflexive predicates, each of them (1,2) (3,4) self-complete, the first being (1) (2) and (3) (4) symmetric, the second (1,2) (3,4) symmetric. Each has $\mathbf{sc} = 1$ and $\mathbf{sy} = 2$; but the first has $\mathbf{cf} = 0$, while the second has $\mathbf{cf} = 1$.

$$\mathbf{v}\,[n\text{-pl. irref., s.c.}] = n$$

and $$\mathbf{v}\,[n\text{-pl. irref., s.c., sym.}] = 1.$$

These formulae cover irreflexive predicates only. In Section 4 we saw how, where an n-place predicate "P" is not irreflexive, to find an equivalent set of irreflexive predicates. This set will contain one k-place irreflexive predicate for each k-variegated prime pattern of "P"; and if "P" is without any reflexive regularity, all possible n-place patterns are prime. Thus the maximum value for a 2-place predicate will be the value of a set consisting of one 2-place irreflexive and one 1-place predicate–that is, 4. For a 3-place predicate, the value will be that of a set consisting of one 3-place irreflexive predicate, three 2-place irreflexive predicates, and one 1-place predicate–that is $5 + 3(3) + 1$, or 15. As the number of places increases, the value rises steeply: $\mathbf{v}\,[4\text{-pl.}] = 59$; $\mathbf{v}\,[5\text{-pl.}] = 250$. The general formula, for which I am indebted to N. J. Fine, is:

3.743 $$\mathbf{v}\,[n\text{-pl.}] = \sum_{k=1}^{n} \sum_{r=0}^{k} (-1)^r \frac{(2k-1)(k-r)^n}{r!\,(k-r)!}.$$

8. SECONDARY COMPLEXITY

According to P3.32, the complexity-value of a basis is the sum of the values of the predicates in it, regardless of any relationships among these predicates. As a result, for example, a basis consisting of two 1-place predicates, "P" and "Q", will have twice the value of a basis consisting of "P" alone, even if "P" and "Q" are coextensive. This is fair enough, since sound policy obviously counsels against using two predicates where one of them will do. But sound policy likewise counsels against using a predicate with more places than necessary; yet our calculus gives an irreflexive, symmetric, self-complete 2-place (or for that matter n-place) predicate the same value as a 1-place predicate. Recognition that the latter is in some sense the simpler must be incorporated in our treatment of complexity. To this end, a subsidiary ordering may be established among bases having the same complexity-value under our six postulates.

The complexity-values dealt with in preceding sections are to be considered *primary* complexity-values. Bases equal in primary complexity may differ in secondary complexity, which is definable in terms of primary complexity. Consider three bases with the (primary) complexity-value 3: b_1 consists of three 1-place predicates; b_2 consists of a 2-place and a 1-place predicate; and b_3 consists of a 3-place predicate. Now the maximum value for bases consisting of three 1-place predicates is 3, for bases consisting of a 2-place and a 1-place predicate is 5, and for bases consisting of a single 3-place predicate is 15. These maxima may be thought of as measures of potential complexity. And the difference between the complexity-potential and the primary complexity of a basis, which thus measures unused complexity-potential, constitutes the secondary complexity-value of the basis. Thus for b_1 the secondary value is 0, for b_2 is 2, and for b_3 is 12.

Putting this more generally, for any n-place predicate "P", whatever its properties, the complexity-potential is **v** [n-pl.] and the secondary value, **s**P, is **v** [n-pl.] $-$**v**P. The secondary values for n-place predicates thus range from 0 to the number given by the Fine formula (Theorem 3.743). The complexity-potential and the secondary value of a basis are respectively the sum of the complexity-potentials and the sum of the secondary values of the predicates in the basis. Hence where K specifies only the number of predicates of each length that b contains, $\mathbf{s}b = \mathbf{v}K - \mathbf{v}b$.

9. EVALUATION OF BASES

The full complexity-index of a basis is thus an ordered pair of integers h, k, where h is the primary and k the secondary value of the basis. And in the overall sense, a basis b with index h, k is structurally more complex than a basis b' with index h', k' if *either* (i) $h > h'$ or (ii) $h = h'$ and $k > k'$. In other words, if $\mathbf{v}b > \mathbf{v}b'$, then b is more complex than b' no matter what $\mathbf{s}b$ and $\mathbf{s}b'$ are; but if $\mathbf{v}b = \mathbf{v}b'$, then b is more complex then b' if and only if $\mathbf{s}b > \mathbf{s}b'$. This may be set forth in a final postulate, with "**V**" abbreviating "the overall structural complexity of":

P3.91 $\quad \mathbf{V}b > \mathbf{V}b' \equiv : \mathbf{v}b > \mathbf{v}b' \;.\mathbf{v}.\; \mathbf{v}b = \mathbf{v}b' \;.\; \mathbf{s}b > \mathbf{s}b'.$

Although 2,3, for example, will accordingly in a sense lie between 2,0 and 3,0, a unit of secondary value is no fraction of a unit of primary value; no sum of secondary values amounts to a unit of primary value.

For bases consisting entirely of irreflexive predicates, all this can be summarized by the injunction: *don't waste places or links*. That is, among such bases with the same primary value and the same number of predicates, choose the one with the lowest aggregate number of places; and among such bases with the same primary value and the same aggregate number of places, choose the one with the lowest aggregate number of links–i.e., with least average length of predicate and hence the most predicates. The corresponding theorems follow readily from P3.91:

3.911 If $\mathbf{v}b = \mathbf{v}b'$, and b and b' consist entirely of irreflexive predicates, and b has the same number of predicates as b' but a greater aggregate number of places, then $\mathbf{V}b > \mathbf{V}b'$.

3.912 If $\mathbf{v}b = \mathbf{v}b'$, and b and b' consist entirely of irreflexive predicates, and b has the same aggregate number of places as b' but distributed over more predicates, then $\mathbf{V}b' > \mathbf{V}b$.

By 3.912, for example, where n is greater than 1 an n-place predicate with a primary value of n is more complex than n 1-place predicates (though by P3.91 simpler than $n+1$ 1-place predicates); for the secondary value of the n-place predicate here is $\mathbf{v}[n\text{-pl.}]-n$ while the secondary value of n 1-place predicates is 0. This does not mean that using an n-place primitive with a value of n is necessarily wasteful; often no simpler basis is adequate for the system in question. Waste consists of choosing any basis when a simpler one will do.

In applying the foregoing formulae, we appraise a basis relative to the available information, whether supplied by the interpretation given for the primitives or by the postulates of the system. The lowest primary value that can be claimed for a basis is that of the least-valued relevant specification that the information in question guarantees the basis to meet. This appraised value, which may be greater than the actual primary value of the basis is, so to speak, the primary value of the set of primitive 'ideas' of the system. For example, a 2-place irreflexive self-complete and symmetric predicate "P" has the actual primary value 1; but if we know only that "P" is 2-place, irreflexive, and symmetric, the primary value of "P" as a primitive 'idea' of a system is 2.

Only where such appraisal gives the same result for alternative bases need we consider secondary values. Here again an appraisal will be relative to the information available. Since we shall rarely be ignorant of the number of places in a primitive predicate, the complexity-potential of the primitive 'idea' will normally be that of the predicate itself. The appraised secondary value will then be the complexity-potential minus the appraised primary value. Thus with appraised primary values equal, the appraised secondary (and combination) values will vary directly with the complexity-potentials.

10. COMPLEXITY OF OTHER PRIMITIVES

All the foregoing treatment of complexity pertains only to bases that contain no extralogical primitives other than predicates of individuals. The postulates as they stand do not apply directly to such other primitives as function-terms, individual-constants, and predicates of classes. The procedure here is rather to correlate the primitive with a set of one or more predicates of individuals and assign to the primitive the complexity-value of this correlated set.

Function-terms are correlated with predicates in the usual and obvious way; e.g., "the father of ..." (or "f_x") amounts to the predicate "____ is the father of ..." (or "Px,y"). Each function of n arguments corresponds to, and takes the complexity-value of, an $n+1$-place predicate with relevant properties derived from properties of the function.

An individual-constant may be correlated with a 1-place predicate applying only to the individual in question. Thus the complexity-value is 1. But to take individual-constants as primitive is surely to admit as relevant the information that the term applies to one and only one individual (though different constants need not apply to different individuals). Using this information, we can replace any number of individual-constants by a sequence, and then replace the sequence by the 2-place predicate of immediate succession in this sequence. Hence no set of individual-constants has a value greater than 3. Admission of individual-constants to the range of our original postulates, together with the consequent admission of their unique application as relevant information, would thus lead immediately to inconsistency; for a set of 4 individual-constants would then have to have the value 4 and the value 3. But individual-constants are not within the range of our postulates, and their unique application is admitted as relevant information only in the process of finding the set of predicates of individuals to be correlated with a set of individual-constants. The original set must always be interreplaceable with the correlated set in the sense of replaceability I have been using throughout, except that certain additional information is admitted as relevant. Nothing requires that the complexity-value of a set of predicates of individuals correlated with a set of individual-constants be the sum of the complexity-values of the predicates correlated with the individual-constants taken severally. A single individual-constant has the value 1; a set of two, the value 2; and a set of three or more, the value 3.

To compare the complexity of bases where one or more is platonistic we need means for evaluating predicates taking classes rather than individuals as arguments. I shall first consider only cases where the argument-classes are thoroughly finite and replete–that is, have finitely many and only non-null members, members of members, and so on. The repleteness condition, but not the finiteness condition, can easily be dropped later.

Here again the procedure is to correlate a predicate of classes with a set of one or more predicates of individuals; and the correlation is by interreplaceability with certain additional information admitted as relevant. For example, information (when available) as to the cardinality of the argument-classes is admitted as relevant; for this affects the centrally relevant factor of the number of places in the correlated predicate.

A 1-place predicate of unit-classes of individuals can be not merely correlated but even identified with a 1-place predicate of individuals if we take advantage of the privilege of identifying individuals with their unit-classes.[18] Such a predicate has the value 1.

A 1-place predicate of two-membered classes of individuals correlates directly with a 2-place irreflexive symmetric predicate of individuals. This correlate has x,y as a pair if and only if x and y make up an argument-class of the original predicate. The complexity-value is 2.

In general, then, a 1-place predicate of k-membered classes of individuals has the complexity-value k.

Now suppose we know of a given 1-place predicate of classes of individuals only that each class has not more than two members. The correlate is then a set consisting of one 2-place irreflexive symmetric predicate of individuals and one 1-place predicate of individuals. Accordingly, the complexity-value is 3. In general the complexity-value of a 1-place predicate of classes of not more than k individuals is the sum of the integers from 1 to k–that is, $\frac{k^2 + k}{2}$.

Where a predicate of classes has more than one place, any information concerning whether or how these classes intersect is admitted as relevant since this affects the reflexivity-properties of the correlated predicate. A 2-place predicate that in every case relates two different (unit-classes of) individuals has the value 3. A 2-place predicate that in every case relates a one-membered to a disjoint two-membered class of individuals correlates with an irreflexive 3-place predicate of individuals, symmetric with respect to its second and third places, and so has the value 4. In general, a 2-place predicate that in every case relates a k_1-membered class of individuals to a disjoint k_2-membered class of individuals correlates with a $k_1 + k_2$-place irreflexive predicate of individuals, symmetric with respect to the first k_1 and also with respect to the remaining k_2 of its places and thus has the value

$$2(k_1 + k_2) - 1 - (k_1 + k_2 - 2), \text{ or } k_1 + k_2 + 1.$$

For some sample values see column A of Table I.

If we know only that a given 2-place predicate in every case relates a class

[18] Following Quine, *Mathematical Logic*, revised edition (Cambridge, Mass.: Harvard University Press, 1955), p. 135.

of not more than k_1 individuals to a disjoint class of not more than k_2 individuals, correlation is with a $k_1 + k_2$-place predicate of individuals that is not irreflexive, and hence with a set of irreflexive predicates of individuals. For any h from 2 through $k_1 + k_2$, the number of h-place predicates in this set will be the least among k_1, k_2, and $h-1$; that is,

$$\min(k_1, k_2, h-1, k_1 + k_2 - h + 1) \qquad (B)^{19}$$

and each such h-place predicate will have the value $h + 1$. The total value of the original predicate will then be

$$\frac{k_1 k_2(4 + k_1 + k_2)}{2}. \qquad (M)$$

For some sample values, see column B of Table I.

k_1	k_2	A	B	C	D
1	1	3	3	4	4
1	2	4	7	7	11
2	1	4	7	7	11
2	2	5	16	12	30
2	3	6	27	16	55

TABLE I

(Sample values, under various requirements, for bases consisting of a single 2-place predicate of classes of individuals. Under "k_1" and "k_2" are listed the maximum number of members in, respectively, the left-hand and the right-hand classes. Column A shows the values where (**i**) these numbers are also minima, and (**ii**) where the two classes related are in every case disjoint. Column B shows the values where **ii** but not **i** is required; column C the values where **i** but not **ii** is required; and column D the values where neither **i** nor **ii** is required.)

A predicate that in every case relates a k_1-membered to a not necessarily disjoint k_2-membered class of individuals will, again, correlate with a set of irreflexive predicates of individuals. Where $k_1 = k_2 = 1$, the complexity-value is 4. For all other cases, if we use "k" for whichever if either of k_1 and k_2 is the lesser and use "m" for the other (the choice being arbitrary where $k_1 = k_2$),

---

[19] Formulae labeled "(*B*)" were supplied by Howard Burdick, that labeled "(*M*)" by David Meredith.

the complexity value (e.g. see column C of Table I) is determined by the formulae

$$\text{if} \quad k \neq m: \quad \frac{k^2 + 5k}{2} + km + m;$$

$$\text{if} \quad k = m: \quad \frac{3k^2 + 7k}{2} - 1. \qquad (B)$$

The correlated set for a predicate that in every case relates a class of not more than k_1 to a not necessarily disjoint class of not more than k_2 individuals will be much larger, except for the degenerate case where $k_1 = k_2 = 1$. The general formula for the complexity of such a predicate, where "k" and "m" are taken as before, is:

$$\frac{km}{12}(25 + 3km + 21k + 9m + 2k^2) - \frac{k}{12}(k^3 + 4k^2 - k + 8). \qquad (B)$$

Sample values are in column D of Table I.

The way of calculating the complexity-value for other specifications of 2-place predicates of individuals–such as that the related classes intersect in a certain one (or more) rather than in all (or no) ways, or that they have one (or more) rather than all (or no) numbers of members up to k_1, k_2, etc.–is by now evident enough, though the calculation may sometimes be laborious.

An n-place predicate that in every case relates disjoint classes of individuals, the first class having k_1 members, and the second k_2 members, and so on, has the value of a $k_1 + k_2 + \ldots + k_n$-place irreflexive predicate of individuals, symmetric with respect to its first k_1 place, also with respect to its next k_2 places, and so on. Its degree of symmetry is thus $k_1 + k_2 + \ldots + k_n - n$; and its value is

$$k_1 + k_2 + \ldots + k_n + n - 1.$$

Symmetry of a predicate of this sort reduces the value only where the related classes have the same number of members. An h-place symmetric predicate of disjoint k-membered classes of individuals has the value hk. For example, while a nonsymmetric 3-place predicate of disjoint 2-membered classes has the value 8, a symmetric predicate of this kind has the value 6.

The complexity-value of a predicate of n disjoint classes having at most $k_1, k_2, \ldots, k_n$ members is

$$\frac{k_1 k_2 \ldots k_n (3n - 2 + k_1 + \ldots + k_n)}{2}. \qquad (B)$$

Since predicates with arguments of still higher type are seldom encountered as primitives of constructional systems, some brief indications concerning their complexity will suffice. A 1-place predicate of unit-classes of unit-classes of individuals identifies, of course, with a 1-place predicate of individuals. Moreover, even though the unit-class of a class of more than one individual cannot be identified with its member, a 1-place predicate of unit-classes (or of unit-classes of unit-classes, etc.) of any classes of individuals correlates with the same predicate of individuals as does a predicate of these latter classes.[20] The correlation established between predicates of non-individuals and predicates of individuals is thus not one-one. Interreplaceability does not require one-one correlation, but requires only that from any basis answering either specification we can always define some second basis, answering the other specification, such that we can redefine the first basis from the second. In general, a 1-place predicate of h-membered classes of disjoint k-membered classes of individuals is correlated with and has the same value, hk, as an hk-place irreflexive predicate of individuals with maximal degree of symmetry. Incidentally, although this is also the value of a 1-place predicate of hk-membered classes of individuals, the correlated hk-place irreflexive predicates of individuals are different in the two cases; both have maximal degree of symmetry, but only the correlate of the lower-type predicate is fully symmetric.

A 2-place predicate that in every case relates an h-membered class of disjoint k-membered classes to a disjoint j-membered class of disjoint m-membered classes of individuals correlates with an $hk + jm$-place irreflexive predicate of individuals that has $\mathbf{sy} = hk + jm - 2$ and consequently the value $hk + jm + 1$.

Our temporary restriction to thoroughly replete predicates can now be dropped; for since stipulation of the nullity of any argument or member (member of member, etc.) of any argument cannot increase complexity, the complexity of a specification when there is no stipulation concerning repleteness remains the same as when repleteness is stipulated.

Often the number of members (members of members, etc.) in the argument-classes of a predicate is unknown; and one must face the problem of comparing the complexity of bases when some or all such information is lacking. By definition, the value assigned to a given specification is the lowest number that is big enough–that is, the lowest number such that the

[20] Thus we can deal wherever necessary with predicates relating arguments of different types; for achieving type-homogeneity by taking the unit-classes of (unit-classes of ..., etc.) the lower-type arguments will leave the complexity-value unchanged.

specification guarantees that no basis answering it has a higher value. Thus if the arguments of a 1-place predicate of classes of individuals are specified to have eight members, the complexity-value is 8, while if these classes are specified only to have some *uniform* number of members not over 18, the value is 18. But suppose that no ceiling is specified, that the number is known only to be constant and finite. Though the value must be finite, no integer is big enough; for whatever integer we might assign, the complexity-value of some basis answering the specification might be higher. In this case, I shall put in for the number of elements in the argument-classes, and hence here for the complexity-value of the predicate in question as well, the arbitrary letter "u" (for "unknown" or "unspecified"). After some further examples of how "u" may enter into certain complexity-indices, I shall consider how it is to be handled mathematically.

The same letter "u" goes in for each utterly unspecified relevant number. Thus every 1-place predicate of classes of individuals that has a constant but unspecified number of members in each argument-class has the complexity-index "u". This does not imply that the constant number in the two cases is actually the same, but follows from the fact that the narrowest relevant specification met by such a predicate is the same in all cases. Where the argument-classes of a 1-place predicate are not even specified to have a constant number of individuals as members, the complexity-value will obviously be $\frac{u^2+u}{2}$. A 1-place predicate having as its arguments classes of some constant but unspecified number of three-membered classes of individuals will have the index "$3u$"; and if the member-classes have rather some unspecified constant number of individuals as members, the index will be "u^2". A 1-place predicate of logical type t taking as arguments classes of some constant but unspecified number of members, themselves in turn having some unspecified constant numbers of members, and so on, will have the index "u^t". If for a 2-place predicate of disjoint classes, the left-hand classes have three individuals as members and the right-hand classes an unspecified constant number, the index is "$3+u+1$", while if the number of members in the left-hand classes is also specified only to be constant, the index is "$2u+1$". As a final example, a symmetric n-place predicate of disjoint classes having an unspecified constant number of individuals as members has the index nu.

Now suppose we want to compare the complexity of some bases such that as least one has a complexity-index containing "u". First, multiply out all such expressions as "$(2u-1)u$" into such expressions as "$2u^2-u$". Second, remove all fractions by multiplying all the indices by their lowest common denominator. Third, remove all minus signs by making an appropriate fixed addition to all the indices.[21] Finally, collect the coefficients in each index and put it in the form:

$$hu^0 + h_1 u^1 + h_2 u^2 + \ldots + h_n u^n.$$

To order these indices according to the complexity they express, we need some evaluation of "u" as it appears in them. We know that "u" must in general be considered as greater than any given integer. Thus in the present context, if we are to avoid risking the anomaly of rating a purely numerical index higher than one containing "u" we must take "u" as greater than any coefficient in any of the indices being compared.[22] But so long as this condition is met, *the resultant ordering will be the same no matter how great "u" is taken to be*. Hence the only evaluation needed for "u" in such a context is that it be greater than any of the coefficients in question. Then from elementary laws of natural arithmetic alone we can derive a ready rule for ordering the indices: one index is higher than another if and only if the first has the greater coefficient where "u" has the greatest exponent such that the two indices have different coefficients for "u" with that exponent. That is, to compare two indices we merely read from right to left until we find parallel coefficients that differ; the higher index is the one with the greater coefficient there. Some consequences are that any index containing "u" will be higher than any purely numerical index, and that any index containing "u" with a given exponent will be higher than any index that does not contain "u" with any exponent at least as great.

In closing this discussion of structural simplicity, I must repeat that many other less tractable factors–some of them mentioned earlier–also enter into the choice of basis for a system.

11. BASIC INDIVIDUALS

In founding a system, we must not only choose the primitives but also deter-

[21] This third step is permissible only because we are here concerned solely with an ordering, not with a quantitative measure.

[22] A purely numerical index is one such that all coefficients other than the first are 0. Every index theoretically contains "u" with each integer as exponent; and wherever "u" with a given exponent does not actually appear, it is understood to have the coefficient 0 in that index. Note that "u" is interpreted only in the context of a comparison of given indices. Otherwise "u" functions, as David Meredith points out, much like ω, the first infinite ordinal. But while ω would indeed be the lowest number that could be assigned as absolute value of a basis with an index containing "u", it is also too high since all bases in our universe of discourse have finite value.

mine the range of the individual-variables–the realm of individuals recognized by the system. This is commonly done in terms of the special primitive predicates chosen. But where a system contains the calculus of individuals, we must bear in mind what we observed earlier (II,5): that the universe of individuals consists of everything that satisfies the predicate "overlaps", and that the universe of individuals *for such a system* consists of everything that satisfies the predicate "overlaps" under the restricted interpretation[23] of it provided for that system. The way the predicate is restricted for a given system generally depends upon the other, special, primitives adopted. The individuals recognized include all that satisfy at least one place of at least one of these primitive predicates, and in addition (by the principle 2.45) all sums of such individuals. Let us distinguish as *basic units* those individuals that satisfy such a special primitive predicate and as *minimal basic units* those basic units that have no others as parts–as "part" is defined in terms of the systematically restricted predicate "overlaps". For all finite systems, and it is these that primarily concern us, there will be atomic individuals–individuals that have no others as systematic parts. But will these atoms and the minial basic units be the same? There is no fixed rule. Most commonly however, in choosing the special primitives of a system we are also selecting the individuals that satisfy those predicates as the elements out of which all others are to be constructed. In general, the minimal basic units are also the atoms of a system.

In any case, it can be shown that the only individuals we can recognize under a finite system are sums of one or more of its atoms. Any other individual would either systematically overlap some individual that is systematically discrete from all the atoms of the system or systematically overlap some atom without containing it. But no individual can be systematically discrete from all the atoms; for every recognized individual x will obviously have as a systematic part some individual y (perhaps identical with x) that has no other as a systematic part and that is therefore an atom. And no individual can systematically overlap an atom without containing it; for in that case the atom would have some other individual as a systematic proper part and would therefore not be an atom.

In brief, then, every admitted value of the individual-variables of a finite system is a sum of, and in this sense reducible to, one or more atoms of the system. And normally the atoms of a system are to be identified with its minimal basic units.

[23] To adopt a "restricted" interpretation, as was made clear in II,5, does not involve any commitment on whether the realm to which the restriction is made is or is not less than the universe.

12. POSTULATES

One might seek to set up for postulational economy standards comparable to those set up for economy of primitives in Sections 3 through 10. The problem is somewhat simplified by the fact that postulates can be transformed by logical operations into some specified standard form in order to facilitate comparison. One must, of course, avoid the pitfall of supposing that postulate set A is simpler than B if and only if A follows from B while B does not follow from A; for then the very important economy effected when one of three postulates is found to be deducible from the other two would be wrongly regarded as no economy.

The whole problem of postulational economy, however, has been complicated by the recent discovery[24] that in a very wide variety of cases all the postulates on a special primitive can be entirely eliminated through purely mechanical procedures. Until some way is found for distinguishing such an apparently superficial procedure from a reduction in postulates that results from some significant and special deductive discovery, adequate standards of postulational economy can hardly be formulated.

As an example of how a postulate upon a special primitive of a system may be eliminated, suppose our primitive is the two-place predicate "intersects" (symbol "Int") applying to lines in the diagram of Chapter I, and suppose the postulate in question is:

$$(x)\,(y)\,(\text{Int}\,x,y \supset \text{Int}\,y,x),$$

which affirms the symmetry of the primitive predicate. Now replace "intersects" as primitive by "crosses" (symbol "C") interpreted in exactly the way that we interpreted "intersects". Then define:

$$\text{Int}\,x,y =_{\text{df}} \text{C}\,x,y\,.\,\text{C}\,y,x.$$

The erstwhile postulate of symmetry can now be proved as a theorem; yet we have not made our basis more complex to accomplish this.

So far as platonistic systems go, the general eliminability of 'extralogical' postulates in virtually all interesting cases–namely, all cases where there is a 'logical' model for the set of postulates in question–has been demonstrated by Professor Quine and the present writer.[25] "Extralogical" here refers to

[24] See W. V. Quine and Nelson Goodman, "Elimination of Extralogical Postulates", *Journal of Symbolic Logic*, 5 (1940), pp. 104–109; reprinted in *P & P*, pp. 325–333. Some more recent ideas in the present section have resulted from further discussions with Professor Quine.

[25] In the article cited in note 24 above, where the proof here outlined is given in much more detail.

everything not contained in the basic logic or the calculus of classes. In very brief outline, the argument is as follows. Let "d" be the sole extralogical primitive term and "Pd" an abbreviation for the conjunction of all the postulates governing this primitive. First find a logical model m such that Pm; for example, the universal class of ordered couples will do if "Pd" affirms just that d is symmetric and transitive. Then replace "d" as primitive by another term "c", explained in the same way that "d" was explained; and define "d" as follows:

$$d =_{\text{df}} (\imath x)(\text{P}c \,.\, x = c \,.\vee.\, {\sim}\text{P}c \,.\, x = m).$$

Now "Pd", instead of being adopted as a postulate, can be proved as a theorem. The method can be extended to cases where there are many extralogical primitives.

Efforts to dismiss or remove this obstacle to the reasonable measurement of postulational economy have been notably unsuccessful. The observation that elimination of postulates in the way described 'removes their empirical content to the semantics of the system'[26] may be true but hardly helps. The plausible idea of counting definitions as postulates in measuring economy is untenable; for since, with a constant total vocabulary, the set of definitions inevitably tends to become more complicated as the set of primitives becomes simpler, we should be working at cross-purposes in trying simultaneously to achieve economy of primitive predicates and of primitive formulae. Genuine postulational economy has to do with formulae that are not mere consequences of definitions; but our result raises the question whether any such distinction can be made to hold. Indeed, after some twenty-five years, no relief has been found; and Quine has recently[27] reaffirmed and extended the result, contending that it shows that for typical platonistic systems, all postulates are implicitly definitions.

In a system that contains the calculus of classes, we can even eliminate the postulates of the calculus of individuals in the way described.[28] But to what extent can postulates be similarly eliminated in a nominalistic system, where classes are not available as models? Our models will then have to be defined in terms of the basic logic and the calculus of individuals. For example, suppose the original primitive predicate is "Q" (given a specific interpretation) and the postulates are:

---

[26] This seems to be the upshot of Langford's "Note on a Device of Quine and Goodman", *Journal of Symbolic Logic*, 6 (1941), pp. 154–155, and Hempel's review of Langford's note, same journal 7 (1942), p. 98.
[27] In "Implicit Definition Sustained", *Journal of Philosophy*, 61 (1964), pp. 71–74.
[28] "Elimination of Extralogical Postulates", p. 108.

(1) $(x)(y)(Q\,x,y \supset Q\,y,x)$,

(2) $(x)(y)(z)(Q\,x,y \,.\, Q\,y,z \,.\!\supset Q\,x,z)$,

affirming that "Q" is symmetric and transitive. Replace "Q" by "R", explained in the same way as "Q". Then, using "F" as an abbreviation for the conjunction of (1) and (2) with "R" written in for "Q" we may define "Q" as follows:

$$Q\,x,y =_{\mathrm{df}} (F) \,.\, R\,x,y \,.\, \mathbf{v} \,.\, (\sim F) \,.\, x = y.$$

The two-place predicate "=" is the model. We can then prove (1) and (2) as theorems, making it unnecessary to adopt them as postulates. Note that we cannot, as would be undesirable, also prove the theorem "$(x)(y)(Q\,x,y \supset x = y)$".

The chief difference between platonistic and nominalistic systems in regard to elimination of postulates concerns postulates with strong existential consequences. We usually have little trouble finding a model for these in platonistic systems, where we can generate entities at will from one or even no individuals (e.g., Λ, $\iota\Lambda$, $\iota\iota\Lambda$, etc.). By comparison, the existential fertility of a nominalistic system is indeed meager. It is perhaps a fair conjecture, however–for which I can offer no proof–that any likely set of postulates on the special primitives of a nominalistic system can be eliminated if these postulates have no existential consequences beyond those of logic and the calculus of individuals alone.

In general, any set of postulates on a given special primitive is eliminable if the analogues of these postulates for any other primitives of the system, or terms defined from other primitives of the system, have already been established by postulates or proof in the system. Accordingly, any postulates on special primitives of a nominalistic system are eliminable if the same statements with some signs definable in terms of "**o**" replacing those special primitives have been established in the calculus of individuals. And postulates on further special primitives are eliminable if their analogues have been established for any terms defined from previously introduced primitives. To put the matter in other words, let us suppose that $S_1 \ldots S_n$ are established statements of a system and contain, besides logical terms, only the predicates "Q_1" ... "Q_m" which are defined in terms of primitives other than "P_1" ... "P_m"; then the statements $S'_1 \ldots S'_n$, which are like $S_1 \ldots S_n$ except that "Q_1" ... "Q_m" are consistently replaced by "P_1" ... "P_m" *need never be adopted as postulates.*

Even this rather broad statement, however, does not fix the outside limits of the use of the general kind of definitional elimination of postulates illustrated in this section. In the first example given, for instance, a postulate of symmetry is eliminated without the help of anything but the postulates of the basic logic. Here the symmetry of conjunction seems to be made to yield the symmetry of the predicate in question; but we can hardly say that the connective of conjunction is used as a model in the same way that "=" is used in the later example. Whether postulates that are analogues of established statements in truth-function theory can in general be eliminated is not at present clear.

In the case of platonistic systems, we have seen that virtually all postulates likely to be needed are readily eliminable. In such systems, the effort to achieve postulational economy by ordinary means therefore becomes pointless unless we find some criterion not now available by which to distinguish trivial from nontrivial methods of reducing postulate sets. In the case of nominalistic systems, where certain postulates at least are likely to resist elimination by the devices above discussed, it is an open question whether we ought to accept these devices as equally legitimate with other ways of reducing postulate sets. All that can be said at present is that if no criterion is found by which some devices are excluded as trivial, the number of postulates required on any primitive will be small and will tend to become progressively smaller for each successive new term that is introduced.

PART TWO

ON QUALITIES AND THE CONCRETE

CHAPTER IV

APPROACH TO THE PROBLEMS

1. THINGS

As an introduction to the detailed study of various systems, I shall in the present chapter attempt to clarify briefly some aspects of their common subject matter, characterize certain of the more important types of basis that may be chosen, and introduce some of the first problems that are to be dealt with. Much of this preliminary discussion will be quite informal, making free use of such everyday language as seems helpful and often dealing rather summarily with terms (such as "thing") whose formal definition we shall not reach at all in this book.

A blue thing sometimes looks green, and may become red. In such a simple statement lie rich opportunities for confusion. If a thing looks green, then how can it be actually blue? How is the *apparent* change that we describe by saying a thing now *looks* different in color to be distinguished from the actual change that we describe by saying the thing *has become* different in color? And what is *it* that retains identity as a thing throughout innumerable apparent and real changes of quality?

The last question is logically the first. If a thing can remain the same while its appearance changes, then clearly the real and the apparent are different. But this means not that the real thing must be something quite separate from its appearances but only that a real thing comprises many appearances. To say that the same thing is twice presented is to say that two presentations–two phenomenal events–are together embraced within a single totality of the sort we call a thing or object.

Although this is obvious enough, our thinking is often muddled with regard to the temporal dimension of experience. We hardly ask for an explanation of the identity of a table throughout its various *spatial* parts; we accept it as a compound of these parts. Because the parts are of certain kinds and are related in certain ways, the compound is called a table. We feel no need to hypostatize an underlying core of individuality to explain how a leg and a top, which differ so drastically, can belong to one table. Yet when we consider the table at different moments, we are sometimes told that we must inquire what it is that persists through these temporally different cross sections.The simple answer is that, as with the leg and the top, the unity overlies rather than underlies the diverse elements: these cross sections, though they happen to be temporally rather than spatially less extensive than the whole object in question, nevertheless stand to it in the same relation of element to a larger totality.[1] And as before, because these elements have certain characteristics and are related in certain ways, the totality they make up is what we call a thing, and more specifically a table. Our tables, steam yachts, and potatoes are events of comparatively small spatial and large temporal dimensions. The eye of a potato is an event temporally coextensive with the whole, but spatially smaller. The steam-yacht-during-an-hour is an event spatially as large as the yacht but temporally smaller. But the steam-yacht-during-an-hour is an element in a larger whole as is the eye of the potato.

We may of course, still ask why certain cross sections, and similarly why certain presentations, belong to a single thing while others do not. The detailed answer would be complicated and difficult to arrive at, but the question is entirely parallel to the question why certain things related in a certain way make a table rather than a desk. The only difference is that in the former case we are asking for a general definition of "thing" rather than for the more specific definition of "table". None of these problems concerns us at present. I simply want to emphasize the point that the identity of a thing at different moments is the identity of a totality embracing different elements.

A presentation, of course, cannot be identified with an entire momentary cross section of a physical object. A visual presentation of a baseball, for example, is spatially bidimensional, whereas the baseball, for example, is spatially tridimensional. To construct the entire physical object or a cross section of it out of presentations means, in effect, bringing in 'possible' as well as actual presentations; for the baseball embraces not only the totality of its presentations when observed but also such presentations as 'would have occurred had it been observed' at other times and under other conditions. This immediately involves us in a difficult set of problems that have in recent years been studied very earnestly but with little success.[2] However, the problem of constructing either the physical object or its cross sections out of presentations (or conversely, the presentations out of physical objects) lies beyond the scope of this book. In dealing with the question immediately at

---

[1] Or we may say, of part to whole–provided we bear in mind that we are using presystematic language that might under different systems be interpreted in terms of "$<$", "$\subset$", "$\in$", etc.

[2] See, for example, Carnap, "Testability and Meaning", *Philosophy of Science*, 3 (1936), pp. 420–471, and 4 (1937), pp. 1–40, and also my "Problem of Counterfactual Conditions", *Journal of Philosophy*, 44 (1947), pp. 113–128, reprinted as Chapter I of *Fact, Fiction, and Forecast* (Indianapolis: Bobbs-Merrill, 3rd ed. 1973).

hand we can safely ignore the distinction between the physical object and the totality of its presentations, as well as the distinction between a cross section and a single presentation; for this will not affect the question of the nature of identity at different moments. As we have seen, the answer to that question is that such identity, like identity in spatially different parts, is an additive identity. An object, or the totality of its presentations, is an event with a relatively long temporal dimension; and parts of it that differ spatially or temporally from one another may differ in other respects as well.

2. PROPERTIES

During its existence, an object ordinarily presents many different qualities of each kind–many different colors, for example. That is to say, even if an object is such as to be regarded as actually constant in color, its presentations are ordinarily of many different colors. In much the same way, an object that is regarded as uniformly colored all over may exhibit many different colors in spatially different parts at a moment when it lies in a nonuniform light. Thus different perceptible parts of an object may be differently colored even if the object itself is uniform and unvarying in color. This is no more paradoxical than the fact that a single object contains spatiotemporally different parts. As the self-identical object is a function of its parts, so the single unchanging color of the object is a function of the colors of its parts. The nature and interrelation of the lesser elements that make up the whole determine what kind of thing the whole is; the kind and arrangement of the colors exhibited by these various parts determine what color the whole is said to have. A table is not made up of tables; not all the presentations of a yellow thing need be yellow.

Roughly, then, to say that a thing *looks* green is to make a statement concerning a presented quality, a color quality of some presentation of the thing, while to say that a thing *is* green is to make a more complex statement concerning the color qualities exhibited by various presentations of the thing. Obviously, the color names are thus used in two different ways in ordinary language: in the one case for presented characters, which I shall hereafter call *qualia*[3]; in the other, for properties of things.

Exactly what is involved in ascribing such a property to a thing has been the subject of some controversy. Some maintain that to say that a thing is carmine is to say that all presentations of the thing that occur under pure

[3] My usage of the terms "quale" and "property", along with much else in this and the following section, is taken from C. I. Lewis's *Mind and the World Order* (New York: C. Scribner's Sons, 1929).

daylight and otherwise optimum conditions exhibit the quale carmine. This *optimum* theory has some initial plausibility since it is under optimum conditions that the most acute distinctions can be made and a property most precisely determined. Furthermore the names of properties are often–as with most colors–the same as those of the relevant qualia of the optimum presentations. But there are serious difficulties. Optimum conditions may not always be uniquely determinable. Again, two objects that persistently present the same color quale under optimum conditions, but contrasting color qualia under another set of conditions, will not usually be said to have the same color property. And the (tridimensional) shape of a spherical object is obviously not the (bidimensional) shape of *any* of its presentations, optimum or otherwise. A better theory has been proposed by C. I. Lewis. He holds that to ascribe a certain property to an object is in effect to describe the complete pattern of qualia (of the kind in question) exhibited under all sorts of conditions. This *pattern* theory meets the difficulties of the optimum theory, but may perhaps err in the opposite direction. Surely we shall usually regard a thing as carmine if it presents the requisite qualia under all more or less 'normal' conditions, regardless of how it looks under very strange or unusual circumstances. Perhaps the truth is that to apply any ordinary property predicate to a thing amounts to describing its appearance under all those sorts of conditions that are regarded as critical or standard. In some cases, optimum conditions alone may be the standard; in other cases, many different conditions may belong to the standard set.

We need not decide finally among these theories. My purpose has been simply to suggest something of the relation between a property of a thing and the qualia of presentations, and to emphasize the difference between the two. It is enough to recognize that to ascribe a property to a thing is in effect to affirm that the qualia it presents under different conditions conform to some more or less fully prescribed pattern. Much as some pieces of the world are regarded as single things while others are regarded as heterogeneous collections of fragments, so some among all possible patterns of qualia presentation have been for convenience dignified as fixed-property patterns while others have not. It is upon this that 'objective' change, or change of property, depends. A thing changes only 'apparently' when all its presentations, however widely varied, are those called for by some one fixed-property pattern. Such change is entirely compatible with 'objective' constancy. A thing is said to change 'objectively', or in a property, only if some two of its presentations conform to no one fixed-property pattern. In some cases, of course, we do not construe such nonconformity as objective change but discard one presentation or both as illusory.

3. QUALIA

Likeness and difference of presentations are thus interrelated with the likeness, difference, constancy, and change of things. Though presentations are momentary and unrecallable, they are nevertheless comparable in that they contain repeatable and recognizable qualia. Yet it is sometimes held that any identification of a quale from one presentation to another is out of the question because it is extremely unreliable and quite untestable. The unreliability seems apparent in our hesitancy to decide that a given presentation has the same quale as an earlier one, and in our readiness to change our minds. The untestability appears guaranteed by the fact that the past presentation cannot be actually revived to stand comparison with a later one. Yet no type of judgment can be *both* unreliable and untestable; for an unreliable judgment, after all, is one that is frequently found to be false *when tested*.

This paradox suggests that we might do well to follow Lewis in explaining the unreliability in question as not genuinely attaching to quale recognition at all but rather as having to do with properties—as attaching, for example, to the prediction that two objects will match in color when both are observed together under some specified conditions. On this view, to say that the color of a thread I see now is the same as the color of a shirt I saw yesterday is to say that if the two objects are placed together in daylight (or other standard illumination) they will present the same quale. It is to say not that the quale now presented by the thread is the same as that yesterday presented by the shirt, but rather that the two objects have identical color properties. Such implicitly predictive judgments are what may be tested and so described as more or less reliable. Quale recognition, on the other hand, Lewis regards as quite untestable; in his own words it is "immediate and indubitable; verification would have no meaning with respect to it".[4]

Now a judgment may be regarded as indubitable and exempt from test on either of two scores: that it is meaningless or that it is certainly true. If one holds that comparison of temporally diverse presentations is meaningless, then under this theory the distinction between properties, and thus between constancy and change of property, also becomes meaningless. We might try to draw these distinctions in a new way by saying that to ascribe a property to an object is simply to affirm that when compared with any of certain standard objects the given object presents a quale that bears a specified relation to (e.g., matches, is complementary to, etc.) the quale presented by the standard object. But the trouble is that there are no 'standard objects', no

---

[4] Lewis, *Mind and the World Order*, p. 125.

objects that we regard as unchanging and unchangeable in color, for example. And we cannot say that any object may be arbitrarily designated as standard; for if after a hot spell the lawn and the cardinal bird contrast less brilliantly, it is the lawn and not the bird that we deem to have changed in color.

The indubitability of a judgment relating a quale now presented to one presented earlier must, then, consist not of the meaninglessness of the judgment but of its certain truth. This amounts, however, to the curious view that while it might be wrong to identify a certain quale of a presentation with a certain quale of a later presentation, no such error is ever made, not even willfully. What Guardian Angel or vestige of Original Virtue keeps us from such mistakes?

These difficulties can be resolved, I think, along the following lines. A comparison of temporally diverse presentations is indeed immune from the direct test of simultaneous comparison. If I say that the green presented by the grass now is the same as the green presented by it at a certain past moment, I cannot verify that statement by reviving the past presentation for fresh inspection. My statement might thus be looked upon as a *decree*. But such decrees are not therefore haphazard. I am not equally inclined to identify the color presented by the grass now with the color presented by the sky an hour ago, even though such a decree would be equally safe from direct test. Moreover, we often withdraw or alter our decrees. A decree, however safe it may be from disproof, is vulnerable to cancellation by another decree. The untestable is not irrevocable. But then the question arises: admitting that we *can* so change our decisions, why *do* we ever change them if we are never faced with negative test results? In part, no doubt, because of a new impulse of the same sort that led to the original decision. But this is not all. More important are the consequences, actual or prospective, of a given decree. When a decree causes us too much trouble, we abandon it; and our decrees can lead us into such serious trouble as outright inconsistency. For example, suppose that I (perversely) decree that the color quale presented by a (red) apple now is the same as that presented by the (blue) sky yesterday noon. Then I cannot very well also maintain both the following (natural) statements: (1) that the quale now presented by the sky is the same as that presented by it yesterday noon, and (2) that the quale presented by the apple now is very different from the quale presented by the sky now. Unless I am ready to abandon the more general and exceedingly useful principle of the transitivity of identity, I must give up one of the three statements.

A decree by itself thus may be unchallengeable; and any decree, however unnatural, can be maintained by giving up enough others. But in practice our choice, when a conflict arises, is influenced by two factors. In the first place, we favor the more 'natural' decree, the one best supported by an instinctive feeling of hitting the mark, as when we select a remembered color. In the second place, we favor the decree that makes necessary the least adjustment in the body of already accepted decrees. Normally, we have not a conflict of two decrees, but a conflict between a new decree and a whole background of accepted decrees. We could uphold the discordant newcomer, but only at the exorbitant price of reconstructing our whole picture of the past.

The unreliability we attribute to quale recognition is then, in so far as it actually pertains to quale recognition at all, chiefly the liability of some of these decrees to be rejected in favor of conflicting decrees that either have stronger psychological support or involve less reorganization. Though quale recognition may not be strictly testable, it is grounded and it is revocable. It is even testable if we take indirect as well as direct testing into account. Any judgment that a quale of one presentation is the same as a quale of another is open to pertinent criticism that may cause it to be abandoned. If it survives because it is psychologically satisfactory and workable, and because it is compatible with the body of other accepted statements, it may be said to be well verified. Indeed, one may question whether this sort of verification can be sharply distinguished from some more direct process; for, as we shall see in a moment, the notion of 'direct' observation is far from clear.

4. PHYSICALISTIC AND PHENOMENALISTIC SYSTEMS

One much-discussed distinction among systems depends upon whether the basic units are phenomenal elements such as qualia, presentations, etc., or physical elements such as things, processes, etc. I illustrate these two types of system rather than define them because a precise and general distinction cannot readily be drawn. In the systems of either type that are in question here, the basic units are 'observable' or 'perceptible' individuals; and, as we shall see, the view that the phenomenal can be distinguished as the cognitively immediate is not easily maintained.

The choice of one rather than the other type of basis often reflects an underlying philosophical attitude: a desire to show, on the one hand, that nothing beyond the phenomenal need be countenanced in order to explain everything, including the physical, or to show, on the other hand, that nothing beyond the physical need be countenanced in order to explain everything, including the phenomenal. But before we consider very briefly some of the arguments in favor of each sort of basis, I must point out that the realization of either program would by no means preclude–and might even assist–realization of the other. Systems of different types, although they may result from opposing philosophical attitudes or convictions, do not themselves necessarily conflict, but may be regarded as answering different problems: in the one case the problems of explaining everything or at least as much as possible in terms of phenomenal elements; in the other, the problems of explaining as much as possible in terms of physical elements.

Choice of a phenomenalistic basis is usually argued for on the ground that since the phenomenal by its very nature comprises the entire content of immediate experience, everything that can be known at all must be eventually explicable in terms of phenomena. A phenomenal system is thus held to constitute a kind of epistemological reduction of the predicates it defines; the definitions indicate the testable, empirical, pragmatic significance of these predicates; and definability in the system provides a criterion of meaningfulness. To the phenomenalist, what cannot be explained in terms of phenomena is unknowable, and words purporting to refer to it are vacuous. The problems that a phenomenalistic system is designed to meet, and the point of view that finds these problems to be of central interest, clearly descend from the tradition of British empiricism that culminated in Hume.

The physicalist usually agrees that it is important that a system be in some sense epistemologically grounded, but argues that physical things, for example, are known more directly than evanescent phenomena and are actually less far removed from 'raw' experience than are the obviously postanalytic atoms of a phenomenalistic system. Verification, experimentation, and pragmatic test, he claims, operate not with phenomena but with things. In addition, he argues that no phenomenalistic basis is adequate for explaining objective and intersubjective fact, while a physicalistic basis can be found that is adequate for a *universal* language.[5]

Let us consider the last point first. I shall return to it again much later (XI,5) and it can be disposed of rather briefly here. The physicalist's charge that phenomenalistic bases are essentially inadequate for a universal language rests chiefly on the admittedly grave difficulties of defining physical things in terms of phenomena. But the physicalist has not proved the problem insoluble; indeed until the distinction between physicalistic and phenomenalistic systems is more accurately defined, the problem will not have been very precisely formulated. Nor has the physicalist constructed, or shown that he can construct, the comprehensive system he claims is possible on his basis. The physicalist is normally unwilling to accept as primitive such predicates of physics as "(is an) electron"; what distinguishes his program from that of

[5] See, for example, Carnap, "Die physikalische Sprache als Universalsprache der Wissenschaft", *Erkenntnis*, 2 (1931), pp. 432–465.

physics itself is that he insists upon beginning with 'observation statements'. Yet if he takes as primitives only predicates that apply to perceptible individuals, and if his claim of universality is to be made good, he will have to explain in terms of these the multitudinous imperceptible particles that the physicist discusses. And the difficulties he will encounter–for example, the problem of defining certain 'disposition' terms–are obviously closely akin to those that arise in the attempt to deal with the physical in terms of the phenomenal. Thus the physicalist has so far come nowhere near substantiating either the claim that phenomenalistic bases are inadequate for a universal language or the claim that some physicalistic basis is adequate. But as a matter of fact, the interest of a system does not depend upon its all-inclusiveness any more than the interest of chemistry depends upon whether it eventually absorbs biology. A partial system may answer many important questions, and partial systems are all we are likely to have for some time to come.

Apart from the argument of universality, we have seen that the physicalist and the phenomenalist each claims the advantage of epistemological priority for his own sort of basis. Actually the argument does not seem to me very sound on either side; for the whole question of epistemological priority is badly confused. The claim is that one basis corresponds more closely than another to what is directly apprehended or immediately given, that one more nearly than the other represents naked experience as it comes to us–prior to analysis, inference, interpretation, conceptualization. Now one may certainly ask whether a given description is *true* of what is experienced; but here the further question is whether one or the other of two *true* descriptions more faithfully describes what is experienced *as it is experienced*–and this I have some difficulty in understanding. What I saw a moment ago might be described as a moving patch of red, as a cardinal bird, or as the 37th bird in the tree this morning; and all these descriptions may be true. But the phenomenalist seems to hold that what I saw I saw *as* a moving patch of red, which I then interpreted as a glimpse of a cardinal bird. The physicalist seems to hold that I saw it *as* a cardinal bird, and only by analysis reached the description of it as a moving patch of red. Both apparently agree (since I made no count) that what I saw I did not see *as* the 37th bird in the tree this morning. Now just what is in question here? Let me try to formulate it.

Perhaps what is claimed is, for example, that I did not know at the time of the visual experience that the bird was the 37th in the tree this morning. Presumably, then, if I had known this at the time, I would have seen what I saw as the 37th bird in the tree this morning. The criterion suggested here apparently is that what I see I see as what I know it to be at the moment I see it. Did I, then, see what I saw just now not only as a cardinal bird but also as more than 5000 miles from China, as weighing less than Aristotle, as containing cells full of protoplasm, etc., etc.? Surely this formulation will not do.

Perhaps, then, the degree to which I strip experience of all interpretation and nullify the effects of analysis is to be judged by the degree to which my description avoids commitment to any doubtful statement about the experience, so that to say I saw a red thing is nearer to my raw experience than to say I saw a red bird, and this in turn is nearer than to say I saw the 8th red bird on the tree this morning. In that case, I am more faithful to my experience if I describe what I saw as a vertebrate rather than as a bird, and faithful to the ultimate degree if I describe it as a cow-or-non-cow. But if we leave such objections aside, does the proposed criterion at least decide between the phenomenalist and the physicalist on the score that I commit myself to less that is dubious if I say that I saw a moving patch of red than if I say that I saw a cardinal bird? It can be argued, on the contrary, that the commitment in the one case is merely different from, not less than, that in the other. For while it is true that if I say I saw a red patch I do not commit myself to having seen a cardinal bird, it is equally true that if I say I saw a cardinal bird I do not commit myself to having seen a red patch. If the light was yellow, I may have seen an orange patch; and if the bird was against a bright sky, I may have seen a black patch. And I may on a given occasion be either more or less uncertain about the presented color than about the kind of bird. Clearly, this criterion of epistemological priority is no better than the first.

I might take up other formulations, but perhaps I have said enough to suggest the difficulty of rating perceptible individuals or predicates of them on a scale of immediacy. My purpose here has not been to show that no criterion of epistemological priority is tenable, but merely to indicate why the claim of greater immediacy for either a physicalistic or a phenomenalistic basis is not easily supported. And as a matter of fact, I think there is no need to make such a claim. An economical and well-constructed system of either sort provides an orderly and connected description of its subject matter in terms of perceptible individuals. It does not have to be further justified by evidence that its orientation reflects some subtle epistemological or metaphysical hierarchy.

It must be borne in mind, therefore, that in applying the terms "phenomenal" and "physical" I am not attempting to distinguish between the immediate and the nonimmediate. For example, the two-dimensional field of vision is clearly different from three-dimensional physical space; a change of position in either may or may not be accompanied by a change of position in the other. But to make this distinction is neither to say that phenomenal space is immediately given and physical space the result of a process of construction nor, on the other hand, to say that physical space is directly apprehended and the visual field the product of a subsequent analysis. I happen to be primarily concerned with problems treated by phenomenalistic systems; and the systems to be considered in this book are all phenomenalistic. But it should be clear by now that I neither make nor recognize any claim that any of these systems has an advantage of epistemological fidelity over physicalistic systems in general or over an alternative phenomenalistic system.

A physicalistic system, I have intimated, is one that takes some perceptible physical individuals as its basic units, while a phenomenalistic system is one that takes some perceptible phenomenal individuals as its basic units. However, since I have not given any precise general definition for either "physical" or "phenomenal" or "perceptible", the terms "physicalistic" and "phenomenalistic" remain useful but somewhat unprecise labels for effecting a rough grouping of systems. Accordingly, a good deal of caution must be exercised in making general pronouncements concerning systems of either kind. A specific system, after all, is of much greater importance than any label that may be attached to it; and it is to specific systems that I shall devote most of my attention. Nevertheless, one or two general points concerning phenomenalistic systems call for brief mention here.

In the first place, a system that is both phenomenalistic and nominalistic can have only a finite ontology. Our powers of perception are not infinite in either scope or discrimination; that is to say, there are only finitely many mininal phenomenal individuals. A phenomenalistic system that is also platonistic may admit an infinite hierarchy of classes as well, but the universe of the nominalistic phenomenalistic system contains at most the finitely many sums of one or more phenomenal individuals. Under such a system whatever is said that ostensibly concerns the infinite, the nonindividual, or the physical must be explained consistently with such a limited ontology.

In the second place, speaking from outside a phenomenalistic system, one may describe its basis as solipsistic, may say that its basic units are comprised within a single stream of experience. But speaking from the point of view of the system itself, this is an anachronism. For the basic units of such a system are not taken as belonging to a subject and representing an object. They are taken as the elements in terms of which must be construed whatever objects, subjects, streams of experience, or other entities the system talks about at all.[6] These basic units are *neutral* material. A presentation, for example, may be at

[6] Cf. Carnap, *Aufbau*, Section 65, "Das Gegebene ist subjektlos".

once part of a stream of phenomena and part of a physical object; but this will depend upon the later constructions of the system. Aside from all such questions, however, it is clear that whatever are the differences among whatever persons there are, the constructions of a phenomenalistic system are discussed and tested quite as intersubjectively as those of any other system.

5. REALISTIC AND PARTICULARISTIC SYSTEMS

Among phenomenalistic systems, perhaps the most important distinction depends upon whether nonconcrete qualitative elements (such as qualia) or concrete spatially or temporally bounded particulars (such as phenomenal events) are taken as basic units. In the former case the system may be called *realistic*, in the latter *particularistic*.

This distinction is independent of the distinction between platonism and nominalism, even if we suppose for the moment that the basic units of a system are its only individuals. Whether a system is platonistic or not depends upon whether it admits any nonindividuals as entities. Whether a system is realistic or not depends upon whether it admits nonparticulars as individuals. A system that admits qualitative nonconcrete individuals is realistic even though it excludes all nonindividuals, while a system that admits no individuals other than particulars is not realistic even though it admits nonindividuals, such as classes of particulars. As emphasized earlier (II,3) the nominalist's decision to exclude nonindividuals does not by itself determine what may properly be regarded as an individual. Thus a system may be platonistic and realistic, or nominalistic and realistic, or platonistic and particularistic, or nominalistic and particularistic.

These are not merely academic alternatives; I shall have occasion to illustrate them all in the course of following chapters. Even though the two distinctions in question indeed stem from the traditional controversy between nominalism and opposing views, they actually cut across one another. The considerations that might lead one to favor, or even insist upon, a nominalistic system are in general not the same as those that might lead one to favor a particularistic system. One may be unwilling to construe repeatable qualities as individuals and yet be willing to construe qualities, or something else, as classes. On the other hand, one who finds the notion of such entities as classes hopelessly elusive may nevertheless feel that a quale, as an identifiable phenomenal element in experience, can be construed as an individual quite as legitimately as can concrete particulars. An inability to understand how a class, which does not differ in content from the sum of its members, can yet be an entity distinct from that sum does not necessarily carry with it an inability to accept as individuals the qualia that, so to speak, result from a latitudinal analysis of the stream of phenomena–as contrasted with the longitudinal analysis that yields particulars. And one may accept qualities as individuals without accepting as individuals those 'attributes' or 'properties' that are regarded as the designata of predicates when predicates are regarded as designating. To regard the color carmine an an individual is not to regard has-the-color-carmine as an individual; for even if both carmine and a particular that is carmine in color are taken as individuals, the statement that the particular has the color carmine requires use of a two-place predicate in addition to the names of the individuals.

It may be argued that the nominalist renounces classes, for example, just because they are abstract, and that he will therefore refuse to admit abstract qualities as individuals. I should say rather that the nominalist renounces classes on grounds of incomprehensibility; that what he refuses to construe as an individual is decided on the same grounds; and that not all nominalists regard as incomprehensible whatever happens to have been called "abstract" (see further VI, 1).

I touch on all these points here in the sketchiest way because at the moment I am not concerned with defending any type of system, but only with indicating that neither logical nor psychological consistency demands that preference for a platonistic or a nominalistic system be accompanied by a preference for, respectively, a realistic or a particularistic system.

One might suppose that at least particularistic systems share with nominalistic systems the advantage, as compared with their respective opposites, of a sparser ontology. Actually, this is not the case. For compare a particularistic system that recognizes as individuals all sums of one or more concrete presentations with a realistic system that also recognizes as individuals certain qualia into which these events may be divided. Assuming that both systems are nominalistic, it appears at first that the second system must have the more comprehensive ontology; but since qualia are the atoms of the second system, and since there are in all probably fewer qualia than presentations, and since the individuals recognized by each system are all and only the sums of one or more atoms of that system, the realistic system in question probably recognizes fewer individuals than the comparable particularistic one (cf. VII, 7).

But all arguments concerning the acceptability and relative merits of systems of these two kinds are best left until we come to consider specific systems in detail. In the end, we may well find that particularistic and realistic systems are about equally reasonable and that each has certain advantages and disadvantages. Indeed, I think that to look upon the choice between the two types of system as the major problem is to hark back to empty and timeworn disputes over whether qualities or particulars have some kind of metaphysical or epistemological priority. A more important problem is to give an intelligible account, in the form of a system of either type, of the relationship between qualities and particulars. Philosophical intuition and personal predilections may perhaps incline some to prefer one type of system even to the point of rejecting the other completely; but it is not necessary to discredit either type of system in order to justify the choice of the other. The two types of system embody alternative treatments of a given subject matter, rather than opposing points of view.

So far, I have considered three principal distinctions among systems. In the case of platonistic and nominalistic systems (Chapter II), I have tried to suggest why I shall avoid the platonistic commitment to nonindividuals if I can. In the case of physicalistic and phenomenalistic systems, I have explained my skepticism concerning the arguments offered for the superiority of either, but have expressed my interest in the phenomenalistic approach and said that I shall consider only phenomenalistic systems in this book. In the case of particularistic and realistic systems, I have indicated that I feel that the honors are about even, but have left the justification of this view to later chapters where systems of both types will be examined in detail.

6. INTRODUCTION TO THE PROBLEMS OF ABSTRACTION AND CONCRETION

Once we have chosen the basis for a system, we turn to problems of defining other wanted predicates. If the system is a particularistic one with concrete individuals of one sort or another as its basic units, the first problems will normally be those of constructing qualities–i.e., of defining "is a quale" and kindred predicates. If, on the other hand, the system is typically realistic, with qualia or other qualitative individuals as basic units, the first problems will ordinarily be those of constructing concrete individuals. The problem of interpreting qualitative terms in a particularistic system, of constructing repeatable 'universal' 'abstract' qualities from concrete particulars, I call the *problem of abstraction*. The problem of defining predicates pertaining to concrete individuals in a typical realistic system, of constructing unrepeatable concrete particulars from qualities, I call the *problem of concretion*. The two problems are so closely parallel that to explain one is to explain a good deal about the other. The parallel between the two, as well as certain interesting differences, will be dealt with in later chapters. In the present section, I shall briefly survey the problem of abstraction only, since the first systems we shall consider are particularistic.

The basic familiar proposal is to construe a term like "white" as ambiguously naming many particulars rather than as naming any individual uniquely. But this is no more than a very general program; it provides no method for defining a specific quality term in a system. If the stock slogan "White is the class of all white things" is intended as a definition, its admissibility depends on the prior introduction into the system of the predicate "(is a) white thing". If that predicate is supposed to be defined, the question how it is defined still faces us and differs little from the original problem. If the predicate is supposed to be primitive, and if the slogan is taken as an instance of a general schema for defining quality terms, then a distinct primitive predicate is required for each quality term. This would give a system so uneconomical as to be virtually no system at all, and would not solve the problem that is really in point: how to define quality terms upon a very narrow particularistic basis, consisting perhaps of a single primitive predicate of particulars.

Now of course we do not expect in practice to give actual definitions for all the specific quality terms. The theoretical procedure for doing this ought eventually to be described, but the immediate problem is rather the definition of such a general predicate as "is a quality". The above slogan suggests only that a quality is a class of all those things that have some one quality in common. But such a definition accomplishes nothing; for it makes use of the predicate of classes "is a class of all those things that have some one quality in common", and the very problem before us is to define some such predicate in terms of a predicate of particulars.

The primitive usually adopted is a two-place predicate of similarity or resemblance, explained as applying between any two things that in ordinary language are said to have some sense quality in common. A quality is then defined as any class of things that resemble each other, provided this class is included in no larger class of this kind.

The charge that such a definition involves a covert circularity is not justified. Even though we explain similarity as applying where identity in some quality obtains, qualitative identity is *systematically* defined in terms of similarity and not vice versa. A quality class of things is defined by means of the primitive similarity predicate; and the statement that two things have an identical quality is then construed as affirming that they belong to the same quality class.

It is about in this stage that the matter was left until quite recently. But the proposed definition is vague, and a satisfactory clarification is not easy to find. Is a 'class of things that resemble each other' a class of things *a* ... *n* such that a chain of similarity relationships runs from *a* to *n*? Any class of things that in fact have a common quality will satisfy this condition; but unfortunately many other classes will also satisfy it. For example, *a* may resemble *b* in color, and *b* resemble *c* in texture, while *a* does not resemble *c* in any quality; hence even though the three had no common quality, they would be members of a single quality class under the proposed interpretation of the vague definition. Accordingly, this formulation must be rejected as too broad. Since there will be more classes satisfying the definiens than there are qualities, the requirement of isomorphism is violated. The practical test of our definiens, of course, is whether it describes precisely those most-inclusive classes of things which, in ordinary language, have some one quality in common; for we rely upon the connection between qualities and such classes as our guarantee of isomorphism.

Perhaps, however, a 'class of things that resemble each other' is a class of which every member resembles every other member. This condition is likewise satisfied by all genuine quality classes. But suppose we have three things describable as follows:

r is white, round, and of maximum hardness;

s is black, square, and of maximum hardness;

t is white, square, and *not* of maximum hardness.

Each of the three things is similar to each of the others; each two have some common quality. Hence, by the proposed definition, all three would belong to the same quality class. But since the three nevertheless have no common quality and so cannot belong to any genuine quality class, the definition is still faulty.

Nor can we correct the trouble be amending the definiens to read "class of things similar to each other in some one respect throughout". This would be circular indeed; for the problem of defining such a class in terms of our primitive is the same as the problem of defining a class of things all of which have a common quality. Similarity is, of course, similarity in *some respect or other*; if two things are similar they are similar in some 'respect'–i.e., they have a common quality. When, however, we say that more than two things are all similar *in one respect* we are in effect saying not only that each two are similar but also that some 'respect' in which any two are similar is the same as that in which any other two are similar. In other words, we are saying that every two members of the class have in common some quality that every other two members have, and this amounts to saying that all members of the class have a common quality. But this is precisely what we are trying to define in terms of our given two-place primitive predicate. And so we find ourselves back where we started.

Although we have not so far found any solution to the problem of abstraction, I shall not for the moment pursue the search further. In this brief survey, my purpose has been merely to introduce the problem and the way of approaching it, as a preliminary to the detailed discussion of a typical particularistic system. Although this problem is not the only one to be dealt with in such a system, it is among the first and most important.

CHAPTER V

THE SYSTEM OF THE *AUFBAU*

1. INTRODUCTION

In *Der logische Aufbau der Welt*, Carnap's purpose is to sketch a particularistic system much more coherent and comprehensive than any proposed before. Starting with the few positive findings of earlier particularistic thought, he avails himself of the methods and the model of symbolic logic in developing his constructions. If the resulting system remains rudimentary and inadequate in many respects, the advance it makes must still not be overlooked.

In discussing Carnap's system, I feel that exposition and criticism are equally important, since few of my readers are likely to be familiar with the system of the *Aufbau* in sufficient detail for the purpose at hand. I shall therefore explain each step of the system before commenting upon it. As for my critical comments, I must point out at once that Carnap does not claim to offer more than an imperfect sketch of the system. He is well aware that many difficulties remain, and he calls attention to some of them himself. The purpose of my critical scrutiny is not to disparage his accomplishment but to determine just where the remaining problems lie and perhaps to pave the way for their solution.

It must be mentioned that the *Aufbau* was published over thirty years ago and that the author's point of view has changed considerably since then. He seems no longer to attach much importance to the approach exemplified by that book. For one thing, the extreme difficulty of dealing with the physical world on a phenomenalistic basis has led him to prefer a physicalistic system. I have already (IV,4) suggested my own views on this point and shall come to it again later (XI,5),[1] but the matter is irrelevant to the purpose of the present chapter.

Since it seems advisable to present the commentary close to the exposition of each step in order to avoid the need for too many long-range references, I shall first explain each section and then usually comment upon it immediately.

---

[1] For recent discussions of Carnap's position and my own concerning the *Aufbau*, see the introduction to the second edition (Hamburg: Felix Meiner Verlag, 1961), and my article "The Significance of *Der logische Aufbau der Welt*", in *The Philosophy of Rudolf Carnap* (La Salle, Ill.: Open Court Publishing Company, 1963), pp. 545–558; (reprinted as "The Revision of Philosophy" in *P & P*, pp. 5–23) and Carnap's reply, pp. 944–947.

The exposition (printed in small type) will seldom be a mere summary; in some cases it will be fuller than the original, and it will often incorporate new devices and diagrams that seem to me illuminating. Nevertheless, I shall try to keep it faithful to the intention of the original, reserving to the commentary (printed in standard type) everything in the nature of criticism, correction, or incidental reflection.

I shall confine my attention to the actual system presented, omitting all introductory and parenthetical material. Moreover, I shall deal only with the first of the three major divisions of the system. The 'lower steps' have to do with the phenomenal, the 'middle steps' with the physical, and the 'upper steps' with the psychical. But only the lower steps are worked out in any detail in the *Aufbau*, and only these fall within the scope of the present book.

Carnap sets forth his system in two parallel series of sections. The first–Sections 67 to 93–contains an informal but careful account of the system; the second–Sections 108 to 120–presents a concise and formal statement of it. The two may conveniently be combined for discussion. Each passage of my exposition will be preceded by the numbers of those sections of the *Aufbau* which it covers; for example, "67 (109)" will indicate that I am dealing with Carnap's Section 67 and his parallel Section 109. These numbers will be immediately followed by a literal translation of his title for the earlier section.

In outline, Carnap's procedure is as follows. He first decides upon his basic units or 'ground elements'. Then he describes the method by which he will seek to construct qualities, and with the requirements of that method in mind then selects his primitive relation. Because he chooses a relation that will also provide him with the means for ordering qualities, he finds it necessary to define quality classes by a somewhat indirect route. That accomplished, he discusses the question how such qualities may be classified into the various sense realms: auditory, visual, tactual, etc. Concentrating next upon the visual qualities, he deals with the problem of separating and distinguishing the spatial qualities, or locations, from the color qualities; this is not another subclassification of the visual qualities but, as we shall see, a further abstraction from the more concrete color-spots he first abstracted from the ground elements. He deals then with the matter of ordering colors and visual places. Finally, he defines 'sensations', which presystematically are certain concrete proper parts of his ground elements.

The general apparatus Carnap uses in constructing his system consists of the ordinary logic of statements together with the full calculus of classes or functions. Thus his system is platonistic (II,3). He does not use the calculus of individuals or any of the predicates belonging to it.

2. THE BASIC UNITS

67 (109). THE CHOICE OF GROUND ELEMENTS: 'ELEMENTS OF EXPERIENCE'

The basic units chosen for the system are called *elementarerlebnisse* [which I shall hereafter abbreviate as *erlebs*]. They are full momentary cross sections of the total stream of experience. They are limited to a least perceivable segment of time, but are otherwise unlimited except by the bounds of immediate experience itself; each includes all the experience at a moment. Their selection as basic units or 'ground elements' does not imply that erlebs are actually separate units marked off in experience, but merely that assertions can be made about, and relating, such places in the stream of experience.

Erlebs are preferred as ground elements because they seem to be the closest practicable approximation to what is given, namely, a single unbroken stream of experience. Such other possible basic units as least perceivable particles of experience are obviously reached only after considerably more analysis. Since we want our system to constitute a 'rational reconstruction' of the knowing process, epistemological primacy is an important factor in the choice of ground elements.

In preceding sections (64, 65),[2] Carnap has given notice that his system would be phenomenalistic and thus in one sense solipsistic; but he has pointed out that strictly the ground elements (i.e., the basic units) are subjectless, since such terms as "subject", "subjective", and "objective" have to be defined at a later stage in the system. Although the system is solipsistic in a loose sense, it by no means embodies a 'solipsism of the present moment'. It commences with a set of momentary erlebs that together exhaust the total temporally long stream of experience. That does not solve the problem of ordering the erlebs in time, nor of distinguishing past, present, and future; but it does make unnecessary, for instance, the construction of past experience solely in terms of memory images and other present experience.

Carnap seems to hold that experience is originally given in a single stream and that lesser elements are known only through subsequent analysis. Others might argue that experience is no more given in one big lump than it is given in very minute particles, and that the single stream is as much the product of an artificial synthesis as the minimal particles are the products of an analysis. To me the debate seems a futile one, for I do not know how one would go about determining what are the originally given lumps. But in any case, despite the emphasis Carnap here lays upon epistemological considerations, the validity and interest of his system do not seem to me to depend at all upon whether it is the sort of epistemological reconstruction he claims;

---

[2] In the present chapter, references to sections of the *Aufbau* can be distinguished from references to sections of the present work by the fact that the latter will always contain roman numerals indicating a chapter.

accordingly, I think it is unnecessary to defend the choice of basic units by maintaining that they are epistemologically primary.

Carnap does not draw the distinction I have drawn (IV, 5) between particularistic and realistic systems, and is not concerned with realistic systems at all. But since his basic units are concrete, each of them containing all the experience at a moment, his system is a typically particularistic one according to my terminology.

In the formal presentation of his system, Carnap of course first introduces his primitive relation and then defines the class of erlebs as the field of that relation.

68.[3] THE ERLEBS ARE INDIVISIBLE UNITS

The ground elements of any system must be treated as indivisible under that system. If conceived of as divisible, they would not be ground elements but rather constructs of the particles into which they are divisible, and hence would be derivative. The particles would then be the true ground elements. Consequently, with erlebs as ground elements we have the problem of constructing what are normally regarded as constituent parts of them (i.e., their various qualities), and indeed everything that is said to be contained within an erleb. To accomplish this a special 'constitutional method' is needed.

Since no erleb is part of another, every erleb is a minimal basic unit. And as we have seen (III,11), the minimal basic units of a system are best treated as the indivisible atoms of that system.

As a matter of fact, the indivisibility of the basic units of Carnap's system follows trivially from the fact that, since he uses no calculus of individuals, no individual is systematically part of any other. As a result, the basic units of any such system are at once the minimal basic units of the system and the atoms of the system. Moreover, no individual can be defined in the system as a sum of individuals. For all practical purposes, indeed, the basic units constitute the range of the individual-variables in such a system.

Carnap next (Sections 69 to 73) investigates the general method for deriving qualities from concrete basic units. Then he chooses a specific primitive relation of erlebs and proceeds to apply the general method in order to define qualities in terms of this primitive.

3. METHODS OF CONSTRUCTION

69. THE TASK OF DEALING WITH INDIVISIBLE UNITS

Given the ground elements and a primitive relation, the problem is to find a way of defining what, for the system, can stand for the qualities or constituent parts of the

[3] Sections 68 to 76 are devoted to matters preparatory to the actual constructions and so have no parallels in the second, formal, statement of the system.

ground elements. Since the ground elements are literally indivisible under the system, this process of definition is called 'quasianalysis'.

All we have to start with in addition to logic is a list of the ground-element pairs that belong to the primitive relation.

In practice, of course, we are never actually given the pair list of the primitive relation of any comprehensive system. We are given rather an explanation designed to enable us to determine whether any given pair belongs to the list. The point is, though, (as I pointed out in III,1) that the list is the *most* that is to be assumed. Given the list, we require no further information.

The present section merely states the problem. In the following section (70), Carnap discusses a process of proper analysis–analysis of wholes regarded as actually divisible–in order to use this as a model for the strictly analogous process of quasianalysis. Section 70 is therefore an extremely important one.

70. THE PROCESS OF PROPER ANALYSIS ON THE BASIS OF A RELATION DESCRIPTION

Assuming that we have a set of divisible individuals, the problem is to analyze them into parts solely upon the basis of a given relation description (i.e., the pair list of a given relation). As an example, suppose there are several things before us, each having one or more of a set of three colors. Without knowing what colors each thing has, but knowing only for which pairs of things the relation of 'color kinship' (i.e., possession of an identical color) holds, we are to define each of the three colors by isolating the several color classes. A color class is the class of all and only those things that have a certain color either alone or in combination with other colors. The color class *red* would include, for instance, all the things of a given set that were entirely red and all that were partially red. Multicolored things would belong to several color classes. Since we have assumed that there are three colors in our example, there will be three color classes. How may these be defined, using only the relation list?

Color classes are those satisfying two requirements. (*A*) Of the members of a color class, each pair must be 'color akin'; that is, each pair must be on the list of pairs for which that relation holds. A color class always conforms to this rule because of the fact that all its members have a color in common. If we knew which classes had a color in common, we could find the pair list of color kinship. Since we are instead supposed to know the pair list, we reverse the rule in order to find color classes. (*B*) The color classes must also be the greatest possible classes satisfying the first requirement; in other words, no thing outside the class may be color akin to all things in the class. Otherwise not only the class of things possessing a common color, but also each subclass included in it, would constitute a separate color class. The second rule serves its purpose, however, only if certain unfavorable circumstances (see below) do not obtain. We assume for the purposes of our example that these do not obtain.

If we have correctly prescribed the rules, we need only apply them to the given relation list in order to derive the three color classes. These we label arbitrarily "k_1", "k_2", and "k_3", since we cannot tell whether one of these classes is the color class of "green" or "red" or "blue" but only that it is some color class. A thing belonging to k_1 will then have the color k_1; a bicolored thing is one that belongs to two classes, say k_2 and k_3, but not to the third. We can accordingly determine how many colors a thing has, and which other things possess the same color; and thus we have analyzed the things into their color components.

All this can be better fixed in mind if we represent each thing, in the example given, by a group of letters in which each letter stands for a color the thing has.[4] For instance, thing 1 may be partly (a specific shade of) red and partly (a specific shade of) green: this we write "1. *rg*". Let us suppose there are just six things, as in Table II.

1. *br*	4. *g*
2. *b*	5. *r*
3. *bg*	6. *bgr*

TABLE II

The conditions of our problem restrict us, however, to much less information than is given in Table II. They do not permit us to know what, or how many, colors each thing has, or what color is possessed in common by two things. We are given simply a list naming each pair of things that possess *some* color (or letter) in common. This relation list is given in Table III.

1:1	2:2	3:3	4:4	5:5	6:6
1:2	2:3	3:4	4:6	5:6	
1:3	2:6	3:6			
1:5					
1:6					

TABLE III

Now the problem is to determine solely on the basis of Table III, without reference to Table II, which classes among the six things are 'color classes'–that is, classes made up of all and only those things whose names in Table II have some one letter in common. We may test whether our effort has been successful by comparing the result with Table II.

The rules for accomplishing the classification are: (*A*) Each class must be such that every pair of members of the class is listed in Table III. (*B*) Each class must be such that no thing that is not a member is paired by Table III with all the members; that is, each class must be a greatest possible class satisfying rule *A*.

Proceeding to apply these proposed rules, we may commence by listing all the things paired with thing 1 by Table III. These are 1, 2, 3, 5, 6. But the class {1 2 3 5 6} breaks rule *A* since pairs 2:5 and 3:5 are not listed in Table III. Hence we must drop either 5 alone or both 2 and 3 if we are to have a color class. Dropping 5, we have left the class {1 2 3 6}. This satisfies the two requirements: (*A*) every pair in it is listed in Table III; (*B*) nothing excluded (i.e., 4 or 5) is paired by the table with every member of the class, since pairs 3:5 and 1:4, for example, are not on the list. The class {1 2 3 6} we therefore label "k_1".

To construct a second color class we may drop from the preliminary class first suggested ({1 2 3 5 6}) numbers 2 and 3 instead of number 5. This leaves us with the class {1 5 6}, which satisfies both rules and which we may call "k_2".

A third color class may be found by listing, say, all the things paired with 3. These are 1, 2, 3, 4, 6. But the pairs 1:4 and 2:4 are not listed in Table III, so that we must

[4] Carnap does not use this device. Here and in later sections I amplify his exposition, in accordance with the policy stated in the introduction to this chapter (Section 1).

drop either 4 or both 1 and 2. If we drop 4 we will have simply class k_1 again, not a new class; but if we drop 1 and 2, we will have the class {3 4 6}, which is a third color class k_3.

Further investigation will reveal no fourth class answering the two requirements. We have then these three classes,

$$k_1\ \{1\ 2\ 3\ 6\};$$
$$k_2\ \{1\ 5\ 6\};$$
$$k_3\ \{3\ 4\ 6\}.$$

Naturally none of these includes another, for the included class would not satisfy the second requirement, would not be a 'greatest possible' class in the intended sense.

Now having constructed these classes solely by the use of Table III and the rules, let us glance back at Table II. We see that k_1 includes all and only the *b*-things; that k_2 includes all and only the *r*-things; and k_3 all and only the *g*-things. Thus our rules have enabled us to discover these true color classes on the basis of a list which told us merely what pairs comprised two things having *some* unit in common.

However, if the things of the original set, listed in Table II, had been such that certain 'unfavorable circumstances' obtained, the method would not have succeeded. For instance, if a certain color, say *r*, happened to occur only in things in which *b* also occurred, separate color classes for the two could not have been constructed. The class for *r* would have to be included in (be part of or equal to) the class for *b*, and this is barred by rule *B*. The following example will illustrate the point. Suppose the things were as in Table IV instead of as in Table II.

1. *br*	4. *g*
2. *b*	5. *bgr*
3. *bg*	

TABLE IV

Then the relation list would be that given in Table V.

1:1	2:2	3:3	4:4	5:5
1:2	2:3	3:4	4:5	
1:3	2:5	3:5		
1:5				

TABLE V

Working as before and listing all things paired with 1, we have the class {1 2 3 5}, which satisfies the two requirements and may be called "k_1". If we look back at Table IV to test this result, we see that the class k_1 is the color class for *b*. Yet we see from Table IV that our rules should be such as to give us also the class {1 5} as one of the color classes, since it is the color class for *r*. This class cannot, however, be derived by the proposed rules since thing number 2, for instance, is paired by Table V with both 1 and 5; and therefore, according to rule *B*, 2 cannot be excluded from the class. The same is true of 3, so that we are brought back to the class for *b*: {1 2 3 5}.

If such an 'unfavorable circumstance' holds–if a certain color occurs only as the companion of another, never appearing except in a thing in which the other appears–the proposed method of analysis will not work. We must therefore assume that the unfavorable condition does not obtain. The assumption is justified by the fact that the probability of the unfavorable condition's holding diminishes as the number of things increases and as the average number of colors to each thing decreases, barring any systematic connection between the two colors. That is, the more things there are, and the fewer the colors most of them possess, the greater will be the likelihood that a given color will occur in some one thing in which some other given color does not occur.

The term "similarity circle" is introduced as a general name for classes constructed by means of the two rules on the basis of the list for any reflexive and symmetric relation. Whether a similarity circle is a quality class will depend on the primitive relation chosen.

Although this section describes only a process to which quasianalysis is analogous, the analogy is so close that we have here the essence of Carnap's method of dealing with the problem of abstraction.

The need for making such an extrasystematic assumption as that indicated, to meet what I shall hereafter call the 'companionship difficulty', obviously constitutes a serious defect in the proposed method. Not only is there no way of determining within the system whether or not the unfavorable circumstance in question obtains in a sphere to which the system is applied, but the assumption that it does not obtain for the class of all erlebs–to which Carnap plans to apply it–is ill-founded. The statement that the probability that the assumption is true increases as the number of particulars involved increases and as the average number of qualities to each decreases, *barring any systematic connection between qualities*, may at first sound plausible. But while it is true that the number of those particulars (i.e., the erlebs) with which Carnap is shortly to deal is very large, it is not true that the average number of qualities to each is small: on the contrary the totality of one's experience at a given moment ordinarily embraces a multitude of different qualities. What is more important, however, is that inclusion of the proviso "barring any systematic connection" seems to make the required assumption circular. Is the excepted connection anything more than that constant co-occurrence of two qualities that constitutes the unfavorable circumstance? If not, then we are only saying that this circumstance is not likely to obtain where it does not obtain. On the other hand, if such a relationship as that between two very similar colors constitutes a systematic connection, it seems that we are arbitrarily ruling out of consideration just the cases where the unfavorable circumstance is most likely to obtain. The probability is by no means negligible that, for example, a certain shade of blue occurs only in erlebs in which another shade just slightly lighter also occurs.

There are other difficulties. Carnap speaks of the unfavorable circumstance described as one among several but defers mention of others until later (Section 72). One of these I must explain here, however, since it is virtually disastrous to the proposed construction. Suppose that the things we start with were not those of Table II but were rather as shown in Table VI. In this group the unfavorable condition above discussed obviously does not obtain; that is, no color occurs only in those things in which some one other color occurs.

1. *bg*	3. *br*	5. *b*
2. *rg*	4. *r*	6. *g*

TABLE VI

The relation list will be that given in Table VII.

1:1	2:2	3:4
1:2	2:3	3:5
1:3	2:4	4:4
1:5	2:6	5:5
1:6	3:3	6:6

TABLE VII

Suppose now someone suggests that things 1, 2, 3 constitute a color class. Let us apply the formal tests: (*A*) All possible pairs (1:2, 1:3, 2:3) are listed in Table VII; hence the first requirement is satisfied. (*B*) No other thing is paired with all these three things (4 is excluded because 1:4 is missing; 5 because 2:5 is missing; 6 because 3:6 is missing); hence the second rule is satisfied. Therefore {1 2 3} must be a color class. But glance back at Table VI. What color is common to 1, 2, and 3? *None at all.* Hence our rules have failed badly even though the unfavorable circumstance first discussed is not present to trouble us.

This failure stems from a serious difficulty. Whenever there are three things such as,

1. *br* 2. *rg* 3. *gb*,

or four such as,

1. *abc* 2. *bcd* 3. *cde* 4. *dea*,

or other groups similarly composed, the whole process collapses *because in such a set, although every pair of things has a quality in common, no quality is common to all the things in the set*. That this can happen invalidates the method of analysis on which the first constructions of the *Aufbau* are based. We cannot even legitimately class this as another 'unfavorable circumstance' that we might–if we were willing to run the risk[5] –assume not to obtain. For

[5] Although the question of the probability of this difficulty's occurring is complex, rather extensive investigations seem to show that the probability in the case of the *Aufbau* system is by no means negligible.

we can hardly state what the requisite assumption is without begging the question. It seems that we should have to assume that no class of things satisfying the two requirements (*A* and *B* above) lack a common quality; but this amounts to assuming outright that the proposed method of analysis will work.

In other words, the very problem of abstraction is to define, in terms of the primitive relation of concrete individuals, the conditions under which a set of such individuals have a common quality. So to formulate the definition that it excludes false quality classes like those just illustrated constitutes the crux of the problem. Until we can do that we have not gone far toward solving our problem, although we may have clarified it.

For purposes of identification, I shall call the difficulty just explained 'the difficulty of imperfect community'. Since the process Carnap finally adopts in his system differs in certain respects from the one here described, we shall have to watch for any modification that may meet this difficulty; but it will become apparent that the only changes have to do with quite different problems, and that this basic difficulty remains unremedied. Carnap promises a later discussion of the various 'unfavorable circumstances'; but as we shall see (81, below), that proves to be rather cursory and quite unsatisfactory.

71. THE PROCESS OF QUASIANALYSIS

Quasianalysis is completely analogous to proper analysis; but since we now regard the elements in question as indivisible, we must regard the constructed quality classes not as means for defining actual constituent parts of these elements but as themselves *quasi-*constituent-parts. So long as the primitive relation used is symmetric and reflexive, like that of Section 70, the process can be carried out in exactly the same way. An indivisible element belonging to a quality class has that quality class as a quasi-constituent-part. Thus quasianalysis does not divide a ground element but correlates it with others in certain classes. The predication of a quality to an element will accordingly be construed as signifying the membership of that element in the class constituting the quality.

The term "*quasi-constituent-part*" ("*Quasibestandteil*") indicates not that Carnap is dealing with pseudo qualities, but that he is construing the relation that qualities bear to erlebs otherwise than as a part-whole relation. Quasi-constituent-parts are not quasi-qualities but they are in a sense quasi-parts. Conformably, the process for deriving them takes the name "quasianalysis". On the other hand, Carnap's use of the prefix in the term "Quasimerkmal" seems uncalled for. Unless "Merkmal" strongly suggests to him a part-whole relationship, there seems no more reason for "*Quasi*merkmal" than for "Quasiqualität", which he seldom if ever uses.

72. QUASIANALYSIS ON THE BASIS OF A PART-SIMILARITY RELATION

The primitive relation used in Section 70 and 71 was the part *agreement* or part *identity* [*Teilgleichheit*] of two things–their possession in common of at least one quality construed as either a genuine part or a quasi-part. We now use instead the relation of part *similarity* between two things–the possession by one of a quality that is like, to at least a certain minimum degree, some quality of the other. For example, two things having no common quality might still stand in a specific part-similarity relationship to each other if some color of one is sufficiently similar to some color of the other. Such a relation is wider than part identity since every part-identical pair is also a part-similar pair.

The explanation of quasianalysis based on a part-similarity relation can conveniently be divided into several sections:

(i) Suppose that our original set of elements is

1. *abd* 2. *cgh* 3. *agh* 4. *irs* 5. *dzy*,

where each letter stands for a specific quality, and relative similarity of qualities is paralleled by relative alphabetical proximity. Let us regard two elements as part similar if and only if some quality of one and some quality of the other are either identical or are such as to be indicated by alphabetically adjacent letters. Then the relation list for the part similarity in question will list not only the part-identical pairs 1:3, 2:3, 1:5; but in addition such pairs as 1:2 (because "*b*", in the description of 1, is alphabetically next to "*c*", in the description of 2), 3:4, 2:5, and so on.

If a complete pair list is given on this basis, the next step is to construct classes conforming to the two rules explained earlier. Then, assuming that our process works properly, for each such similarity circle of the relation in question, there will be some two adjacent letters such that each member of the similarity circle will have at least one of the two qualities indicated by these letters. One such class, not selected from the set of five elements above listed, but from some much larger set, might contain the following things,

axy	*btr*
aqr	*bsf.*
akl	
amn	

Included within this class are two quality classes; the problem now is formally to isolate such quality classes.

(ii) The procedure to be employed for this task depends upon the (classial) overlapping of the similarity circles constructed. For if each class has as members elements containing at least one of two qualities indicated by adjacent letters, we shall obtain such classes as are shown in Figure 2

a . . .	b . . .	and	b . . .	c . . .
a . . .	b . . .		b . . .	c . . .
a . . .	b . . .		b . . .	c . . .

Figure 2. Similarity circles containing common quality class.

(where the strings of dots stand for other qualities that do not here need to be specified). Thus a class of elements having a single quality in common can be isolated by defining it as the overlap of two such similarity circles. Of course, the original part-similarity relation might have been broader, obtaining between any two things having qualities indicated by letters not more than two places from each other in the alphabet. Each member of a similarity circle would then have some one of three qualities; and each similarity circle would embrace three quality classes. The rule for isolating a quality class thus needs to be generalized to read: a quality class is any most comprehensive class left undivided by the overlapping of all the similarity circles. This applies whatever the original span of similarity is taken to be. However, it is not quite accurate and will need some revision later (see 81(112)).

(iii) The point of introducing this roundabout way of defining qualities is that it provides us with the means for *ordering* them. Two similarity circles *A* and *B* constructed on the basis of a part-similarity relation are 'nearer together' than *A* and *C* if the overlap of *A* and *C* is included within the overlap of *A* and *B*. This suggests a method for ordering the contained quality classes and thus for constructing such orders as the color sphere. Because the analysis based on part identity gave isolated quality classes instead of overlapping similarity circles, it supplied no corresponding principle of order.

(iv) We must still assume that certain unfavorable circumstances do not obtain. Suppose, for example, that the given set of elements is as follows,

1. *axy*	4. *bxr*	7. *xrg*
2. *ars*	5. *bgm*	8. *zqf*,
3. *ahg*	6. *btr*	

and that the part-similarity relation in question is represented by alphabetical identity or adjacency. Then elements 1 through 6 make up the class we want to isolate first. But application of our two rules will give us the class of elements 1 through 8, for each member of this class is part similar to every other. Thus our rules do not work properly; they do not give us a class all the members of which have at least one of just two qualities indicated by alphabetically adjacent letters. We must assume that the given set of individuals will not be such as to give rise to difficulties of this sort.

The shift to a part-similarity relation, occasioned solely by exigencies of the problem of order, ameliorates none of the troubles already encountered. Indeed, they become even more serious. The difficulty explained in (iv) above is essentially the difficulty of imperfect community discussed in my comment on Section 70. Now, however, the probability of the difficulty's arising is increased, since (as the example shows) a similarity circle of the wrong kind will result whenever every member of the class we want first to isolate happens to have some quality that is similar to, not necessarily identical with, some 'foreign' element. Again, in order to meet a broadened form of the companionship difficulty, we have to assume not merely that no quality is a constant companion of another, but also that no quality occurs only in elements that have at least one of two (or more if the span of similarity taken is greater) other qualities similar to each other.

It is the process of quasianalysis on the basis of a part-similarity relation that Carnap will use in his system. The process to be described in the following section is not a further revision of this, but a supplementary process that he will use later in the system for another purpose.

73. QUASIANALYSIS ON THE BASIS OF A TRANSITIVE RELATION

The relations so far employed have been reflexive and symmetric but neither transitive nor intransitive. Quasianalysis based on a relation that is transitive as well as symmetric and reflexive is useful for many constructions and has great formal simplicity. No nonmember of a similarity circle of such a relation will bear that relation even to *one* member of the circle; for if it were so related to one it would, because the relation is transitive, be so related to all, and therefore be a member of the class. These similarity circles will consequently by mutually exclusive. A similarity circle of a transitive relation is called an 'abstraction class'; it can be defined simply as a not-null class of all elements bearing the given relation to a given element, since with a transitive relation this will give the same result as the two rules needed in earlier sections.

The transitive relation we are to use will be such that its abstraction classes will themselves be the quasi-constituent-parts we want; they will not need to be subdivided like the similarity circles we get in earlier constructions.

Carnap will select a nontransitive primitive relation, then construct certain quasi-constituent-parts (e.g., certain color-spots) by applying the method of Section 72; he will then define a transitive relation holding between the qualities so constructed; and then (see 90(118)), using that transitive relation, will derive other qualities (e.g., colors) by means of the process mentioned in this section. The *members* of the classes to be constructed by this second method will thus be *themselves classes*; so that the quality classes constructed on the basis of the transitive relation will be classes of classes, and qualities of qualities. They bear to the quality classes first constructed a relation like that which the latter bear to the ground elements. They represent a further stage in the analysis, an abstraction from the qualities first abstracted from concrete elements.

The process of quasianalysis on the basis of a transitive relation is beset by all the difficulties discussed above. In some of the examples used above in explaining these difficulties the basic relations are transitive, though alternative examples with nontransitive basic relations can be readily given.

4. THE CHOICE OF A PRIMITIVE

The sections so far discussed have dealt with the choice of basic units and with general methods for quasianalyzing indivisible elements. Carnap now goes about choosing a primitive relation to use in effecting a quasianalysis of erlebs.

I omit Sections 74 and 75. In the following two sections, Carnap considers specific relations of part identity and part similarity among erlebs before finally choosing his primitive relation, in Section 78.

76. PART IDENTITY

To say that two erlebs x and y are part identical, or in symbols that x Gl y means that x and y have a quasi-constituent-part in common. Were this relation taken as primitive we could define *sense qualities* by means of it; but, for reasons to be explained, we shall choose a different primitive relation, then define sense qualities in terms of it, and only then define part identity, in terms of sense qualities.

The term "sense quality" is used in a special way that will become clear as the discussion proceeds. It refers to certain grosser qualities which are first derived and from which such ordinary qualities as colors are abstracted in turn.

77. PART SIMILARITY

As noted earlier, qualities cannot be ordered in their several realms (e.g., the color sphere, the tone scale) on the basis of a part-identity relation between erlebs. Such orders depend upon 'neighborhood' relations that do not reduce to part identity. Therefore we need a different relation; for example, part similarity, symbolized by "Ae" and obtaining between every two erlebs that either have a common quality or are such that a quality of one is similar to a quality of the other.

By "similarity" (symbol "Aq") as distinct from part similarity is meant the relation of likeness between sense qualities. Two color sensations are similar if they are either identical in or alike to a certain degree in hue *and* chroma[6] *and* brightness *and* location in the visual field. Aq, the similarity of qualities, will be systematically defined in terms of Ae, the part similarity of erlebs, after qualities have been derived. Both Ae and Aq are reflexive.

It is to be especially noticed that what Carnap here calls a 'color sensation' is a complex quality determined in both color and place but not in time; in other words, it is a *repeatable color-spot*. It is a sensation character, not a sensation event. Two erlebs are considered to be part similar (in their visual qualities) only if a color-spot in one is similar in all respects to a color-spot in the other. This of course is a very special kind of part similarity, for which the necessary and sufficient condition is the possession by the two erlebs in question of similar colors *at* similar places.

As for the similarity involved in part similarity, one naturally asks how similar two color-spots must be for the erlebs possessing them to be considered as standing in the part-similarity relation in question. Carnap would perhaps say that it makes no difference what degree of similarity is required so long as it is fixed. One might thus regard two color-spots as similar if and only if they are at least as much alike as some arbitrarily designated pair. A standard pair of qualities of each other kind would also have to be designated. Such a method of arriving at a relation list seems rather awkward and unreliable; but this, of course, does not affect the actual constructions once a relation list is available.

[6] Color terminology varies a good deal. I use "chroma" throughout for what is sometimes called "saturation" or "purity"; the latter terms carry a suggestion of actual mixture that is preferably avoided.

78(108). RECOGNITION-OF-SIMILARITY AS PRIMITIVE RELATION

Even the relation described in Section 77 is unsuitable as our primitive because it provides no means for defining the *temporal* order of erlebs. Consequently we choose recognition-of-similarity–that is, recognition of part similarity–which is also epistemologically more fundamental. Since an erleb contains all the experience at a moment, each erleb has a distinct position in time. To know that two erlebs are part similar, we must compare a memory image or afterimage of the earlier with the later; and a recognition is thereby involved. Recognition-of-similarity (symbol "Er") differs logically from simple part similarity in being asymmetric "x Er y" means that a memory image or an afterimage of x is part similar to y, and therefore that x is earlier than y. This supplies us with that power of determining temporal direction that we need for constructing temporal order.

In terms of Er. we can easily define the symmetric relation Ae (part similarity); for x Ae y if and only if either x Er y, or y Er x, or x belongs to the field of Er and is identical with y. But had Ae been chosen as ground relation, Er could not have been defined in terms of it.

The supposed ability to construct temporal order is not gained without some expense. Our study in Chapter III has shown that an irreflexive asymmetric predicate like "Er" has a complexity-value of 3 as against a value of 2 for a join-reflexive symmetric predicate like "Ae". It would have been desirable, of course, to find a way of constructing temporal order, as well as spatial order and other quality orders, without having to make any such special provision for it. But the problem is difficult; and it is questionable whether even the means with which Carnap here takes care to provide himself actually make possible a satisfactory construction of temporal order (see 87(120)).

Carnap's argument that recognition-of-similarity is 'epistemologically more fundamental' than Ae would seem to assume that memory images and afterimages are epistemologically as fundamental as erlebs. Indeed, if we have to explain how two erlebs are known to be part similar, why do we not have to explain in a similar way how past erlebs are known at all? That would mean that our basic units, if they are to be epistemologically fundamental, would have to be the present erleb and various memory images and afterimages, rather than a series of erlebs exhausting the complete stream of experience.

Moreover, Carnap's epistemological argument covers only cases where the present erleb and some past one are in question. It does not account for our apprehension of part similarity between two past erlebs or between a future one and any other. If Er obtains only between the present erleb and past ones, it does not obtain between every pair of part-similar erlebs; accordingly, aside from the fact that the Er-list would thus change completely at every moment, the relation would hardly be adequate for the construction by Carnap's methods of either temporal order or qualities.

The best course is simply to admit that the whole epistemological argument in terms of memory images and the nature of recognition is irrelevant, and adopt Er as the relation that obtains between every two erlebs *x* and *y* if *x* is part similar to and earlier than *y*. This in effect is Carnap's procedure throughout most of the book (cf. 101).

5. DEFINITION OF QUALITIES

With his primitive relation chosen, Carnap is now almost ready to apply the general method of quasianalysis described in Section 72. He first warns against a likely error.

79. THE POSSIBILITY OF FURTHER DERIVATIONS

We might think we could define the part identity (Gl) of erlebs as follows: *x* Gl *y* if and only if *x* is part similar (Ae) to the same erlebs as *y*. But if there were three erlebs,

1. *abc* 2. *cde* 3. *axy*,

then 3 would be part similar to 1 but not to 2. Thus although 1 and 2 are actually part identical, since both have the quality *c*, they would not be part identical by this definition.

This shows that we must remember that such a relation as Gl or Ae obtains between two erlebs if *any* quality of one is identical with or similar to any quality of the other; and that where different pairs of erlebs are concerned, the relationship may result from agreement in different qualities. If we ignore this, our constructions may easily be faulty.

We cannot, then, define Gl directly from Ae, but must first construct qualities by quasianalysis and then define Gl as the relation holding between any two erlebs that have (membership in) a common quality (class).

The defect in the rejected definition is obviously akin to the difficulty of imperfect community, which Carnap's constructions do not succeed in meeting.

80(111). SIMILARITY CIRCLES

Part similarity (Ae) of erlebs has already been defined; and on the basis of Ae, similarity circles may be constructed as explained in Section 72.

Each erleb, having many qualities, will belong to several similarity circles. Each similarity circle will thus overlap others. The overlapping will be called *essential* if the classial product of the two circles consists of one or more complete quality classes; it will be called *accidental* when the product is less than a complete quality class.

This is a preliminary explanation of the distinction between essential and accidental overlapping. Since in the system the distinction is to be used in the definition of quality classes, a way must be found of drawing the distinction without reference to quality classes. This Carnap attempts in the course of the following section.

Essential overlapping has already been illustrated in section (ii) of the exposition of Section 72. Accidental overlapping results from the fact that any erleb having any two different qualities will belong to two similarity circles that contain no quality class in common. Both kinds of overlapping will be further explained and discussed below.

81(112). QUALITY CLASSES

Two quality classes overlap accidentally if any erlebs have both the qualities in question. For example, the quality class of all erlebs having the color-spot *a* and the quality class of all erlebs having the color-spot *b* will overlap through having as common members all erlebs having both color-spot *a* and color-spot *b*.

Similarity circles may overlap either essentially or accidentally. They overlap essentially if they contain one or more complete quality classes in common. Suppose a similarity circle *S* of Ae consists of all erlebs that have color-spot *a* or *b* or *c*, while *S'* consists of all erlebs that have color-spot *b* or *c* or *d*. Then *S* and *S'* overlap essentially, since they both include the quality class of all erlebs having the color-spot *b* (and also the quality class of all erlebs having the color-spot *c*). On the other hand, two similarity circles that contain no complete quality class in common will still overlap accidentally if either contains a quality class that accidentally overlaps a quality class contained in the other.

How the occurrence of accidental overlapping affects the definition of quality classes may be made clear with the help of an example. Suppose first that we have four similarity circles, as in Figure 3.

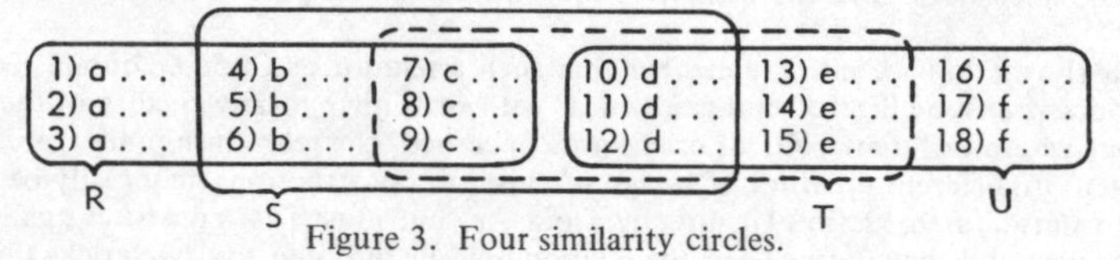

Figure 3. Four similarity circles.

Now the tentative definition of a quality class runs: a class *k* of erlebs is a quality class if (I) *k* is wholly included in every similarity circle to which any member of *k* belongs, and if (II) for every erleb *x* that does not belong to *k*, there is at least one similarity circle to which *x* does not belong but which entirely contains *k*. Let us test whether any other than a true quality class, like that composed of erlebs 7, 8, and 9, can satisfy this definition. Proviso I effectively excludes classes like {4 7 8 9} and {3 6 7 8 9} and {2 5 8}, and indeed any class that cuts across the boundary line of any similarity circle. Proviso II bars any class properly included in the class {7 8 9}; the class {7 8}, for example, violates the proviso because 9, though not belonging to this class, belongs to every similarity circle that includes the class.

Suppose, however, there should also be the erleb

19) *afh*.

It would belong to similarity circles *R* and *U* both (but not to *S* or *T*). Thus *R* and *U* overlap accidentally, with the result that the class {1 2 3 19}, although plainly a quality class, would not be one by definition; for 19 belongs to a similarity circle (*U*) that does not contain the whole class {1 2 3 19}. Proviso I must consequently be altered.

In seeking the appropriate revision, we note that an accidental overlap usually has few members. More particularly, the number of erlebs having a given quality *x* will be much greater than the number that have both that quality and one or another of the qualities within any one 'foreign' similarity circle (i.e., similarity circle that does not have as members all erlebs having the quality *x*).[7] Thus no class is a quality class if any

[7] This assumption is stronger than the one Carnap sets forth here, but is the one actually required for his purpose.

considerable number of its members belong to any similarity circle that does not contain the whole class. To avoid the erroneous results of our tentative definition, we may accordingly revise Proviso I to read: "if *k* is wholly included in every similarity circle that contains at least half of *k*".

How the revised definition works can be illustrated by means of our example above. We must remember that 19 belongs to *R* and *U* but not to *S* or *T*. The new proviso still bars such classes as

(i) {4 7 8 9} because the subclass {7 8 9}–more than half the class–is included in *T*, but the whole is not;

(ii) {3 6 7 8 9} for the same reason;

(iii) {2 5 8} because {5 8}–more than half–is included in *S*, but the whole is not;

(iv) {7 8 9 10 11 12} because {10 11 12}–half the class–is included in *U*, but the whole is not.

But now the proviso no longer excludes the genuine quality class {1 2 3 19}; for while 19 belongs to *U*, which does not include the whole quality class, the erleb 19 is less than the class {1 2 3 19}.

Of course, if there should be two more erlebs, 20 and 21, both having the two qualities *a* and *f*, we should still be in trouble, for then {19 20 21}–half the class {1 2 3 19 20 21}–would be included in *U*, which does not include the whole class. Hence {1 2 3 19 20 21} would be excluded as a quality class although it would actually be one. Such a contingency can never arise, however, under our assumption that as many as half the members of a given quality class can never belong to any similarity circle that does not contain the whole class.

Thus again we see that our method works only under assumptions to the effect that various unfavorable circumstances do not obtain. Further investigation would show, however, that the conditions necessary for the proper working of our formal method of quasianalysis are also necessary for the proper working of the actual knowing process itself; that is, if the unfavorable conditions in question obtained, then the quasianalysis that is intuitively accomplished in actual cognition would not lead to the normal result.

Carnap does not try to dispose of accidental overlapping by assuming that it never occurs, for that would amount to assuming that no two different qualities ever occur in one erleb. But in seeking to frame a definition that will work despite accidental overlapping, he finds it necessary to make a certain assumption. The earlier assumption that no quality always accompanies another has to be broadened; we must now assume that fewer than half the erlebs having any given quality *q* also have one or another of any group of mutually similar qualities other than *q*. By thus presupposing that accidental overlapping is always numerically 'trivial', Carnap is able to differentiate it in his system from the essential overlapping that he wants to use for isolating quality classes.

The need to resort repeatedly to such dubious extrasystematic assumptions is obviously objectionable. Carnap seeks in this section to justify the assumptions he needs by arguing that if they were not true, certain qualities could not in fact be abstracted from experience; in other words, that the assumptions must hold for all known qualities. Were he dealing with such qualities as the aspects of colors and were the only required assumption to the effect that no two of these qualities always occur together, the argument might have a certain plausibility. Perhaps if a certain hue and a certain chroma never occurred except in combination with each other, they would never be abstracted from the unitary color and known as two. However, Carnap is not dealing with intermingling qualities like hues and chromas but with color-spots. While it might seem reasonable to suppose that a hue and a chroma always occurring together would never be known as separate, it seems very much less reasonable to suppose that a certain red color-spot near the top of the visual field and a certain green color-spot near the bottom would never be known as separate if the two occurred in all the same erlebs. And no trace of plausibility remains if one seeks to stretch the argument to justify all the varied and much stronger assumptions found to be required in the sections we have studied and those that are to follow.

A minor revision is required in the definition of quality classes explained in this section. As the definition stands, the unit class of erleb 19 in our example would be a quality class. It obviously satisfies the revised Proviso I; and it also satisfies Proviso II, since every erleb other than 19 lies outside of some similarity circle that includes this class. To correct this difficulty, Proviso II should be changed to read: "if *k* is included in some similarity circle but not in any class except itself that also satisfies Proviso I".

6. FURTHER CONSTRUCTIONS

82(113). DOES ONE PRIMITIVE RELATION SUFFICE?

Part identity may now be easily defined: two erlebs *x* and *y* are part identical (*x* Gl *y*) if and only if they are both members of some one quality class.

The tasks remaining in the 'lower steps' of the system are to construct temporal order, spatial order, and such qualitative orders as the color sphere. For these constructions, as well as for those in the middle and upper steps of the system, it seems likely that Er will be the sole primitive required.

I am concerned solely with the lower steps of the system. The problems to be dealt with in the next few sections are these: to group the qualities into their several sense realms (visual, auditory, etc.); to distinguish each of these realms from the others; to order the erlebs in time; to analyze the visual qualities further, i.e., to abstract colors and places from the grosser color-spots; and to order these colors and places.

I omit the nonessential Sections 83 and 84.

85(114, 115). SENSE CLASSES

Two qualities are similar (Aq) if and only if every member of one is part similar (Ae) to every member of the other. That is to say, two qualities are similar if and only if every erleb in which one occurs is part similar to every erleb in which the other occurs.

Sense classes of quality classes may now be defined in terms of Aq. Two qualities belong to the same sense class if they are similar (Aq) or are connected by a chain of Aq pairs. Any two tones, for example, are connected by such a chain, but a tone and a color, or a tone and an odor, are not. A sense class consists of any most comprehensive class of qualities every two of which are so connected; it is a class of all the qualities of one sense or feeling.

Even assuming that quality classes have been properly defined, the success of the definition of Aq depends upon certain favorable conditions similar to those that had to be assumed earlier. Otherwise, for example, a pair of two *dis*similar quality classes would satisfy the definition of Aq if every member of each happened to be similar to every member of the other through other qualities these erlebs possessed.

86(115). DEFINITION OF THE VISUAL SENSE

The qualities in each sense class may next be ordered by means of the relation Aq. Although dimensionality will not be defined here, each such order will have a determinate number of dimensions. Tone qualities form a two-dimensional array, color-spots a five-dimensional one. All the others are two-, three-, or four-dimensional.

Since the visual sense class alone is five-dimensional, it can be defined as that sense class which has five dimensions. Any objection to defining it through so nonessential a feature mistakes the nature of a constructional definition, which is based upon logical but not necessarily psychological equivalence.

Until the orders of the other sense classes are better known, one cannot feel entirely confident that none is five-dimensional. But what is more serious is that according to ordinary mathematical definitions of dimensionality–and Carnap's definition[8] of dimension number (DZ) does not appear to differ in this respect–any finite array is zero-dimensional. Since we are here concerned with a finite set of erlebs and therefore a finite set of quality classes, a sense class can thus hardly be multi-dimensional according to ordinary mathematical usage. One might, of course, define a different sort of dimensionality, pertaining to finite arrays; but Carnap does not discuss the rather difficult problem of formulating an appropriate definition.

87(120). TEMPORAL ORDER

Erlebs are orderable in time by means of the primitive relation Er; for, as explained in Section 78, x is earlier than y if $x\,\mathrm{Er}\,y$. This fixes the order of precedence only for part-similar erlebs, since Er holds only where part similarity holds. However, the precedence of erlebs near together in time will usually be determinable since such erlebs will usually be part similar, possessing in common some persisting quality. And because precedence is transitive, the precedence of erlebs that are temporally distant and wholly dissimilar

---

[8] For which he refers us to his *Abriss der Logistik* (Vienna: J. Springer 1928), Section 33. A revised and enlarged version in English has been published (New York: Dover Publications, 1958) under the title *Introduction to Symbolic Logic and its Applications.*

will then be mediately determinable in many cases. Yet there may be exceptions. If so, only a preliminary and incomplete temporal order can be constructed; but this may form the basis for constructing later the full temporal order of physical things. In the actual process of cognition, the order that rests on 'temporal perception' is incomplete and needs to be filled in with the help of established physical and psychological laws.

Among all the aspects in which particulars may be like or unlike, temporal position acquires a peculiar status in Carnap's system through the kind of basic units chosen. Comprising all the experience of a single moment, each erleb has but one time; and at each time there can be but one erleb. Many erlebs may have a color or place in common, and one erleb may have many colors and places; but no two erlebs are simultaneous, and no one erleb occurs at or occupies more than one moment.

This one-to-one correlation makes it unnecessary to define times as classes of erlebs (as color-spots have been defined) or as classes of classes of erlebs (as colors and places will be defined). Instead, each erleb unambiguously stands for a time; and temporal order is an order not of classes, but of erlebs. The means for determining which of two part-similar erlebs is earlier has been consciously incorporated in the ground relation.

With respect to the possible incompleteness of the time order based on Er, Carnap argues that actual psychological time order also is incomplete relative to physical time order. This is beside the point; for the incompleteness of the temporal Er-order is an incompleteness of temporal order among erlebs themselves, not merely relative to a physical time order. Perhaps psychological time order, even considered by itself, is actually broken and ambiguous; but there is no evidence that this incompleteness is in any way parallel to that of the time order based on Er (see 78 (108), above).

This section on time order stands somewhat apart from the main systematic development. Carnap now turns to the problem of defining and ordering places and colors.

88(117). DERIVATION OF VISUAL-FIELD PLACES

We have not yet abstracted colors and places from color-spots. Two color-spots stand equally in the relation Aq whether they are identical in color and near in place or identical in place and similar in color. We might try first to distinguish place identity from color identity in the following way. No two different color-spots that have the same place will occur in any one erleb. In order for a single erleb to belong, for instance, to the quality class of erlebs having color 1 at place p, and also to the quality class of erlebs having color 2 at p, that erleb would have to have two colors in the same place. All place-identical color-spot classes are therefore separate from one another; and this suggests defining as place identical any two color-spot classes that have no common member. But that would be inaccurate because, while all place-identical color-spot classes are thus separate, not all pairs of separate color-spot classes are place identical; it may just happen that two differently placed color-spots do not occur in any one erleb.

A more satisfactory method is to define visual-field places directly, as 'abstraction classes' of the relation of separateness (symbol "Fre") between quality classes; that is, to define a place class as a greatest possible class of discrete color-spot classes. This radically diminishes the probability of trouble like that met above, for the following reason: a color-spot class not really belonging to a class of place-identical color-spot classes could not accidentally become a member of a place class as defined unless that color-spot class were discrete not only from one other color-spot class but from all the color-spot classes belonging to the proposed place class. In other words, a color-spot not place identical with the color-spots forming a place class must, to belong accidentally to that place class, happen never to occur in any erleb in which any of the color-spots properly belonging to that place class occurs. Such a color-spot would thus have to occur solely in erlebs in which the place in question was entirely missing. This is very unlikely to happen. Furthermore, the class of such a troublesome color-spot would probably be separate from other member classes of other place classes as well and therefore would belong to several place classes. This multiple relationship would betray the class as but a false member of at least some of these place classes.

An example may help to clarify some of these points. Let us suppose that there are but fifteen erlebs, represented by columns 1 to 15 in Table VIII (16 being ignored for the moment).

	1	2	3	4	5	6	7	8	9	10	11	12	13	14	15	16
x	*R*	*R*	*G*	*BK*	*G*	*G*	*R*	*BK*	*BK*	*G*	*R*	*G*	*G*	*G*	*R*	–
z																
w			*BU*			*BU*					*BU*			*BU*		
t																*Y*

TABLE VIII

The lower-case letters stand for places; the capital letters for specific shades of color: "*R*" for a specific shade of red, "*G*" for a specific shade of green, "*BU*" for a specific shade of blue, and "*BK*" for black. Thus erleb 1 has the color *R* at place *x*, erleb 6 has color *G* at place *x* and color *BU* at place *w*, and so on. The blank intersections may be regarded as occupied by *other*, unspecified letters.

Now the following classes of erlebs constitute the indicated color-spots:

The class {1 2 7 11 15}, the color-spot *xR*;

The class {3 5 6 10 12 13 14}, the color-spot *xG*;

The class {4 8 9}, the color-spot *xBK*.

The three classes are of course separate from each other. The class of the three classes is the visual-field place *x*. It is a greatest-possible class of mutually separate classes. Whatever colors may occur in the other places in the erlebs, each color-spot class having one of the other places will have some member in common with at least one color-spot class belonging to *x*, and will consequently not form with them a class of mutually disjoint classes. For example, the color-spot *wBU*, which is the class {3 6 11 14} has in common with the color-spot *xG* the members 3, 6, and 14; and has in common with the color-spot *xR* the erleb 11.

Note that the distinctness of the several places will not be obliterated by the fact that the member classes of *each* place class exhaust *all* the erlebs; for places are defined as classes of *classes*, not of erlebs.

If now, however, there is an erleb like 16 in which the visual-field place *x* does not occur at all, and if in that erleb alone a certain shade of yellow (*Y*) occurs at place *t*, then the color-spot *tY* is the unit class of erleb 16. This class is separate from all the other member classes of place *x* and hence would belong to place *x* according to the definition of a place class. This would give us a false place class, instead of the true place class for x. However, since tY would also be separate from all other color-spot classes belonging to place class t, it would belong to place t as well as to place x according to the definition. It might therefore be possible to eliminate the unwanted cases on the basis of this multiple membership; but further investigation would be needed to discover the precise rule for ascertaining to *which* of the place classes such a color-spot as tY properly belongs.

By "visual-field place" is meant literally "minimal perceived part of the visual field". As I move my eyes in looking at a fixed dot, that dot–though remaining in one physical place–is seen successively at several different visual-field places. The construction of *physical* places belongs to the much later 'middle steps' of the system.

The sort of difficulty that Carnap points out in his own definition will arise under other circumstances than those he mentions. Suppose there are just the erlebs 1 through 15. The color-spots *wBU* and *xBK* are separate classes and none of the other specified color-spots are separate from both. Let us assume also that the unspecified colors are so distributed that no color-spot class is separate from both these two classes. Then the class having as its members the quality classes *wBU* and *xBK* will satisfy the definition of a place class, although it is obviously not a true place class. Since *xBK* belongs to two different classes satisfying the definition, we shall know that something is wrong. But for such cases, as well as for the more unusual ones that Carnap mentions, some way must eventually be found for picking out the true place classes from among those that satisfy the incomplete definition proposed.

Incidentally, Carnap's use of the term "abstraction class" in the present section is inconsistent with the explanation he gave earlier (73); for the relation of class separateness (Fre) is certainly not transitive. Only in Section 90 is the process that was described in Section 73 actually applied.

89(117). THE SPATIAL ORDER OF THE VISUAL FIELD

Two color-spot classes are *identical in place* (Glstell) if they belong to the same place class.

Two place classes are *near* (Nbst) each other if some color-spot class belonging to one is similar (Aq) to some color-spot class belonging to the other. The order of places may be constructed on the basis of Nbst.

Two actually near places fail to be so according to this definition only if no color that ever occurs at one is similar to any color that ever occurs at the other. This is extremely unlikely to happen.

Since Carnap does not go on to construct the order of visual-field places, it remains an open question whether Nbst is indeed an adequate tool for the purpose.

90(118). THE ORDER OF COLORS

With colors, the sequence of construction is a little different. Whereas place classes were first defined, then place identity, then place nearness, in the case of color we define color identity, then color classes, then 'color nearness' or rather color likeness.

As a preliminary to defining color identity, we first define *color identity in near places*. Two color-spots that are near in place stand in this relation if they are similar (Aq) to exactly the same color-spots among those having some third place near to both. More precisely, color-spots *x* and *y* are color identical in near places if and only if (1) the place class *r* to which *x* belongs and the place class *s* to which *y* belongs are near places, and (2) for some place class *t*, near to both *r* and *s*, every color-spot *z* belonging to *t* is similar to *x* if and only if *z* is similar to *y*.

Color identity (Glfarb) is then easily defined: two color-spots *x* and *y* are color identical if and only if they are connected by a chain of color-identical-in-near-place pairs. Two color-identical color-spots having widely separated visual-field places will reveal their color identity by the existence of such a connecting chain.

A *color*, or color class, is then definable as an abstraction class of Glfarb, i.e. as a most-comprehensive class of color-identical quality classes. The members of a color *x* are the color-spot classes *x*-at-place-1, *x*-at-place-2, and so on for all the places at which *x* occurs in any erleb.

The definition of likeness between colors is entirely analogous to that of nearness of places, running: two colors are alike (Nbfarb) if and only if some color-spot belonging to one is similar to some color-spot belonging to the other. The familiar order of the color sphere may be constructed by means of this relation.

The present constructions depend rather heavily upon the occurrence in the total set of erlebs of a sufficient variety of combinations of colors and places. Each of these definitions would fail to operate properly if certain combinations happened to be missing. However, the more nearly complete the variety of combinations exhibited, the greater is the probability that the difficulty of imperfect community will arise to disrupt earlier constructions. As the assumptions required pile up, some seem to conflict with others; however, see my Section 7 below.

Carnap does not go on to construct the order of the color sphere, or to analyze colors into their constituents–hues, chromas, and brightnesses. Except for incidental remarks in Sections 91 and 92, which I shall omit, he here concludes his treatment of phenomenal qualities. Before going on to the 'middle steps', however, he deals with the definitions of 'sensations'–the phenomenally minimal concrete particles of experience.

93(116). 'SENSATIONS' AS PARTICULAR CONSTITUENT PARTS OF EXPERIENCE

We have seen that two erlebs that extrasystematically have a common quality are systematically said to belong to some one quality class. We now have to distinguish the occurrence of a quality in one erleb from its occurrence in another. Such an occurrence or 'sensation' is defined as the ordered couple of an erleb and a quality class to which the erleb belongs. The ordered pair is an unrepeatable quasi-constituent-part of the erleb.

Two sensations are simultaneous if they are such quasi-constituent-parts of the same erleb; or in other words, if the first components of the couples defining them are identical.

It is to be noted that sensations, like qualities, are defined not by a literal analysis of erlebs but by a synthesis that effects a quasianalysis. Erlebs, as ground elements, are no more divisible into sensations than into qualities.

In the formal statement of the system, the definition of sensations precedes those of colors and places. Only the quality classes first derived, such as color-spots, are coupled with erlebs to define sensations; for a sensation is determinate in all respects (including time) appropriate to a given sense.

What Carnap calls 'sensations' are presystematically those events that are the minimal perceivable concrete parts of the stream of experience. But even though they are concrete, they must of course be defined in this system by a process of abstraction, for the system recognizes no proper parts of erlebs as individuals.

7. CONCLUSION

Carnap's sketch of the lower steps of his system makes progress toward clarifying certain problems, and illustrates methods for attacking them. It is nevertheless incomplete and frequently defective. The most serious problem left unsolved is perhaps what I have called 'the difficulty of imperfect community'. In an unpublished manuscript of 1923, *Die Quasizerlegung,* which Carnap kindly lent me, he further investigates the problem of defining quality classes. He shows how his constructions might be improved so that the assumptions required to exclude the difficulty of imperfect community no longer run counter to those needed for later constructions. But the treatment, as he recognizes, is unsuccessful in avoiding the need for doubtful extrasystematic assumptions. Until such assumptions are rendered unnecessary, the project of defining qualities in terms of some similarity relation of concrete elements remains unrealized.

I shall not at present try to find other means of meeting these difficulties, but turn at once to the construction of a realistic system. If in the course of that construction the problem of concretion is solved, the question may then be raised whether a parallel procedure will help in solving the parallel problem of abstraction.

CHAPTER VI

FOUNDATIONS OF A REALISTIC SYSTEM

1. QUALIA AS ATOMS

If we divide the stream of experience into its smallest concrete parts and then go on to divide these concreta into sense qualia, we arrive at entities suitable as atoms for a realistic system. A visual concretum might be divided, for example, into three constituent parts: a time, a visual-field place, and a color. The stream of phenomena can be exhausted by division into these and other qualia as well as by division into erlebs or concreta. Other entities, both qualitative and concrete, may indeed be present; but all are to be explained in terms of those taken as basic for the given system. Just as the atoms for a particularistic system may be chosen from among the various sorts of concrete parts of experience, so the atoms for a realistic system may be chosen from among various sorts of qualitative parts. A final choice may be left until the following section. First I want to discuss certain general aspects of the proposal to treat qualia of any kind as atomic individuals. Let me emphasize here that I am not discussing whether qualia are in fact individuals (whatever that might mean) or justifying taking them as individuals (for anything may be so taken), but rather considering advantages and disadvantages of taking qualia as atomic individuals for the system to be described.

In some ways it is psychologically more natural to begin with qualia and construct concrete individuals out of them than to take concrete individuals as indivisible and construe qualities in terms of these. The habit of looking upon qualities as constituent parts of concreta is deep-seated enough that we have to make a conscious and sustained effort to treat concreta as not analyzable into qualitative components. On the other hand, of course, we are equally unaccustomed to regarding qualia consistently and literally as individuals. The truth is that any systematization calls for departures from the circular ruts worn by common sense; and there is no need to show that any proposed system is the most natural in all respects. It is important, however, to remove certain possible misunderstandings of what is involved in adopting qualia as atomic individuals for a system.

In the first place, to say that qualia are phenomenal individuals discoverable within experience is not to say that any quale is literally separable from the rest of experience. A color quale cannot, indeed, be actually lifted out of the stream of experience, but neither can a complete visual concretum (a color-spot-moment) or a patch consisting of several of these. Concrete phenomenal individuals have as fixed relations to the totality of experience as do qualia; but in neither case does this fact preclude the analysis of experience into such elements. If it be argued that nevertheless a color-spot-moment *could* exist by itself while a color could not, I shall have to look at the evidence for this statement before I understand what is meant. The sort of evidence usually claimed for it is that we can get an image of a concrete individual but not of a quale; but obviously we cannot get an image of a color-spot-moment unaccompanied by others that surround it any more than we can get an image of a color unaccompanied by a place and a time. Analysis indeed, requires that the elements in question, whether concrete or qualitative, be found within the whole in question, and be distinguished from one another, but not that each or any of them can somehow be enthroned in splendid isolation.

In the second place, division of a concretum into qualia is of course not a spatial division. A visual concretum is already a spatially smallest discernible particle of phenomena, and the further analysis into its three component qualia leaves its space undivided. This analysis consists simply of distinguishing the place from both the time and color that, together with the place, make up the concretum. To say that these qualitative elements are individuals is to say that they lie within the field of application of the general predicate "overlaps" and some of its derivatives; each quale overlaps itself, is discrete from other qualia, is part of certain sums of qualia, and so on. But these terms, although their most familiar illustrations are spatial or spatiotemporal, have not been narrowly construed (in II,4) as spatial or spatiotemporal predicates. The spatial and spatiotemporal applications are but some among many; for example, the overlapping of two individuals is spatial or spatiotemporal just in case the two contain a place or place-time in common. All this will be dealt with more fully in due course. At the moment, I want only to point out that not every individual need be or contain a portion of space or space-time.

In the third place, the relation of an indivisible quale to its many instances is sometimes thought to be essentially incomprehensible. As a result, qualia themselves are regarded as incomprehensible and thus as appropriately definable only as classes, which are likewise incomprehensible. I hope to show later that the relation of an atomic quale to its instances is capable of very straightforward interpretation involving no resort to a realm of Ideas or to subtle distinctions of grades of being.

In the fourth place, as pointed out earlier (IV,5), acceptance of qualia as individuals is not inconsistent with the refusal to countenance classes and other nonindividuals; for the views I have called 'nominalism' and 'particularism' are quite independent of each other. The particularist remains a particularist, no matter what classes or other nonindividuals he recognizes as entities, so long as the only *individuals* he recognizes are concrete. The nominalist remains a nominalist, no matter what individuals–qualitative or concrete–he recognizes, so long as the only entities he recognizes are individuals. Nominalism, in other words, excludes all except individuals but does not decide what individuals there are. The nominalist who is also a phenomenalist will refuse to admit that there are any such individuals as electrons or magnetic fields or even *things*, if he cannot construe them as made up of phenomenal individuals. He will then have to construe the terms "thing", "electron", and "magnetic field" syncategorematically. On the other hand, the nominalist who is a physicalist will refuse to admit that concreta or presentations or erlebs are individuals if he cannot define them as made up of physical individuals. Thus two nominalists may end with completely separate ontologies–nominalism is neutral between them. Similarly, whether the color quale white is admitted as a value of a variable in a nominalistic system will depend upon whether it is construed as an individual or as a class of concrete individuals; and this is a question that nominalism alone does not decide. The mere fact that white *can* be construed as a class of individuals does not render it inadmissible; for anything, even the most concrete entity, can be construed as a class of other individuals.

It might be argued that the rejection of classes naturally carries with it the rejection of qualities because both classes and qualities are abstract. If this means only that classes and qualities are alike in not being concrete individuals, one could equally well argue that rejection of classes ought to carry with it rejection of concrete individuals because classes and concrete individuals are alike in not being qualitative individuals. If the argument is based rather on the analogy between the relationship of a class to its many members and that of a quality to its many instances, an equally good analogy can be drawn between the relationship of a class to its many members and that of a concrete individual to its many qualities.

Yet it may be feared that if we admit some qualities as individuals, then we are obliged to admit that there are qualitative individuals corresponding to all the applicable 'attributive' terms of our natural language; that if we admit the quale white as an individual, then we must go on to admit horseness and politeness as qualitative individuals. But we have just seen that acceptance of some concrete individuals does not require one to suppose that all 'substantive' terms refer to individuals; and similarly, recognition of "white" as the name of an individual carries with it no obligation to recognize such terms as "horseness" or "politeness" as names of individuals. Phenomenalists, for example, will admit only phenomenal qualities; and any given philosopher will usually impose much more complex and often not easily formulated conditions. The mere admission of an individual designated by some term belonging to a broad syntactical category by no means obliges us to populate our world with individuals designated by every term belonging to that category.

In the fifth place, and finally, to treat qualia as atoms for a system is not to suppose that they are the units in which experience is originally given. Since presumably there are no qualia floating free of concreta, and all concreta, and all concreta contain qualities, and all erlebs contain concreta, and all concreta are contained in erlebs, and so on, it is pointless to ask which sort of unit comes first. And if the question is thought to be rather in which way experience is packaged on original delivery, I have no idea what criteria would be applied in seeking an answer (cf. IV,4). Moreover, even should it be shown that certain units of experience are in some sense apprehended prior to any operation of analysis or synthesis, this would not at all preclude the selection of other units as atoms, for a constructional system is not necessarily intended as an epistemological history.

All I have tried to do in these preliminary remarks is to prevent certain misunderstandings that might give rise to unfounded objections against taking qualia as atomic individuals for a system. My purpose is not to show that a realistic basis is superior to a particularistic one, for the two seem to me equally legitimate alternatives. Such advantages as a system of one sort may have over a system of the other sort are to be found, I think, by examining their consequences rather than their foundations (e.g., see Section 6 below). The study of systems of both types is therefore in point even for those who may decide in the end that one is definitely preferable to the other.

2. ATOMS OF THE SYSTEM

Not all the qualitative parts of experience are to be taken as atoms for a realistic system any more than all the concrete parts are taken as atoms for a particularistic system. For a particularistic system we divide the stream of phenomena into certain concrete units–e.g., concreta or erlebs–and eventually define in terms of these such others as we want to discuss. Likewise, for a realistic system we divide the stream into certain qualitative units, and eventually define in terms of these such others as we want to discuss. Thus even after the general type of system has been determined, a choice of atoms has to be made from among a considerable number of alternatives. I shall choose as atoms: colors, times, visual-field places, and various nonvisual qualia. Three aspects of this selection call for special comment: first, the inclusion of places and times among qualia; second, the considerations deciding the general level of analysis at which the atoms are chosen; and third, the exclusion of certain familiar qualities such as phenomenal shapes and sizes.

To argue over whether places and times ought to be called qualities would of course be fruitless. But the perverted tradition that exalts places and times as 'primary' must not be allowed to obscure the fact that phenomena, not only physical things, have spatial and temporal aspects. As we have already seen, the distinction that must be drawn between the color qualia of phenomena and the color properties of things must also be drawn between locations in the visual field and physical locations. As I move my eyes past a physically fixed object, I have a series of presentations of it that differ in phenomenal position. Location in the visual field may change when physical position is constant, or remain constant when physical position changes. Moreover, visual-field places bear to one another certain important relations–of the sort Carnap calls 'neighborhood' relations–that the qualia of each of the other categories bear to one another. Also the association between visual-field places and qualia of alien categories is symmetric. A grammatical convention requires us, indeed, to say of a color *h* and a place *k* that *h* occurs at *k* and not that *k* occurs at *h*; but this can be construed as a condensed way of saying that the two occur together and that *k* is a place or time, just as to say that *s* is the husband of *t* is to say that *s* and *t* are married and that *s* is a man. On all these counts, visual-field places seem to have equal status with colors as components of visual concreta.

An analysis that gives us a color and a place as components of a concretum gives us also a time as the remaining component, and much of what I have said of places applies also to phenomenal times. There are admittedly special problems about time, and I shall deal with these in a separate chapter (XI). But I think that in the end none of them offers any obstacles to treating times much as we treat colors and places.

Arguments urging that places and times differ essentially from other qualities are often based on distinctions that apply solely to properties, not to qualia. It is sometimes argued that space is infinite and location purely relative. But place *qualia* are finite in number; and to say that a concretum is located at the center of the visual field is quite as positive as to say that it has a certain red color; arbitrary referents are no more needed in the former case than in the latter. It is true that while we have a number of different color names we have few nonrelative terms for spatial position. But this is largely because the phenomenal positions of ordinary things change so rapidly and so continually that we are seldom practically concerned with other than relative objective location. With color, on the other hand, the change is less rapid and less frequent, so that color qualia are of greater interest. Furthermore, the realm of objective colors is itself finite, and the names of color properties are often used also for color qualia.

At the same time, I do not want to obscure genuine differences among the several kinds of qualia that I propose to treat as atoms of a system. That colors, sounds, places, times etc. are qualia and are to be treated as atoms does not in any way conflict, for example, with the fact that these several categories are ordered in arrays of different shape, or with the fact that only one color can occur at one place at one time though several places can have one color at one time. These distinctions will be recognized and expressly formulated in our system,[1] and some may be actually used in its construction. Whether these differences are in some sense of greater or of less importance than the likenesses among our atoms is a futile question. The ultimate justification for choosing certain atoms has to lie in the overall effectiveness of the system based upon them. I am not trying to prove that the chosen set of atoms is united by some fundamental bond, but only calling attention to certain similarities that make this set much more homogeneous than might be supposed offhand.

The question may be asked why places and times rather than place-times are chosen as atoms. A sufficient reason is that time qualia appear in all the various sense realms, while visual-field places appear only in the visual realm. There may be aspects appropriately called spatial in other realms, but they are quite distinct from locations in the visual field; for example, the sound of a bell is not at any visual-field place. Time qualia, in contrast, are the same in all realms; a toot and a flash may be phenomenally simultaneous.

As a matter of fact, the choice of places and times instead of the more nearly concrete place-times as atoms for a realistic system hardly calls for explanation. Rather, reasons must be given for choosing places instead of visual-field longitudes and latitudes, and colors instead of hues, chromas, and brightnesses; for any choice of more nearly concrete over less nearly concrete individuals as atoms for a realistic system must be regarded with suspicion. In so far as a solution of a general constructional problem of realistic systems depends upon thus not carrying the qualitative analysis of experience below a given level, that solution is *ad hoc* and incomplete. But where no evasion of any important problem results, there may sometimes be

---

[1] With the result that many loose statements like "No two colors can occur at one place and at one time" are soon seen to be false.

good reason for such a choice. One important general criterion for determining the level at which atoms are best taken is a formal one that cannot well be explained until the problems of quality order have been investigated (X,13). But there are often special considerations as well; for example, colors are preferable to their components as atoms on the score that the mode of analyzing colors into components is obscure enough to be still a matter of some dispute.

Furthermore, we shall see later that while the choice of place-times would have resulted in partially evading a crucial problem, the use of colors and places rather than their components actually throws certain problems into relief. The presence of 'multi-dimensional' units among our atoms calls for the development of a theory of quality order applicable to arrays of any complexity; and a more adequate understanding of the whole problem of order is thus achieved than would be the case if we had to deal with linear arrays only (see X,8). Also we are forced to devise a kind of quasianalysis for defining the components of colors, for example; and the method devised gives promise of wider usefulness (X,13).

Since we are to take care that our constructions do not depend for their workability upon the choice of atoms at a particular level of qualitative analysis, the official choice of qualities at a given level is less consequential than it might be. Thus while places, times, colors, etc. are officially taken as atoms, I shall often illustrate how a given construction works if components of these qualia are chosen instead. This will serve as some insurance against dodging important questions and give the formal structure something of the character of a schema of systems. The transformation of many constructions of the official system into constructions of a system based upon atoms chosen at a different level would require only reinterpretation of the primitives and occasional mechanical adjustments.

The decision to choose our atoms from among units at a certain level of analysis does not automatically determine how comprehensive will be the units chosen at that level. However, as I said at the beginning, I here take qualities of concreta as atoms. That is, the colors taken as atoms are single phenomenal shades; the places, times, and atoms of other kinds are likewise minimal distinguished (commonly called 'least-discernible') qualia. Not even presystematically is such a color or place divisible into other phenomenal colors or places. Certain apparent advantages of choosing *more* comprehensive units as atoms turn out to be illusory (IX,3).

Sizes and shapes are not included among the atoms of our system, since they are not respects in which concreta differ. They are of course comparable in many ways to the qualities chosen as atoms. The quale-property distinction is clearly applicable to shapes and sizes; and phenomenal shapes, as well as phenomenal sizes, bear to one another certain similarity relations like those obtaining among qualities of other kinds. Nor does the decision not to take shapes and sizes as atoms have anything to do with the fact that shape and size predicates are ordinarily construed as having spatial or spatiotemporal applications only; for I shall show in Chapter VIII that analogous predicates apply with respect to other kinds of qualities as well. The pertinent difference between shapes and sizes on the one hand and the qualities we take as atoms on the other lies in the way shapes and sizes are related to those other qualities and thus to concrete individuals. Every place, time, and color is a quality of some concretum, but no concretum is oblong, for example; only more comprehensive concrete individuals can be oblong. And whereas a comprehensive concrete individual may still be uniform in color or place or time in that all concrete phenomenal parts of it have the same color or place or time, no individual can be uniformly square or of a given size in that every concrete phenomenal part of it is square or of that size. It is differences of this sort, which will later be dealt with more fully, that lead me to treat shapes and sizes not as atoms–or even as individuals at all–but in quite a different way (VIII).

Our atoms then–to which I shall hereafter confine the term "qualia"–are to be such qualities as single phenomenal colors, sounds, degrees of warmth, moments, and visual locations. While some of these qualities may be presystematically divisible into components, none is presystematically divisible into other phenomenal colors, sounds, etc. As atoms of a system, they are of course not divisible at all in that system and are thus systematically discrete from each other. As finite parts of a finite stream of experience, they are finite in number.

While I have not yet indicated what the basic units of the system–that is to say, the members of the field of the primitive relation–are to be, they must obviously be among those individuals that are sums of one or more of the chosen atoms, for no other individuals are recognized by a system. However, a choice of atoms does not involve any commitment as to whether qualities that are not sums of one or more such atoms might properly be construed as individuals for some other system.

Since a realistic system treats all individuals as made up of qualities, it might be regarded as founded upon the principle that all differences are intensive. The principle is not peculiar to the realistic approach, for the notion of sheer undifferentiated quantity seems hardly intelligible from any point of view. But a realistic system, by reducing everything to qualia, gives special emphasis to the principle. No indissoluble particularity is admitted, and no variation in the quantity of a single quale. Quantities are always correlated with numbers of qualia; and even the most concrete individuals are to be defined in terms of qualia.

3. TOGETHERNESS

In order to draw a close parallel with the system of the *Aufbau*, let us suppose for the present that the general apparatus of our system is platonistic, and that no calculus of individuals is used.[2] The chosen atoms themselves will then be the only individuals recognized by the system.

These individuals are repeatable universals. Each has many instances. We shall in our system have to reconcile the identity of such a quale with the multiplicity of its instances. In a particularistic system, this problem took the form: given particulars, define qualities–each of which may be common to many particulars. In a realistic system, conversely, it takes the form: given qualities, define particulars–many of which may be instances of a single quality.

It should be clear at the outset that, quite apart from systematic restrictions, a nonconcrete single discernible quale is not a whole divisible into several particles that occur in different particulars. Literally, there is no greater quantity of a color or time or place in a thousand concreta than in one; the significant difference is in the number of qualia of some other kind to which a given quale bears a certain relation. If two concreta have the same location, they are diverse in time and perhaps in color. Two simultaneous concreta together endure no longer than one; their diversity is diversity of place and possibly of color but not of time. If concreta *a* and *b* are of color *c*, than "the color of *a*" and "the color of *b*" are merely different descriptions of the single quale *c*. The same can be said of such descriptions as "that portion of color *c* which is in *a*" and "that portion of color *c* which is in *b*"; for that portion of any quale which occurs in a concretum is the whole quale.

But if the whole quale appears in every concretum in which it appears at all, perhaps concreta may be defined as certain classes of qualia. Two concreta that are exactly alike in color will then be certain classes having a color quale as a common member but having different places or times as members. And the unity of a quale will be reconciled with the multiplicity of its instances by construing these instances not as particles of it but as concreta

[2] The present chapter is a preliminary informal discussion of problems to be dealt with formally in Chapter VII. Since in the present chapter both platonistic and nominalistic approaches are considered, platonistic language will be freely used. The development in Chapter VII, however, will be strictly nominalistic.

(i.e., concretum classes) to which it belongs. The construction of concrete particulars as classes of qualities is the exact converse of the particularistic construction of qualities as classes of concrete particulars.

We need, then, a primitive relation that will enable us to differentiate systematically between those classes of qualia that form concreta and those that do not. We saw that the abstracting relation used to define quality classes in a particularistic system is typically a relation of similarity of particulars. The basic concreting relation is a symmetric relation of togetherness among qualia. Such a relation obtains, for example, between a color and a time or place at which the color occurs; likewise between a place and a time at which the place is presented.[3] In general, it obtains between any two atomic qualia belonging to some one concretum. If this seems circular when the problem at hand is to define concreta, we must remember that all explanation of a primitive is outside the system. Definitions of the system make no use of the terms in which such an explanation is framed but only of the primitive itself. In a particularistic system, as we saw, similarity may be described as obtaining between any two particulars having a common quality, even though the problem is to define qualities in terms of that relation.

No two qualia of one category are together; no place is at another place; no two colors appear in a single concretum. Also primitive togetherness may conveniently be construed as irreflexive, so that no quale is said to occur with itself. Thus if x and y are together, x and y are distinct qualia of different kinds.

However, two qualia that are together always belong to the same sense realm, even though some qualia–notably times–may belong to several sense realms. A color and a sound may occur at the same time, but they are not therefore together any more than two places or two colors that occur at the same time are together. As noted earlier, phenomenal sounds are not located at visual-field places. In cases where sounds seem to be rather definitely at places, objective rather than phenomenal places are involved. Qualia that do not belong to the same sense realm never occur together.

A quale ordinarily occurs with numerous different qualia of each of certain other categories. Many colors occur at one place (at various times), a place occurs at many times, many places occur at one time, and so on. From this and the symmetry and other characteristics of togetherness already described, it will be obvious that the relation is not transitive. We have just noticed that a color and a sound that occur at the same time are not together and that two

[3] It is quite possible that a visual-field place may not be presented at certain times, as for instance when the field of vision is narrowed by the closing of one eye.

places at which a single color occurs are not together. And color *c* may occur at place *p*, and *p* may occur at time *t*, without *c*'s occurring at *t*; for *c* may occur at *p* at some other time and not occur at any place at time *t*.

Togetherness must not be narrowly conceived as location in space or time. It is true that in our system at least one member of every togetherness pair in the visual realm will be a place or a time (likewise, incidentally, at least one member of every such pair will be a color or a place, and at least one will be a color or a time). But this is an accident of the level of analysis at which our atoms are chosen; under a system for which hues, chromas, and brightnesses, rather than complete colors, are atoms, a hue and a chroma–as well as a hue and a brightness, and a chroma and a brightness–occur together in every visual concretum. More important is the point that, even with our chosen atoms, togetherness may obtain in other sense realms between qualia neither of which is a place or a time. The location of a quale at a place or a time is thus a special case of the togetherness of two qualia in a single concretum.

By saying that two qualia are together, I shall always intend to make an assertion having the same nontemporal character as a mathematical statement. I shall say that a color is at a place if the color was, is now, or ever will be at that place. And I shall say that a color occurs at a time even though the time be past or future. Thus such expressions as "occurs at" or "is together with", like "intersects" in mathematics, are here to be taken as without effective tense. The whole problem of tense and words like "past", "present", and "future" will be discussed in Chapter XI.

For the togetherness relation described, which is tentatively taken as primitive, I shall use the symbol "Wh" (from "with"). As noted, the relation obtains between qualia only; and it is symmetric, irreflexive, and nontransitive.

4. THE PROBLEM OF CONCRETION

A concretum is a *fully* concrete entity in that it has among its qualities at least one member of every category within some sense realm. It is a *minimal* concrete entity in that it contains nothing more than one quale from each such category. Other concrete entities comprise more than one concretum; and anything less than a concretum is not fully concrete. In the visual realm, for example, a concretum is a color-spot-moment, consisting of a time and a place and a color that are together; and any entity made up of but two of these three is not fully concrete, since it lacks a quality from one of the relevant categories. The problem of concretion as it now presents itself is to define concreta in terms of the relation Wh.

It is convenient to use the auxiliary term "*complex*" to distinguish those entities that are comprised within concreta from those that are not. Complexes thus include all concreta, and (as degenerate cases) all qualia, and also whatever consists of more than one but fewer than all of the qualia of any one concretum. Obviously, if we can define complexes in general we can readily define concreta, which are the most comprehensive of complexes.

Every two qualia that belong to one complex are together. Moreover, since every two qualia that are together are comprised within some one concretum, any two such qualia make up a complex. Thus a complex consisting of two qualia–a bicomplex–may easily be defined as a class of two qualia that are together (Wh). Color-spots and place-times are examples of bicomplexes. Note especially, however, that two qualia make up a bicomplex *only* if they are together. A place and a color that never occurs at that place do not form a color-spot.

If a bicomplex is the sum of two qualia that are together, one might naturally suppose that a tricomplex would result from adding to these two a third quale that occurs with both, and that complexes consisting of more qualia would result from continuing this process. The principles underlying these suppositions would be two already noted: (1) that every two qualia belonging to a single complex are together; and (2) that every two qualia that are together belong to a single complex. A concretum would then be defined as a most-inclusive class of qualia that are with each other. A visual concretum according to this definition would be a class having as members a color, a place, and a time such that the color is at the place, the color is at the time, and the place is at the time; no other visual quale is with all three.

The proposed method of definition is analogous to Carnap's method of defining a quality in the *Aufbau*. If we disregard those peculiarities of his primitive relation that he introduces expressly for meeting problems of order, his definition says in effect that a quality is a class of erlebs such that every two members are similar and such that no erleb outside the class is similar to every member of the class. Our tentative definition of a concretum makes it a class of qualia such that every two members are together and such that no quale outside the class is with every member of it. Thus the two definitions are parallel. But unfortunately both are faulty. Our tentative definition of a concretum suffers from the same central defect as Carnap's definition of a quality, and is thus quite inadequate.

The proposed definition of a bicomplex is unobjectionable, as are the principles that every two qualia that are together belong to a single concretum and that any two that belong to a single concretum are together. But a serious difficulty arises whenever more than two qualia are involved. The notion of the all-togetherness of three or more qualia was tentatively construed as meaning that each one is with each of the others, but the simple and troublesome fact is that a group of qualia so related may still fail to belong to any one concretum. Suppose that color *c* occurs at place *d*, that *c* occurs at time *f*, and that *d* occurs at *f*; suppose, that is, that

$$\text{Wh } c,d \,.\, \text{Wh } c,f \,.\, \text{Wh } d,f.$$

Still it may be that *c* and *d* and *f* belong to no one concretum; for color *c* may occur at place *d* at some other time than *f*, and *c* may occur at time *f* at some other place than *d*. Plainly the fact that every two qualia among a given group form a togetherness pair is not a sufficient guarantee that some one concretum contains them all. The difficulty here confronting us is by this time an old acquaintance: it is the same difficulty of imperfect community that wrecks Carnap's definition of qualities.

To discover the error, however, is easier than to rectify it. Obviously, we were wrong in supposing that for three or more qualities to be together means simply for each two of them to be together. What then *does* "being all together" mean? The failure of our tentative definition shows that *something* more is involved than the togetherness of every pair of the set; but how are we to express systematically just what more is involved?

Since the problem appears as soon as more than two qualia are concerned, let us concentrate for the present upon the case of three qualia; for example, a color, a place, and a time. What we want to express is not only that the color is at the place and at the time, and the place with the time, but also that the color is *at the place at the time*. Common language makes here an acute but insufficiently stressed distinction that has a definite bearing on our problem. To say that a color is at a place at a time is to say more than that the color is at a place *and* at a time; for a color *c* may be at place *d* and time *f* and yet not be *at place d at time f*.[4] The difference is easily overlooked in ordinary discourse because we seldom have occasion to make the weaker assertion except where the stronger one is also true. But the sentence "*c* is at *d* and at *f*" affirms the togetherness of certain pairs only, while "*c* is at *d* at *f*" affirms the all-togetherness of the three. The former sentence is readily translated as "Wh *c*,*d* . Wh *c*,*f*"; but what the latter sentence asserts is not so easily translated, for it is not exhausted even by "Wh *c*,*d* . Wh *c*,*f* . Wh *d*,*f*".

Since the place *d* may occur at many times, we might be tempted to suggest

[4] Analogously, the veteran who exclaims "I was with the U.S. Army in Pearl Harbor *and* at the time of the Japanese attack!" is not lying even if he was with the U.S. Army in Pocatello at the time of the Pearl Harbor attack and was with the U.S. Army in Pearl Harbor in 1945.

that for color *c* to be at place *d* at time *f* is for *c* to be at that part of *d* which is at *f*. But we have seen that a quale cannot be divided into parts that occur severally with different other qualia. Rather if *c* is at *d* at *f*, then not only is *d* at *f* but *c* is with that entity which comprises both *d* and *f*. In other words, for a color to be at a certain place at a certain time is for the color to be at a certain place-time. Similarly, to affirm that a certain color-spot occurs at a certain time is to affirm not only that the color and place occur together and that each occurs at the time, but also that the color-spot consisting of the color and the place is at that time. Likewise for a color-moment and a place. The 'more' that is involved in each case, beyond the togetherness of each of the qualia in question with each other, is the togetherness of a quale with the entity consisting of the other two. But place-times, color-spots, color-moments, do not belong to the field of our tentative primitive relation, which obtains only between qualia. Some way must be found of bringing the needed additional togetherness relationships within the scope of our system. This we might seek to do in any of the following ways.

(a) By trying to define in terms of Wh a broader relation having certain entities other than qualia in its field.–But our investigation of the problem in the present section should disclose the hopelessness of trying to define the togetherness of a color and a place-time, for example, in terms of togetherness pairs of qualia.

(b) By discarding places and times as atoms in favor of place-times, and revising our primitive relation accordingly.–But we have already noted objections to this course (Section 2 above). It provides at best an *ad hoc* way of defining visual concreta on the basis of a certain special choice of atoms. It gives us no general method applicable in other sense realms, or even in the visual realm if slightly different atoms are chosen. Inapplicability in other sense realms is of course the more serious deficiency; but also, we resolved that our methods of construction shall so far as possible be applicable no matter at what level of analysis qualitative atoms are chosen.

(c) By revising our primitive in some way that is free of these objections and that violates no other conditions of our problem. Let us explore this course.

5. A REVISION AND ITS CONSEQUENCES

Among several different possible revisions, the best is perhaps also the most obvious. The choice of atoms need not be changed, but all sums of two or more atoms are likewise admitted as individuals, and some of these are included as basic units. In particular, primitive togetherness is construed as obtaining not merely between qualia, but between any two separate sums of one or more qualia contained in a single concretum. Whereas Wh obtained between every two distinct atomic qualia in a concretum, the new primitive, W, obtains between every two discrete parts of a concretum.[5] This involves no departure from the ordinary notion of togetherness, but merely interprets it systematically by a less restricted primitive. A color may quite as naturally be said to occur at a place-time, or a color-spot at a time, or a color-moment at a place, as a color at a place or a time. The relation W obtains in all these cases. And if a concretum in some other sense realm contains four or more qualia, W obtains between each two, between each one and the sum of any two or three of the others, and between the sum of any two and the sum of the other two. The field of W consists of all complexes except concreta.

The relation W, like Wh, is symmetric, irreflexive, and nontransitive. And just as the two qualia forming a Wh-pair always belong to different categories of some one sense realm, so likewise two qualia contained in the sum of two individuals forming a W-pair always belong to different categories of some one sense realm.

If we are to admit individuals that are sums of atoms and are to deal with relationships of overlapping part-whole, etc. that obtain among these individuals, we shall need to incorporate as part of the general apparatus of our system such a calculus of individuals as was described in II,4. So to adopt the relation of overlapping as a primitive is by no means a net expense, for we shall then drop the calculus of classes, eliminating membership as a primitive. And this exchange had been independently planned anyway, in order to render the system nominalistic. The new approach, then, consists of replacing "$\in$" by "**o**", and "Wh" by "W".

As a result of the change from "$\in$" to "**o**", a bicomplex is to be construed as the sum rather than the class of two qualia that are together; and the togetherness relationship between a color and a place-time, or a color-spot and a time, will be simply a W-relationship between two individuals. If color c is at place-time k, their relationship is expressed in the system by "W c,k". If d is the place and f the time comprised in k, then this same relationship may be expressed by "W $c,d+f$". Now "W $x,y+z$" is stronger than "W x,y . W x,z". A color r may occur at a place s and a time t even though s and t are not together; and if s and t are not together, then the individual $s+t$–since it is contained in no one concretum–does not bear the relation W to r or to anything else. Moreover, even if the place and time do occur together and the

[5] Or, more accurately, between every two individuals that are sums of qualia, that are systematically discrete, and that are parts of one concretum. Similar condensations are to be taken as understood in the following discussion.

color occurs at each, still "W $r,s + t$" will be false unless r occurs at s *at* t. Thus "W $x,y + z$" is stronger even than "W x,y . W x,z . W y,z"; the difference in strength is precisely that illustrated by the difference between saying that a color is at a place at a time and saying that a color is at a place *and* a time that are together. In shifting from Wh to W we have not replaced a nonagglomerative[6] primitive by an agglomerative one; we can no more infer the togetherness of a color and a place-time (or of a color-moment and a place) from W-relationships among the qualia than from Wh-relationships. But because other individuals than qualia may belong to W-pairs, we are now able to express in our system togetherness relationships we could not define in terms of togetherness pairs of qualia.

Complexes are now readily definable. An individual is a complex if and only if every two discrete parts of it bear the relation W to each other. Obviously, a quale, since it has no two discrete parts, satisfies this definition, and a bicomplex, consisting of two qualia that are together, likewise satisfies it. Obviously also, any individual that contains two qualia that are not together fails to satisfy the definition. So far the new definition coincides with the old one. But according to the new definition, an individual composed of three qualia, even if each of the three is with every other, will not be a complex *unless* each is also with the sum of the other two. More generally, an individual is not a complex unless every sum of one or more of the qualia it contains is with the sum of the rest. It is just under this condition that the qualia in question are 'all together'. Thus our new definition succeeds in imposing the requisite condition of all-togetherness, which we were unable to formulate in terms of Wh. As a result, the troublesome cases that arose before are excluded.

Since two individuals are together if and only if they are discrete parts of some one concretum, any sum of two individuals that are together is a complex. A complex might thus alternatively have been defined as any individual that is either a quale or the sum of two individuals that are together. But this is somewhat more cumbersome than our adopted definition when both are expanded in terms of primitives.

Once complexes are defined, a concretum is definable as any complex that bears the relation W to no individual. No concretum x, then, is a proper part of any complex y; for then x, contrary to definition, would bear the relation W to $y-x$.

---

[6] For definitions of "agglomerative" and "nonagglomerative", see II,4.

6. RECTIFICATION OF PARTICULARISM

So closely does the problem of concretion parallel the problem of abstraction that the question naturally arises whether the method above devised for meeting the difficulty of imperfect community in a realistic system might not be used to meet the same difficulty in a particularistic system.

Suppose we have, to begin with, a typical particularistic system in which the atoms are erlebs, and the primitive is the similarity relation that obtains between two erlebs if and only if they have some quality in common. This system differs from the *Aufbau* system only in features that are irrelevant to the problem of abstraction. We have seen that the attempt to define qualities on such a basis ends in failure.

Now application of the general method of the preceding section will involve broadening our primitive relation so that not only erlebs but also certain sums of erlebs are included among our basic units. Let us, therefore, drop the calculus of classes in favor of the calculus of individuals. A quality for this system, then, will not be a class but a whole–the sum of all the individuals that, in ordinary language, have the quality in common. The term "quality stretch" may be used for any sum of one or more erlebs all of which have a quality in common. Quality stretches–which are to play in the present system a role like that of complexes in realistic systems–may be continuous or discontinuous and they vary in size from a single erleb up to a complete quality whole.

Let us now take as our new primitive the similarity relation L that obtains between every two discrete parts of a quality whole. A quality stretch may now be defined as any individual of which every two discrete parts form an L-pair. A quality whole is then any quality stretch that is a proper part of no other. This method of definition meets the difficulty of imperfect community here as well as in a realistic system.

In view of the results of our examination of the criteria of definition in Chapter I, there is no objection to defining a quality as a quality whole, even if a quality is presystematically a proper part of such a quality whole, so long as care is taken to preserve the isomorphism of the entire set of definientia of the system with the entire set of definienda. Indeed, if qualities are presystematically proper parts of concrete individuals, then nothing more than isomorphism can be claimed for the definition of qualities as classes of concrete individuals. However, if one prefers the definition of qualities as classes strongly enough to add the calculus of classes to a general apparatus that already contains the calculus of individuals, a quality may be defined as a class of all the erlebs that are parts of a single quality whole. In such a case it is clear that the calculus of individuals is used as an auxiliary tool in reaching a classial definition in a platonistic system. Concreta might, of course, be similarly defined as classes of qualia in a platonistic realistic system, though the advantages of such a definition are difficult to perceive.

Thus the expression defining qualities in the particularistic system described is identical with the expression defining concreta in our realistic system except that "L" appears in the one where "W" appears in the other.

However, we saw in Chapter V that the particularistic definition of qualities in terms of erlebs is beset not only by the difficulty of imperfect community but also by the companionship difficulty. The proposed revision does not meet the latter difficulty. If, for example, a quality *a* happens to occur only in erlebs in which another quality *b* also occurs, while *b* occurs in some erlebs in which *a* does not occur, then the sum of all the erlebs in which *a* occurs will fail to satisfy the definition of a quality whole. If neither *a* nor *b* occurs in any erleb in which the other does not occur, then the two distinct qualities will be erroneously defined as identical.

Some of the trouble can be blamed on taking too-comprehensive entities as atoms. If concreta are chosen rather than erlebs, one may argue with at least some show of plausibility that two qualities occurring in exactly the same atoms will in actual fact never be distinguished. But still it is hardly plausible to argue that if a color *d*, for example, never occurs except at one certain place *f*, while other colors occur at place *f* at other times, then color *d* and place *f* will in actual fact never be distinguished. Thus the companionship difficulty remains unresolved in particularistic systems.

Now the parallel to the companionship difficulty would arise in a realistic system if any concretum were composed entirely of qualities contained in some other concretum. But no two distinct concreta have exactly the same qualities. And, since the number of qualities in a concretum is the number of quale catagories in the sense realm in which the concretum lies, no concretum is composed of some but not all of the qualities of another concretum. The situation is quite different from that in a particularistic system; for since a quality may occur in any number of concreta, one quality easily occur only in some of the concreta in which another occurs.

Hence, despite the close parallel between the two systems, the realistic system seems to have as a definite point in its favor its freedom from anything comparable to the companionship difficulty. I do not say this constitutes a decisive advantage; the point is merely to be noted along with any that may appear later in favor of either type of system.

7. ALTERNATIVE TREATMENTS OF THE PROBLEM OF CONCRETION

Among various possible alternatives to the procedure outlined in Section 5 for dealing with the problem of concretion, a few deserve consideration here for some advantage of convenience or economy they seem to offer.

(i) The first alternative differs from our adopted method only in ways consequent upon construing concreta and complexes as classes rather than sums of qualia. The togetherness relation, Wt, taken as primitive obtains between each two separate subclasses of a concretum. A complex is then defined as any class of which every two separate subclasses form a Wt-pair. The calculus of individuals is therefore not needed.

Obviously such a course is open only to platonists. Moreover, one who is a platonist to the extent of accepting the calculus of classes as part of his general apparatus might still balk at taking a relation of classes as an irreducible special primitive for a system. In *A Study of Qualities*, for instance, platonistic apparatus was used; but only relations, or predicates, of individuals were admitted as special primitives.

Suppose, however, all scruples of this sort are set aside. How does this alternative basis compare with our adopted one on the score of economy alone? I have remarked that addition of the calculus of individuals involves no sacrifice of economy if the calculus of classes can then be dropped.[7] But I cannot prove that the calculus of individuals by itself will serve to replace the calculus of classes for all eventual purposes of our system as well as for our immediate purposes. Let us therefore look upon "**o**" not as replacing "∈" but as an additional primitive. The question then is whether it is more economical to take both "**o**" and "W" as primitive, as in our adopted procedure, or to take "Wt" alone as primitive, as in the alternative now under discussion.

According to the calculus of complexity developed in Chapter III, each of the 2-place symmetric predicates "**o**" and "W"–the one totally reflexive and the other irreflexive–has the complexity-value 2; and a basis consisting of both thus has the total value 4. The complexity of "Wt", a 2-place disjoint symmetric predicate of classes of individuals, will depend upon the maximum number of qualia contained in any complex in the field of Wt. What this num-

[7] This, indeed, is a rank understatement. The predicate "**o**" has the complexity-value 2. We cannot directly apply our calculus of complexity to "∈" since "∈" has a non-finite extension and furthermore must itself be incorporated in the basic apparatus for effecting the correlations enabling us to evaluate predicates of classes. But this much can be said: a thoroughly finite 2-place predicate that like "∈" relates elements of each type to elements of next higher type up to an unspecified level will have a complexity-index containing "u^u", and will count as more complex than any set of finite predicates of specified types.

ber is will depend upon how the atoms are chosen in each sense realm. If they are so chosen that there are not more than three atoms in any concretum, with the result that complexes in the field of Wt comprise at most two qualia, "Wt" can be correlated with a basis consisting of one 3-place irreflexive symmetric and one 2-place irreflexive symmetric predicate of individuals.[8] Thus "Wt" alone will have the complexity-value 5 as against a value of 4 for "**o**" and "W" together. While I have indicated earlier (Section 2) a choice of visual atoms, the atoms in the other sense realms have been left unspecified. Furthermore, neither the adopted method of constructing concreta nor the present alternative depends upon the choosing of atoms at one level of qualitative analysis rather than another, and we have seen that such dependence is to be avoided. The relevant comparison of complexity values, therefore, is that made upon the assumption that the maximum number of atoms in a concretum is undetermined. In this case, the total complexity of "**o**" and "W" together remains at 4. "Wt", however, now correlates with a basis consisting of irreflexive symmetric predicates of individuals, one predicate for each number of places from two up to an unspecified number, and so has the complexity-index "$\frac{u^2+u}{2}-1$", which counts as higher than any specified integer. *Accordingly, a basis consisting of the single predicate* "Wt" *is more complex than a basis consisting of the two predicates* "**o**" *and "W"*. And furthermore, the primitive "**o**" is of course used for many purposes other than the construction of concreta.

On every count, then, the method adopted in Section 5 is preferable to the present alternative.

(ii) A second proposal might be to make W reflexive and include concreta in its field, so that they are immediately definable as the most inclusive members of the field. A suggestion that carries the same idea even further and uses a monadic rather than a dyadic primitive is simply to drop W entirely and take as primitive the class of complexes.

The trouble is that both proposals violate the conditions of our problem.

---

[8] In effecting this correlation, I use not only the information that "Wt" under the stated assumptions is a 2-place disjoint symmetric predicate of classes having at most two individuals as members, but also the further information that it never relates two 2-membered classes, and that if it relates (in either order) a 1-membered and a 2-membered class, then it relates the class of any one of the three members to the class of the other two. Although I have not drawn very explicitly the line between general or relevant and special or non-relevant information for the replaceability involved in the correlation of predicates of classes with predicates of individuals, some of the information just detailed may very well be excluded as special. In that case, the complexity-value of "Wt" will be even higher.

The problem of concretion is to construct concreta from qualities; or more precisely, to define concreta in terms of primitives such that only nonconcrete qualities are basic units.[9] We do not solve that problem if we include concreta among the basic units. The problem of abstraction would be similarly evaded in a particularistic system for which quality classes or quality wholes were basic.

The fact that these question-begging proposals are not far removed from the legitimate method adopted suggests how far the problem of concretion in its initial version resists solution. The construction of concreta in our system is not–despite the trouble we had in arriving at it–a very long step. But then, a constructional system tends to be a succession of short steps taken with great difficulty.

(iii) A quite different sort of proposal is founded on the idea of treating qualities as quality wholes and making these basic for a system. They are regarded either as divisible into atomic concreta, or as themselves atomic. In the former case, the class of all quality wholes and their concrete parts is taken as primitive; in the latter case, the primitive is the relation obtaining between any two quality wholes that presystematically contain a concretum in common.

In neither version does this proposal give us a realistic system; for it involves taking concrete individuals, not nonconcrete qualities, as basic units. Incidentally, as a proposal for founding a particularistic system, it is question-begging in either version, since it takes quality wholes as basic units.

For varying reasons, therefore, all these alternatives are rejected. The method proposed in Section 5 for dealing with the problem of concretion will be used in the formal constructions now to be presented.

[9] Or basic classes if the special primitive is a class of classes or a relation of classes.

CHAPTER VII

CONCRETA AND QUALIFICATION

1. THE INDIVIDUALS OF THE SYSTEM

The special primitive of the present calculus is the two-place predicate "W", which has already been explained as applying between two individuals if and only if they are discrete complexes comprised in a single concretum. "W" is read "is with", "is at", "occurs with", "occurs at", or "are together"; and it applies, for example, between a color and a place at which the color occurs, between a place and a time at which the place occurs, and between a color-spot and a time at which the color-spot occurs.

The general apparatus of the system consists of the usual truth-functional signs, individual-variables, the signs of quantification and punctuation, and the predicate "**o**" together with the calculus of individuals (II,4) based upon it. The term "∈" and the others of the calculus of classes are omitted; and no variables are used that call for classes or other nonindividuals as values. The identity sign is introduced by definition in terms of "**o**".

The range of the individual-variables, and therefore the range of application of "**o**", are restricted so that only sums of one or more qualia are admitted. As a matter of fact, this restriction is theoretically dispensable, since it is not needed if we include appropriate modifying clauses in each definition, postulate, and theorem of the system. Practically, however, an initial general restriction results in a great saving of effort. The restriction is effected by a postulate (7.1 below) stipulating that every admitted individual has some quale as a part. First, a quale is defined as a basic unit (i.e., an individual to which "W" applies) that has no other basic unit as a part:

D7.01 $\quad \text{Qu}\, x = (\exists y)\, (\text{W}\, x,y)\, .\, (z)\, (t)\, (\text{W}z,t \supset {\sim} z \ll x).$

Then the postulate mentioned is introduced:

7.1 $\quad (\exists y)\, (\text{Qu}\, y\, .\, y < x).$[1]

By adopting this postulate we do not commit ourselves on the question whether there are individuals that have no quale as a part, but merely exclude such individuals–if there are any–as values of our variables. The situation

[1] Initial universal quantifiers of unconfined scope will usually be omitted.

is much the same as that in the calculus of individuals, where by adopting postulates that give the theorem "$(x)(x \mathbf{o} x)$" we do not commit ourselves as to whether there are any nonindividuals but merely exclude such entities–if there are any–as values of our variables.

Now it must be especially borne in mind that the postulates of the calculus of individuals *remain in force*, and that the postulate 7.1 is in addition to them. As one result, two individuals overlap under this system only if they are individuals recognized by the system and have as a common part some individual recognized by the system (cf. II,5). Indeed, it can be shown that they overlap if and only if they have some quale as a common part:

7.11 $\quad x \mathbf{o} y \equiv (\exists z)(\mathrm{Qu}\, z \,.\, z < x \,.\, z < y)$.

Proof:

1. $x \mathbf{o} y \equiv \mathrm{E!}xy$	C.I.[2]
2. $\mathrm{E!}xy \equiv (\exists z)(\mathrm{Qu}\, z \,.\, z < xy)$	7.1, C.I.
3. $\mathrm{E!}xy \equiv (\exists z)(\mathrm{Qu}\, z \,.\, z < x \,.\, z < y)$	2, C.I.
4. *Q.E.D.*	1 and 3

Of course the predicates ("$\wr$", "<", etc.) that are defined in terms of "**o**" are consequently restricted in easily discoverable ways.

It is now easy to show that 7.1 has the effect of restricting the individuals of the system to those that are sums of one or more qualia; in other words,[3] all and only those individuals that overlap y overlap some quale that overlaps y.

7.12 $\quad x \mathbf{o} y \equiv (\exists z)(\mathrm{Qu}\, z \,.\, z \mathbf{o} x \,.\, z \mathbf{o} y)$.

Proof:

1. $x \mathbf{o} y \supset (\exists z)(\mathrm{Qu}\, z \,.\, z \mathbf{o} x \,.\, z \mathbf{o} y)$	7.11, C.I.
2. $z \mathbf{o} x \,.\, {\sim} z < x \,.\supset (\exists y)(\mathrm{Qu}\, y \,.\, y < z - x)$	7.1, C.I.
3. $\qquad \supset (\exists y)(\mathrm{Qu}\, y \,.\, y \ll z)$	2, C.I.
4. $\qquad \supset (\exists y)(\exists t)(W y,t \,.\, y \ll z)$	3, 7.01
5. $\qquad \supset {\sim}\mathrm{Qu}\, z$	4, 7.01
6. $\mathrm{Qu}\, z \,.\, z \mathbf{o} x \,.\supset z < x$	5

[2] In the annotation of proofs, "C.I." indicates an application of the calculus of individuals. A line without annotation is logically valid by itself.

[3] Cf. the schema for the definition of the sum of the individuals satisfying a given predicate, in II,4.

7. $(z)(\mathrm{Qu}\, z\, .\, z \circ x\, .\, z \circ y\, .\supset .\, \mathrm{Qu}\, z\, .\, z < x\, .\, z < y)$ 6
8. $(\exists z)(\mathrm{Qu}\, z\, .\, z \circ x\, .\, z \circ y) \supset (\exists z)(\mathrm{Qu}\, z\, .\, z < x\, .\, z < y)$ 7
9. $(\exists z)(\mathrm{Qu}\, z\, .\, z < x\, .\, z < y) \supset x \circ y$ 7.11
10. $(\exists z)(\mathrm{Qu}\, z\, .\, z \circ x\, .\, z \circ y) \supset x \circ y$ 8 and 9
11. *Q.E.D.* 1 and 10

Then two individuals are identical if and only if one is the sum of all the qualia that overlap the other:

7.13 $x = y \equiv (z)\, \{z \circ x \equiv (\exists t)(\mathrm{Qu}\, t\, .\, t \circ y\, .\, t \circ z)\}.$

Proof:

1. $(z)\, \{z \circ y \equiv (\exists t)(\mathrm{Qu}\, t\, .\, t \circ y\, .\, t \circ z)\}\, .\, (z)\, \{z \circ x \equiv (\exists t)(\mathrm{Qu}\, t\, .\, t \circ y\, .\, t \circ z)\}\, .\supset (z)(z \circ y \equiv z \circ x)$
2. $(z)\, \{z \circ y \equiv (\exists t)(\mathrm{Qu}\, t\, .\, t \circ y\, .\, t \circ z)\}$ 7.12
3. $(z)\, \{z \circ x \equiv (\exists t)(\mathrm{Qu}\, t\, .\, t \circ y\, .\, t \circ z)\} \supset (z)(z \circ y \equiv z \circ x)$ 1 and 2
4. $(z)(z \circ y \equiv z \circ x) \supset x = y$ C.I.
5. $(z)\, \{z \circ x\, . \equiv (\exists t)(\mathrm{Qu}\, t\, .\, t \circ y\, .\, t \circ z)\} \supset x = y$ 3 and 4
6. $x = y \supset (z)\, \{z \circ x \equiv (\exists t)(\mathrm{Qu}\, t\, .\, t \circ y\, .\, t \circ z)\}$ 2
7. *Q.E.D.* 5 and 6

Logical transformation of 7.13 then shows that every individual is identical with the sum of the qualia that overlap it; it exhausts and is exhausted by such qualia:

7.14 $x = (\imath y)(z)\, \{z \circ y \equiv (\exists t)(\mathrm{Qu}\, t\, .\, t \circ x\, .\, t \circ z)\}.$

A number of theorems closely related to the foregoing ones may be proved along similar lines. For example, two individuals are identical if and only if they have the same qualia as parts; and any individual is the sum of the qualia that it has as parts.

So much for the reducibility of all individuals of the system to sums of qualia. I want now to show that qualia are the atoms of the system. It is helpful first to have a preliminary theorem stating that every quale is a basic unit:

7.15 $\mathrm{Qu}\, x \supset (\exists y)(\mathrm{W}\, x, y).$

This follows directly from definition 7.01. The next step is to show that all qualia are atoms of the system, that is, that under the system, no quale has a proper part.

7.16 $\quad \text{Qu}\, x \supset (y)(\sim y \ll x)$.

Proof:

1. $(\exists y)(y \ll x) \supset (\exists z)(\text{Qu}\, z \,.\, z \ll x)$ — 7.1, C.I.
2. $(\exists z)(\text{Qu}\, z \,.\, z \ll x) \supset (\exists z)(\exists t)(\text{W}\, t, z \,.\, z \ll x)$ — 7.15
3. $(\exists y)(y \ll x) \supset (\exists z)(\exists t)(\text{W}\, t, z \,.\, z \ll x)$ — 1 and 2
4. $(t)(z)(\text{W}\, t, z \supset .\, \sim z \ll x) \supset (y)(\sim y \ll x)$ — 3
5. $\text{Qu}\, x \supset (t)(z)(\text{W}\, t, z \supset \sim z \ll x)$ — D7.01
6. $\text{Qu}\, x \supset (y)(\sim y \ll x)$. — 4 and 5

Conversely, all the atoms of the system are qualia:

7.17 $\quad (y)(\sim y \ll x) \supset \text{Qu}\, x$.

Proof:

1. $(y)(y < x \supset y = x) \,.\, (\exists y)(\text{Qu}\, y \,.\, y < x) \,. \supset (\exists y)(\text{Qu}\, y \,.\, y = x)$
2. $(\exists y)(\text{Qu}\, y \,.\, y < x)$ — 7.1
3. $(\exists y)(\text{Qu}\, y \,.\, y = x) \equiv \text{Qu}\, x$ — C.I.
4. $(y)(y < x \supset y = x) \equiv (y)(\sim y \ll x)$ — C.I.
5. $(y)(\sim y \ll x) \supset \text{Qu}\, x$. — 1, 2, 3, 4

The atoms of the system are then the same as qualia, for 7.16 and 7.17 conjoined give:

7.18 $\quad \text{Qu}\, x \equiv (y)(\sim y \ll x)$.

As a result we have analogues, in terms of atoms, of 7.1 through 7.15. That is, every individual has an atom as a part; two individuals overlap if and only if they have an atom as a common part; every individual is a sum of atoms; two individuals are identical if and only if one overlaps all and only those atoms that overlap the other; every individual is the sum of the atoms it overlaps; and every atom is a basic unit.

Every individual, then, is exhaustively divisible into indivisible individuals. It does not follow from this that there are only finitely many atoms. Consequently it does not follow that no individual is infinitely divisible, for the sum of an infinite number of atoms would be infinitely divisible. The fact that there are only finitely many atoms for this (as for any phenomenalistic) system need not be stated or used in the system, but is of course reflected in the avoidance of postulates that imply infinity.

The predicate "Qu" is rather trivially dissective in that every part of a quale x is a quale because identical with x:

7.19 $\text{Qu}\, x \,.\, y < x \,.\, \supset .\, x = y \,.\, \text{Qu}\, y.$

But "Qu" is not expansive; for if x is a quale, no other individual of which x is a part is also a quale:

7.191 $\text{Qu}\, x \,.\, x \ll y \,.\, \supset \sim \text{Qu}\, y.$

Finally, whatever overlaps a quale has it as a part:

7.192 $\text{Qu}\, x \,.\, y \circ x \,.\, \supset x < y;$

and two qualia are always either identical or discrete:

7.193 $\text{Qu}\, x \,.\, \text{Qu}\, y \,.\, \supset .\, x = y \vee x \wr y.$

2. PRINCIPLES OF TOGETHERNESS[4]

Since our problems are primarily definitional, I shall not lay much emphasis on deductive development in this chapter. The selection of postulates, where it is indicated at all, is to be regarded as very tentative and made with little thought of postulational economy–whatever that may prove to be (see III, 12). Proofs will be omitted or only briefly outlined.

An important feature of the primitive predicate "W" is that it is *pervasive* in the weaker sense described in II,4. That is, if "W" applies between two individuals, it applies between every two discrete parts of their sum; for the sum is then part of a concretum and every two discrete parts of it are therefore together.[5] Thus, the statement:

7.2 $\text{W}\, x,y \,.\, t < x + y \,.\, z < x + y \,.\, t \wr z \,.\, \supset \text{W}\, t,z$

belongs to the system and might be taken as a postulate. The predicate "W"

[4] Unavoidably, this and the following two sections for the most part duplicate material covered informally in Chapter VI.

[5] This argument is, of course, not even an outline of a proof, since it makes use of terms and principles that have yet to be introduced into the system. It is merely a substantiation based upon the interpretation of "W". Such substantiation of postulates has the same sort of unofficial status as the explanation of primitive terms.

has been so explained that it is external; and the following statement might thus be taken as a second postulate:

7.21 $\mathrm{W}x,y \supset x \wr y$.

Thus no complex is with any quale that it contains. Although a color-spot, for example, is sometimes said to be at its place, the systematic predicate that applies in such a case is not "W" but another to be defined later (see Section 5).

It follows that if x is with y then every part of x is with y:

7.211 $\mathrm{W}x,y \,.\, z < x \,.\supset \mathrm{W}z,y$.

For example, if a color-spot occurs at a certain time, then the color and the place occur at that time. Furthermore, every two discrete parts of an individual that is with another are together:

7.22 $\mathrm{W}x,y \,.\, t+z < x \,.\, t \wr z \,.\supset \mathrm{W}t,z$.

If a color occurs with the sum of a place and a time, for example, the place and the time are together.

It likewise follows from 7.2 that if a color is with the sum of a place and a time, then the place is with the sum of the color and the time, and the time is with the sum of the color and the place. This is a case of the theorem:

7.23 $\mathrm{W}x,y+z \,.\, y \wr z \,.\supset.\, \mathrm{W}y,x+z \,.\, \mathrm{W}z,x+y$.

Put in a different way: if three individuals are discrete, then one is with the sum of the other two if and only if each of the three is with the sum of the remaining two. In general:

7.24 $x \wr y \,.\, x \wr z \,.\, z \wr y \,.\supset.\, \mathrm{W}x,y+z \equiv \mathrm{W}y,x+z \,.\, \mathrm{W}x,y+z \equiv \mathrm{W}z,x+y \,.\, \mathrm{W}y,x+z \equiv \mathrm{W}z,x+y$.

To say that any one of three discrete individuals is with the sum of the other two is to say that the three are 'all together'.

That "W" is irreflexive follows from 7.21 and C.I.:

7.26 $\sim \mathrm{W}x,x$.

Not even a place or time, for example, is at (in the sense of with) itself. And from 7.2 and 7.21, it can be shown that "W" is symmetric:

7.27 $\mathrm{W}x,y \equiv \mathrm{W}y,x$.

As was emphasized earlier, to say that a color is at a place is to say that the place is with the color, or that the two are together.

Quite as important as the postulates and theorems listed is the absence from the system of certain other statements. In the first place, "W" is neither one-one, one-many, nor many-one; for example, a color may occur at many places, and many colors may occur (at different times) at one place. Again, "W" is not transitive; a color may occur at a place that occurs at a time, and yet the color may happen not to occur anywhere at that time. Moreover, as emphasized in Chapter VI, "W" is not cumulative nor even agglomerative; if a color c occurs at a time t, and c occurs also at place p, and p occurs at t, still it may be that c does not occur at $p + t$. From "W x,y . W x,z . W y,z", we cannot infer "W $x,y + z$".

3. COMPLEXES

In Section 1, it was shown that every individual (recognized by the system) is a sum of one or more qualia. We have now to differentiate those sums of qualia that are complexes from those that are not. As was explained in Chapter VI, a complex is distinguished by the fact that every two discrete parts of it are together. In symbols the definition runs:

D7.03 $\quad \mathrm{Cm}\, x = (y)\,(z)\,(y + z < x \,.\, y \,\imath\, z \,.\supset \mathrm{W}\, y,z).$

Obvious consequences are that every two discrete parts of a complex are together, and that every proper part of a complex is with the rest of it:

7.31 $\quad \mathrm{Cm}\, x \,.\, y + z < x \,.\, y \,\imath\, z \,.\supset .\, \mathrm{W}\, y,z;$

7.32 $\quad \mathrm{Cm}\, x \,.\, y \ll x \,.\supset \mathrm{W}\, y,x - y.$

From 7.22, it is clear that "W" applies to complexes only:

7.33 $\quad \mathrm{W}\, x,y \supset .\, \mathrm{Cm}\, x \,.\, \mathrm{Cm}\, y;$

for example, if a color is with the sum of a place and a time, that sum is a place-time. It is not conversely true that "W" applies to all complexes; a color-spot-moment, for example, is with no individual (see further 7.41 below).

Since a quale is atomic (7.16), it has (by C.I.) no two discrete parts, and thus every two discrete parts of it are together. Hence all qualia are complexes:

7.34 $\quad \mathrm{Qu}\, x \supset \mathrm{Cm}\, x.$

As was remarked earlier, application of the term "complex" to atomic individuals appears somewhat anomalous; but convenience is served by treating qualia as degenerate cases of complexes.

Definition 7.03 makes every part of a complex also a complex:

7.35 $\text{Cm}\, x \,.\, y < x \,.\supset \text{Cm}\, y$;

for example, if a color and a time are contained in a complex, the sum of the color and the time is a color-moment. In other words, the predicate "Cm" is dissective. It of course follows immediately that "$\sim$Cm" is expansive, that no individual is a complex if it has any noncomplex as a part:

7.36 $y < x \,.\sim \text{Cm}\, y \,.\supset .\sim \text{Cm}\, x$.

For example, no individual is a complex if it contains a color and a place that are not together, or two colors. But it may happen that all parts of a noncomplex *except itself* are complexes. This is obviously true of any sum of two qualia that are not together. It is also true of any sum of three qualia that are not all together but of which each two are together; for example, each of the proper parts of the sum of a color, a place, and a time that are so related is a quale or a bicomplex, but the whole is not a complex.

I shall hereafter sometimes use "compound" as an alternative to "noncomplex" in reading verbally the symbol "$\sim$Cm".

The sum of any two individuals that are together is a complex:

7.37 $\text{W}\, x{,}y \supset \text{Cm}\, x + y$;

for by 7.2, if W x,y then every two discrete parts of $x + y$ are together and hence, by D7.03, $x + y$ is a complex. If, for example, a color is at a place, their sum is a complex–a color-spot; if a color is at a time, their sum is a complex–a color-moment; if a place is at a time, their sum is a complex–a place-time; if a color-spot is at a time, their sum is a complex–a color-spot-moment.

The exact converse of 7.37 cannot be affirmed, since the sum of two individuals that overlap and are therefore not together may be a complex; for example, a color-spot and a place-time contained in one concretum are not together, since they have a place as a common part, but their sum is nevertheless a complex. However, if the sum of two *discrete* individuals is a complex, they are together:

7.38 $\text{Cm}\, x + y \,.\, x \wr y \,.\supset \text{W}\, x{,}y$.

It can now be shown that every quale is a proper part of some complex:

7.39 $\text{Qu}\, x \supset (\exists y)\, (\text{Cm}\, y \,.\, x \ll y)$.

For by 7.15 and 7.21 any quale x is with some invividual z that is discrete from x; by C.I., then, x is a proper part of $x + z$; and by 7.37, $x + z$ is a complex.

For some purposes it may be convenient to have definitions classifying complexes according to the number of qualia they have as parts. The definitions of bicomplexes and tricomplexes follow:

D7.031 $\quad \mathrm{Cm}_2 x = \mathrm{Cm}\, x \,.\, (\exists y)(\exists z)(\mathrm{Qu}\, y \,.\, \mathrm{Qu}\, z \,.\, y \neq z \,.\, y + z = x);$

D7.032 $\quad \mathrm{Cm}_3 x = \mathrm{Cm}\, x \,.\, (\exists y)(\exists z)(\exists t)(\mathrm{Qu}\, y \,.\, \mathrm{Qu}\, z \,.\, \mathrm{Qu}\, t \,.\, y \neq z \,.\, y \neq t \,.\, z \neq t \,.\, y + z + t = x).$

On our chosen basis, place-times and color-spots are examples of bicomplexes; and color-spot-moments are examples of tricomplexes. If hues, chromas, and brightnesses instead of colors are taken as basic units, then colors are tricomplexes. Definitions of complexes of higher degrees can obviously be readily supplied if wanted. From its degree alone, however, we cannot tell whether a complex is concrete, for concreta in different sense realms may contain different numbers of qualia.

4. CONCRETA

A concretum is distinguished from all nonconcrete complexes by the fact that it is with no individual, and from all compounds by the fact that it is a complex. Concreta are defined as complexes that are not basic units:

D7.04 $\quad ¢\, x = \mathrm{Cm}\, x \,.\, (y)(\sim \mathrm{W}\, x,y).$

A concretum is thus an individual of which every two discrete parts are together but which is not itself with any individual. Parts of it, of course, may be with individuals that are discrete from it; normally, for example, each of the qualia in a concretum is with other qualia not contained in the concretum.

Definition 7.04 is the means by which minimal concrete individuals are constructed from nonconcrete individuals, i.e., by which concreta are defined in terms of the general apparatus of the system and a special predicate that applies only to nonconcrete individuals. If qualia are the atoms of the system, concreta might in many ways be likened to molecules. As no fragment of a molecule of water is water, so no fragment of a concretum is concrete; the aqueous molecule is made up of what is nonaqueous as the concretum is made up of what is nonconcrete. And just as a molecule of water is as truly water as is a glassful, so a color-spot-moment, for example, is quite as concrete as is the sum of a thousand color-spot-moments.[6]

---

[6] The analogy must not be carried too far: for example, a molecule of water, unlike a concretum, contains two atoms of one kind. I shall later (Section 8) discuss in more detail the use of "concrete" and related terms.

From 7.04 it is clear that an individual is a basic unit if and only if it is a subconcrete complex:

7.41 $(\exists y)(\mathrm{W}\, x,y) \equiv .\, \mathrm{Cm}\, x\, .\, \sim \not{c}\, x.$

In other words "W" applies to all and only those complexes that are not concreta. Since (by 7.15) all qualia are basic units, no quale is a concretum:

7.42 $\mathrm{Qu}\, x \supset \sim \not{c}\, x.$

Concreta are just those complexes that are proper parts of no complex, and might alternatively have been so defined.

7.43 $\not{c}\, x \equiv .\, \mathrm{Cm}\, x\, .\, (y)(\mathrm{Cm}\, y \supset \sim x \ll y).$

The proof in outline is as follows. (1) If there is some complex y of which x is a proper part then (by 7.32) x is with $y-x$, and therefore (by 7.41) x is not a concretum. Thus if x is a concretum it is a proper part of no complex. (2) If x is with any individual z, then (by C.I.) x is a proper part of $x+z$, which (by 7.37) is a complex. Thus if x is a proper part of no complex, x is with no individual; and therefore (by 7.04) if x is a complex and a proper part of no complex, x is a concretum.

It is clear from 7.43 that no individual having a concretum as a proper part is a complex:

7.44 $\not{c}\, x\, .\, x \ll y\, . \supset \sim \mathrm{Cm}\, y.$

Add anything to a color-spot-moment, for instance, and you get a noncomplex; for what is added will not be with the color-spot-moment.

On the other hand, all the parts of a concretum are of course complexes:

7.45 $\not{c}\, x\, .\, y < x\, . \supset \mathrm{Cm}\, y,$

as may be shown from 7.04 and 7.35. For example, the parts of a color-spot-moment are: itself, which is a concretum and therefore a complex; three qualia, which are complexes; and a color-spot, a place-time, and a color-moment, all of them bicomplexes.

But no individual is a concretum if it is a proper part of any concretum or has any concretum as a proper part.

7.46 $\not{c}\, x\, .\, x \ll y \vee y \ll x\, . \supset \sim \not{c}\, y.$

The proof is from 7.04, 7.43, and 7.44. However, while the proper parts of a concretum are nonconcrete as well as not concreta, individuals of which a concretum is a proper part may be concrete even though not concreta–much as a glassful of water is water but is not a molecule of water (cf. Section 8 below).

Neither the product nor the sum of two different concreta is a concretum:

7.47 $\not{c}\, x \,.\, \not{c}\, y \,.\, x \neq y \,.\supset .\sim \not{c}\, xy \,.\sim \not{c}\, x + y.$

For by C.I., if two different individuals overlap, their product is a proper part of at least one of them; and if the individuals in question are concreta, then (by 7.46) the product is not a concretum. And by C.I., at least one of any two different individuals is a proper part of their sum; thus if both individuals are concreta, the sum (by 7.46) is not a concretum. Note, however, that while the sum of two different concreta is not even a complex, the product of two different overlapping concreta is always a complex.

5. ELEMENTARY QUALIFICATION

Different statements containing such a predicate of ordinary language as "is a quality of" may call for interpretation in our system in quite different ways, depending upon the other terms these statements contain. In some cases, to be dealt with later (Chapter VIII), the quality terms in such statements may not even be construed as designating individuals but rather as syncategorematic terms. In the present chapter, however, I am concerned only with those cases where both the quality and what is qualified are individuals. In the present section, I begin by defining an elementary qualification predicate that applies between complexes only; it applies between every proper part of a complex and the whole.

D7.05 $\mathrm{K}\, x,y = x \ll y \,.\, \mathrm{Cm}\, y.$

In the present section, "K" may be read "is a quality of", but later we shall often have to distinguish it carefully from other qualification predicates by reading, for example, "is a K-quality of". The converse of "K" may, with a similar reservation, be read "has as a quality" or "is an instance of". Any complex, then, is an instance of any of its proper parts. A color-spot is an instance of or has as qualities a color and a place. The qualities of a color-spot-moment are the three qualia and the three bicomplexes it contains.

To be a quality of an individual, then, is not merely to be a quale and to be a part or proper part of that individual; for the quality may be a complex other than a quale, and the qualified individual, or instance, must be a complex. Indeed, since (7.43) every proper part of a complex is a non-concrete complex, qualities are just those complexes that are not concreta:

7.51 $(\exists y)(\mathrm{K}\,x,y) \equiv .\, \mathrm{Cm}\,x \,.\, {\sim}\not{c}\,x.$

On the other hand, since no quale has a proper part (7.18), instances are just those complexes which are not qualia:

7.511 $(\exists x)(\mathrm{K}\,x,y) \equiv .\, \mathrm{Cm}\,y \,.\, {\sim}\mathrm{Qu}\,y.$

Thus concreta are instances but never qualities, while qualia are qualities but never instances. Complexes that are neither qualia nor concreta, then, are both instances (of their complex proper parts) and qualities (of more comprehensive complexes that contain them):

7.512 $\mathrm{Cm}\,x \,.\, {\sim}\mathrm{Qu}\,x \,.\, {\sim}\not{c}\,x \,. \equiv (\exists y)(\exists z)(\mathrm{K}\,x,y \,.\, \mathrm{K}\,z,x).$

A color-spot, for example, is both an instance of its place and its color, and also a quality of any color-spot-moment that contains it.

Inasmuch as "K" is an internal predicate,

7.52 $\mathrm{K}\,x,y \supset x \circ y,$

while W is external (7.21), no case of either predicate is a case of the other:

7.521 $\mathrm{K}\,x,y \supset {\sim}\mathrm{W}\,x,y.$

This theorem should be especially noted because words like "at", as was remarked earlier, are often used ambiguously. We say that a color is at a place when the two are together (i.e., when "W" applies between them); but we also sometimes say that a complex is at a place when the place is contained in the complex. Here it is not "W" that applies but "K". Confusion on this score is fostered by the fact that a color is a quality of a color-spot if and only if the color and place in question are together. More generally, when two individuals are together each is a quality of their sum:

7.522 $\mathrm{W}\,x,y \supset \mathrm{K}\,x,x+y$

(proof by C.I., 7.05, 7.37). Nevertheless, as 7.521 shows, the predicates are mutually exclusive, and must be carefully differentiated in exact discourse. "K" applies between any proper part of a complex and the complex itself; "W" applies between any two discrete parts of a complex. The several individuals with which a given quality may occur, then, are not instances of it, for it is discrete from these but is a part of each of its instances.

"K", like "$\ll$", is asymmetric:

7.53 $\mathrm{K}\,x,y \supset {\sim}\mathrm{K}\,y,x.$

Although in some cases an individual may be a quality of a second and an instance of a third (7.512), no individual is both a quality and an instance of any one individual. Accordingly, nothing is a quality of itself:

7.531 $\sim K\, x,x$.

Again like "$\ll$", "K" is transitive:

7.532 $K\, x,y \,.\, K\, y,z \,.\supset K\, x,z$.

A quality of any quality of an individual is a quality of that individual; hence, of course, an instance of any instance of an individual is itself an instance of that individual.

This latter theorem, like some earlier ones, reminds us that the words "quality" and "instance" have the flexibility of relative terms. Whether we regard a color-spot as an instance or a quality depends upon whether we are concerned with its relationship to its color and place or to a color-spot-moment. Whether we think of color-spots or color-spot-moments as the instances of a color depends likewise upon our interests in a given discourse. Our systematic interpretation of "quality" and "instance" merely makes explicit and extends a latitude exhibited in ordinary usage.

Every part of a quality of an individual is also a quality of the whole:

7.54 $x < y \,.\, K\, y,z \,.\supset K\, x,z$,

as is readily shown from C.I., 7.05 and 7.35. It is not true that a quality of an individual is a quality of every whole containing that individual, for the qualified individual must (by 7.511) always be a complex; we thus have, from C.I. and 7.05, only the more limited theorem:

7.541 $K\, x,y \,.\, y < z \,.\, \mathrm{Cm}\, z \,.\supset K\, x,z$.

For example, if a color is a quality of a given color-spot, the color is also a quality of every color-spot-moment containing that color-spot. Both these theorems are closely akin to but slightly stronger than 7.532; the consequent in each case is the same as in 7.532, but the antecedent is weaker.

Two qualities that have a common instance either overlap or are together:

7.55 $K\, x,y \,.\, K\, z,y \,.\supset .\, x \circ z \vee W\, x,z$.

The color and the color-spot qualifying a given color-spot-moment overlap, as do the color-moment and the color-spot qualifying it; but the color and the place, being discrete, are together. Any quality of an individual, indeed, is with any part of the rest of that individual:

7.551 $K\, x,y \,.\, z < y - x \,.\supset W\, x,z$.

The color of a color-spot-moment, for example, is with the place-time, the place, and the time. The latter are thus also qualities of the color-spot-moment; for we have the theorem:

7.552 $\quad \mathrm{K}\, x,y \,.\, z < y - x \,.\supset \mathrm{K}\, z,y.$

Whether two qualities of x overlap or are discrete, their sum is part of x; but we do not have as a theorem the statement that the sum of any two qualities of x is a quality of x; for the sum may be identical with x, and nothing (7.531) is a quality of itself. Our theorem then affirms only that the sum of two qualities of an instance is either a quality of or identical with that instance:

7.56 $\quad \mathrm{K}\, y,x \,.\, \mathrm{K}\, z,x \,.\supset .\, y + z = x \vee \mathrm{K}\, y + z,x.$

Obviously, a quality of two individuals is a quality of their sum only if that sum is a complex; hence

7.561 $\quad \mathrm{K}\, x,y \,.\, \mathrm{K}\, x,z \,.\, \mathrm{Cm}\, y + z \,.\supset \mathrm{K}\, x,y + z.$

A color that is a quality of a given color-spot and a given color-moment is a quality of their sum only if that sum is a color-spot-moment.

Since (by 7.05) an individual completely contains each of its qualities, two instances of a single quality overlap:

7.57 $\quad \mathrm{K}\, x,y \,.\, \mathrm{K}\, x,z \,.\supset y \circ z;$

for the quality itself is a common part of the two. In some cases, as we have seen from 7.532, the two instances are parts of a single complex; but the cases to be especially noted here are those where the sum of the instances is a compound–for example, the case of two different color-spot-moments that have the same color. It is clear from such cases that the old question how two entirely separate concreta may 'participate in' or 'partake of' a single quality is to be answered under our system not by maintaining that the quality is a sum of various particles that occur in the several instances, but by denying that any two instances of a quality can be entirely separate. Instances of a color may be discrete in time or space or both, but they still have the color as a common part. The similarity of these instances to one another is thus construed as involving literal part identity, i.e., overlapping. In the course of the following section (see D7.061), I shall define a similarity predicate that applies in just such cases of overlapping.

6. COMPOUND QUALIFICATION

By no means every statement affirming that an individual is a quality of another can be construed as saying simply that the one K-qualifies the other. Only subconcrete complexes K-qualify, and only complexes consisting of two or more qualia are K-qualified; but often what we take as an instance or even as a quality is a compound rather than a complex. For example, if I say that a certain spatiotemporally extended presentation *b* has a certain color *c*, my statement cannot be systematically interpreted by "K *c*, *b*"; for "K" does not apply between any individual and *b*, which is not a complex. We need to extend our systematic treatment of qualification by defining in terms of "K" other predicates that apply in this and other cases.

If presentation *b* is only partly of color *c*, and partly of other colors, then *c* partially qualifies *b*. "Partially qualifies" is readily defined as applying between *x* and *y* if and only if *x* K-qualifies some part of *y*:

D7.06 $\mathrm{Kp}\, x,y = (\exists z)\, (z < y \,.\, \mathrm{K}\, x,z)$.

While only subconcrete complexes Kp-qualify, every compound containing a complex of two or more qualia is a Kp-instance. A sum of several color-spots is Kp-qualified by each of the colors and places in question, but not by any of the color-spots. The sum of several colors or places is not Kp-qualified by any of them. For *x* to be a Kp-quality of *y* it is not enough that *x* be a part of *y*; *x* must be a proper part of some complex contained in *y*.

In the present section, I shall in general omit all theorems except those relating the several qualification predicates to one another. If *x* is a K-quality of *y*, then *x* is a Kp-quality of *y*:

7.61 $\mathrm{K}\, x,y \supset \mathrm{Kp}\, x,y$.

The converse is obviously not a theorem.

In terms of "Kp" one important similarity predicate can be defined. Two individuals are 'part-identically' similar if they are Kp-instances of some one quality:

D7.061 $\mathrm{Ps}\, x,y = (\exists z)\, (\mathrm{Kp}\, z,x \,.\, \mathrm{Kp}\, z,y)$.

Every two individuals that are thus similar overlap. The predicate "Ps" applies between every two individuals such that each contains a complex of more than one quale that overlaps a complex of more than one quale contained in the other. It does not apply, for example, between a sum of several qualia of one kind and any other individual, or between any individual and the sum of a color and a place that are not together. Thus two overlapping individuals are not always part-identically similar. An individual is part-identically similar to itself only if it contains at least two qualia that are with each other.

I must emphasize here that I do *not* hold that all similarity is part-identity similarity. Indeed, I shall later (IX,2) dispute that contention and shall deal at some length with similarity predicates that apply between discrete individuals.

If an extended visual presentation b has some one color c throughout, then c *uniformly qualifies* b, but "throughout" is a rather vague term. One might suppose offhand that "uniformly qualifies" could be readily defined as applying between x and y if and only if x is a K-quality of every part of y. But obviously this will not do, for no two individuals are so related. To appreciate how badly this proposal fails, and how much it must be altered to be made correct, consider the following difficulties, each of them fatal by itself:

(1) If y is a compound (and it is just such cases that make "uniformly qualifies" broader than "K-qualifies"), then there is at least one part of y (namely y itself) that nothing K-qualifies.

(2) Every individual has one or more qualia as parts; but nothing K-qualifies any quale.

(3) If x K-qualifies some part of y, then there is some part of y (namely x itself) that x does not K-qualify.

(4) If x K-qualifies a part z of y, then some part (namely $z-x$) of y is discrete from and therefore not K-qualified by x.

These difficulties seem to be removed if we change the definition so that x is required to K-qualify not every part of y but only every major complex of y, where a major complex or *section* of an individual is any complex contained in that individual, but not in any other complex contained in that individual. There are cases where an individual K-qualifies every *such* part of a compound even though no individual K-qualifies any compound or any quale or itself or anything discrete from itself. The predicate so defined quite properly applies, for example, between a color and the sum of several color-spots having that color; and it applies also wherever "K" applies. Furthermore, it quite properly fails to apply, for example, between a color and a presentation that is partially of any other color, or between a place and any individual that contains some other place.

And yet for a new and curious reason, this definition also breaks down. Suppose that $p_1 + t_1$ and $p_2 + t_2$ and $p_1 + t_2$ and $p_2 + t_1$ are place-times, and that color c_1 occurs at $p_1 + t_1$ and at $p_2 + t_2$ but not at the other two place-times. Let y be the sum of color-spot-moments $c_1 + p_1 + t_1$ and $c_1 + p_2 + t_2$. Clearly c_1 uniformly qualifies y; but does the predicate "uniformly qualifies", if defined as suggested in the preceding paragraph, apply between c_1 and y? Certainly y contains no other tricomplexes, for no other sum of three of the five qualia contained in y is a complex. But, strangely, y has other sections. Place-times $p_1 + t_2$ and $p_2 + t_1$ are complexes, and are parts of y since they are made up of qualia contained in y; and neither is part of another complex contained in y. Hence each is a section of y. Yet of course neither of them is K-qualified by c_1. Thus, although c_1 uniformly qualifies y, it does not K-qualify all sections of y; and so our proposed definition is shown to be unsatisfactory.

To require only that x K-qualify every concretum in y takes care of the case illustrated but is much too weak; for the predicate so defined applies between any individual and any other that contains no concreta. On the other hand, if we require in addition that the qualified individual be a sum of concreta, the predicate is too narrowly defined; for it then does not apply in perfectly genuine cases of the uniform qualification of certain sums of subconcrete complexes.

Nevertheless, all these difficulties can be met by a rather simple definition. An individual x uniformly qualifies another individual y if and only if x is a proper part of y and is with every quale in the rest of y:[7]

D7.062 $\quad \text{Ku}\, x,y = x \ll y \,.\, (z)\, (\text{Qu}\, z \,.\, z < y - x \,.\supset \text{W}\, x,z).$

So defined, "Ku" gives the wanted results in the cases discussed above. Furthermore, while a uniformly qualified individual may contain two or more qualia that are not together, a uniform qualifier may contain no individual that is not with every quale in the qualified whole.

Uniform qualification obtains just where the qualified individual y is exhaustively divisible into parts that are K-qualified by the qualifying individual x–that is, where every quale in y is contained in some complex in y that is K-qualified by x. The theorem,

7.611 $\quad \text{Ku}\, x,y \equiv (z)\, \{\text{Qu}\, z \,.\, z < y \,.\supset (\exists t)\, (z < t \,.\, t < y \,.\, \text{K}\, x,t)\},$

is easily proved. The parts that jointly exhaust y and that are K-qualified by x will of course not be mutually exclusive, since they will have x, at least, as a common part.

[7] In the first edition, the biconditional now entered as Theorem 7.611 was used to define uniform qualification. The simpler definition now adopted is the result of a suggestion by a student, Alan Haussman.

The interesting cases of uniform qualification are those where y is divisible into complexes that are all of the *same category* (e.g., all concreta, all color-spots, all place-times, etc.) and each is K-qualified by x. But D7.062 embraces less interesting cases also; e.g., if c_1 is at p_1 and at t_1, even if not at $p_1 + t_1$, then c_1 Ku-qualifies $c_1 + p_1 + t_1$. We can easily define the special variety of uniform qualification (call it "Kt") where y is divisible into concreta K-qualified by x; but the other interesting cases can be formally differentiated from less important ones only after the notion of categories (see D9.052) has been defined.

Although x Ku-qualifies y only if x is with every quale in the rest of y, notice that x K-qualifies y only if x is with the rest of y as a unit:

7.612 $\quad \mathrm{K}\,x,y \equiv . x \ll y . \mathrm{W}\,x,y - x.$

Thus it is clear that while Ku-qualification does not imply K-qualification, K-qualification does imply Ku-qualification:

7.62 $\quad \mathrm{K}\,x,y \supset \mathrm{Ku}\,x,y.$

And Ku-qualification implies Kp-qualification:

7.63 $\quad \mathrm{Ku}\,x,y \supset \mathrm{Kp}\,x,y.$

Where we want to say that x only partially qualifies y, we may write: "$\mathrm{Kp}\,x,y . \sim\mathrm{Ku}\,x,y$".

Incidentally, erlebs are most conveniently defined with the help of "Ku". A phenomenal event occupying a single moment is uniformly qualified by a time quale. An erleb is simply any most-comprehensive individual of this kind. Thus after "T" ("is a time quale") has been defined (see IX,5; XI,2), erlebs can be defined as follows: "$\mathrm{Erl}\,x =_{\mathrm{df}} (\exists y)((\mathrm{T}y . \mathrm{Ku}\,y,x) . (z)(\mathrm{Ku}\,y,z \supset z < x))$".

In all cases of qualification so far considered, the qualifier is a subconcrete complex, even though in some cases the qualified individual, or instance, may be a concretum or a compound. Now clearly, to say that a given extended presentation y is red is not to say that "Kp" (and therefore not to say that "Ku" or "K") applies between the sum of all red color qualia and y, but rather to say that "Kp" (and perhaps one of the narrower predicates) applies between some red color quale and y; or–to give an equally good alternative interpretation–it is to say that some section of the compound sum of all red color qualia Kp-qualifies y. In other cases, we may want to say that some section of an individual Ku-qualifies or K-qualifies another individual. Special qualification predicates for each of these three kinds of cases, and for those that follow, are easily defined; but such a multiplication of definitions and symbols seems hardly likely to be helpful at present.

Cases different from any of those already mentioned are quite common. For example, when we say that an extended presentation y is all red we are not saying that y has some one red quale throughout, but only that y everywhere has *some red quale or other*. In view of Theorem 7.611, we are saying that y is exhaustively divisible into parts each of which is K-qualified by some red quale, i.e., by some section of the compound red. Another example of the same sort occurs when we say that a presentation y occurs in the right-hand half of the visual field; here we are saying that y is exhaustively divisible into parts each of which is K-qualified by some place in (i.e., by some section of) the right-hand half of the visual field.

When we say, however, that a presentation y covers–without saying that it covers only–the right-hand half of the visual field, we are saying that every section of (i.e., place in) the right-hand half of the visual field Kp-qualifies y. If we say that y *exactly* covers the right-hand half of the visual field, then systematically speaking we are saying *both* that every section of the right-hand half of the visual field Kp-qualifies y *and* that y is exhaustively divisible into parts each of which is K-qualified by some section of the right-hand half of the visual field. Another example of this same sort occurs when we say that a presentation y of a Dalmatian is black-and-white; the quale black and the quale white (hence every section of the sum of the two) Kp-qualifies y, and y is exhaustively divisible into parts each of which is K-qualified by either black or white.

Thus a considerable variety of sentences of ordinary language can be interpreted in terms of the elementary qualification predicate "K". Certain qualification sentences that require different treatment will be dealt with in Chapter VIII.

7. A PARADOX AND ITS LESSON

Examination of a rather interesting paradox may provide a reminder of certain important points already noted as well as a warning against some easily committed errors.

Suppose that each of the two places p and q occurs at each of the two times r and s. There are then four place-times $p + r$ and $p + s$ and $q + s$ and $q + r$, which may be called respectively a, b, c, and d.[8] Consider now two

[8] Nothing in the following depends upon the fact that our example is in terms of places, times, and place-times. We might equally well let p and q be colors, r and s places, and a, b, c, and d color-spots; or let p and q be colors, r and s times, and a, b, c, and d color-

simple systems. For System I, the atoms are the qualia *p, q, r*, and *s*; and the universe of individuals consists of sums of one or more of these. For System II, the atoms are the place-times *a, b, c*, and *d*; and the universe of individuals consists of sums of one or more of these. Now each of the place-times is a sum of two atoms of System I; thus all the atoms of System II and therefore all the individuals of System II are individuals of System I. In addition, *p, q, r*, and *s*–which are presystematically parts of the atoms of System II–are individuals of System I. If the individuals of System I thus include all the individuals of System II and at least four others, then there are more individuals for System I than for System II. But on the other hand, for each of the two systems there are just four atoms, and therefore just fifteen individuals. Accordingly, it appears that System I must have at once more individuals than System II and exactly the same number as System II. How can this conflict be reconciled?

The individuals of System II are

1. a	6. $a+c$	11. $a+b+c$
2. b	7. $a+d$	12. $a+b+d$
3. c	8. $b+c$	13. $a+c+d$
4. d	9. $b+d$	14. $b+c+d$
5. $a+b$	10. $c+d$	15. $a+b+c+d$.

Let us now list the same individuals in the same order, but describe each as a sum of atoms of System I.

1. $p+r$	6. $p+r+q+s$	11. $p+r+p+s+q+s$
2. $p+s$	7. $p+r+q+r$	12. $p+r+p+s+q+r$
3. $q+s$	8. $p+s+q+s$	13. $p+r+q+s+q+r$
4. $q+r$	9. $p+s+q+r$	14. $p+s+q+s+q+r$
5. $p+r+p+s$	10. $q+s+q+r$	15. $p+r+p+s+q+s+q+r$.

Since "+" is commutative and $x+x=x$, the individuals listed here as 6, 9, 11, 12, 13, 14, and 15 are seen to be identical with one another. The second list actually lists only nine individuals of System I. Even though every individual of System II is an individual of System I, not every two individuals that are distinct for System II are distinct for System I. This coalescence invalidates

moments. Or, taking less comprehensive qualities as atoms, we might let *p* and *q* be visual-field elevations, *r* and *s* visual-field azimuths, and *a, b, c*, and *d* visual-field places; and so on.

the argument that System I has more individuals than System II. For each system, there are just fifteen individuals. The first list above names the fifteen individuals of System II. Besides the nine individuals of System I named in the second list there are six more: the atoms p, q, r, and s; the sum of the two places, $p + q$; and the sum of the two times, $r + s$.

But now we seem to face another difficulty. How can two individuals admitted under each of two legitimate systems be distinct for one and identical for the other? According to System I, for example, $a + c = b + d$, while according to System II, $\sim(a + c = b + d)$. Must not one of two systems that thus contradict each other be simply wrong?

The contradiction is only apparent. As a result of the different restrictions of the variables of the two systems, the cases in which the predicate "**o**" applies in the two are not exactly the same (see II,5); and "+" and "=" are defined in each system in terms of "**o**". The statement "$a + c = b + d$" in System I affirms in effect that those individuals (of System I) that System-I-overlap a or c are just those that System-I-overlap b or d. This is true: the individuals that System-I-overlap a or c are all the individuals of System I, and all of these also System-I-overlap b or d. On the other hand, the statement "$\sim(a + c = b + d)$" in System II affirms in effect that those individuals (of System II) that System-II-overlap a or c are *not* just those that System-II-overlap b or d. This is also true: the individuals that System-II-overlap a or c are a, c, and $a + c$, while the individuals that System-II-overlap b or d are b, d, and $b + d$. There is thus no conflict.

Resolution of the paradox, however, leaves us with an awareness of the need for caution in describing certain facts within a given system. For example, if white occurs at place-times a and c of the above example, it may or may not be the case under our present system that white occurs at every section of $a + c$. For the sum of the place-times a and c also has as sections the place-times b and d. And if white occurs at a and c while black occurs at b and d, then $a + c$ is exhaustively divisible into place-times at which white occurs and also exhaustively divisible into place-times at which black occurs.[9] Thus we must distinguish very carefully between saying that a color occurs at each of several place-times that make up a spatio-temporal path or region and saying that the color occurs at all place-times contained in that path or region.

The fact that discourse within our present system is subject to such pitfalls

[9] However, although white uniformly qualifies the sum of white and the four place-times, and black uniformly qualifies the sum of black and the four place-times, neither white nor black nor anything else uniformly qualifies the sum of all the concreta involved here–a sum identical with the sum of white and black and the four place-times.

might seem to constitute one good reason for choosing more nearly concrete or even fully concrete individuals as atoms. For System II, as we saw, $a + c$ contains no other place-times than a and c, and $a + c$ is discrete from $b + d$. Again, if concreta are taken as atoms, the sum of the concreta lying in the diagonal path of the presentation above described will contain no concreta other than those lying in that path, and will be a proper part of the whole presentation. But thus to choose different atoms creates pitfalls parallel to those encountered under our present system. Look again at Systems I and II. Even though p, q, r, and s are, presystematically, proper parts of the atoms of System II, still each of these qualia may be defined in System II– quite in conformity with the criterion of isomorphism–as the sum of two of the atoms of System II. For instance, the definition may be such that each quale is the sum of those two place-times that presystematically contain that quale, so that p is defined as $a + b$, r as $a + d$, s as $b + c$, and q as $d + c$. Now certain individuals that are distinct for System I will be identical for System II; i.e., for System II, the sum of the two places will be identical with the sum of the two times, with the sum of any three of the four qualia (p, q, r and s), and with the sum of all four. And to say that a color occurs at each of two places will not be the same as to say that the color occurs with every quale contained in the sum of the two places, for that sum will contain the two times as well. To say that the color is with every quale in the sum of the two places is thus to say, as one might not want to say, that the color occurs with each of the two times as well. Again, if concreta are taken as atoms, the sum of all the places, the sum of all the times, and the sum of all the colors in any presentation will be identical with each other and with the whole presentation.

Thus, whichever sort of atoms we choose, we have to be careful if we are to succeed in saying in the language of a system just what we want to say and no more. Accuracy seems to require special vigilance against natural confusion in about as many cases under one sort of system as another.

8. A NOTE ON ABSTRACT, CONCRETE, UNIVERSAL, AND PARTICULAR INDIVIDUALS

It became clear earlier (Chapter IV) that while nominalistic systems admit no nonindividuals, they may admit individuals of widely varying kinds. Some individuals, such as the atomic qualia of our present system, are appropriately termed abstract; others, such as concreta, are naturally called concrete. This usage is quite in keeping with tradition and ordinary practice, and is entirely compatible with the individuality of the entities to which the predicates in question are applied. I now want to suggest precise general definitions to govern the use of the terms "abstract", "concrete", "universal", and "particular" as predicates of individuals. This need involve no decision on the use of these terms in application to nonindividuals in platonistic systems.

As a beginning, we have already observed that an individual is concrete if it is fully determinate, i.e., if it is K-qualified by some quale from each of the categories within some one sense realm.[10] A concretum obviously answers this description, while a proper part of a concretum does not. A quale, for example, has no K-qualities whatever; and a visual bicomplex has either no color or no place or no time as a K-quality. Moreover, the sum of a color, a place, and a time that are not all together is not concrete; for this compound sum, although it consists of one quale from each category in the visual realm, is not K-qualified by anything. But of course we do not want to confine the term "concrete" to concreta alone; certain compounds, sums of two or more concreta, are no less concrete–although no more so–than a single concretum. A compound is concrete if it is exhaustively divisible into fully determinate parts. In general, *an individual is 'concrete' if and only if it is exhaustively divisible into concreta.* Thus, any sum of concreta is concrete, whether it be an erleb, a presentation, or made up of scattered concreta from one or many sense realms. On the other hand, for example, the sum of a concretum and a color not contained in it is *non*concrete, as in any proper part of a concretum and any sum of several qualities of one kind.

Every individual is exhaustively divisible into abstract parts (7.14), but abstract individuals are those that have no concrete part. Qualia and all other proper parts of concreta are of course abstract, but certain compounds also are abstract–for example, the sum of several colors or times or of a place, time, and color that are not all together. Since every individual containing a concretum has some concrete part, *an individual is 'abstract' if and only if it contains no concretum*.

These definitions are purposely so formulated as to leave a middle ground. Some individuals are neither concrete nor abstract. One example is the sum of a concretum and a color it does not contain. Others are easily discovered.

The terms "universal" and "particular" as predicates of individuals coincide rather closely, but not exactly, with "abstract" and "concrete", respectively. Whereas abstractness is a matter of partial indeterminateness, universality is rather a matter of multiplicity of instances. For example, a color or a color-spot is abstract through lacking a time quality, but is

[10] The terms "category" and "sense realm" have been explained earlier but not yet systematically defined. However, their definitions (in IX,5) make no use of the terms explained in the present section.

universal through having many color-spot-moments as instances. In other words, universality depends on repeatability, where repeatability is explained as follows. A complex is *repeated* if it occurs with some two individuals of one category (see IX,5). A color, for example, is repeated if it occurs at two places, even at the same time; and a time is repeated if two qualia of some one kind occur at that time. Any complex that is thus repeated will have two concreta as instances. A complex, whether repeated or not, is *repeatable* if some complex of the category to which it belongs is repeated. Note that the disposition term "repeatable" is here defined without use of modal terms or counterfactual locutions.

A complex is universal if repeatable, particular if unrepeatable. Whether a compound is universal or particular depends upon its parts. In general, *an individual is 'particular' if and only if it is exhaustively divisible into unrepeatable complexes;* while *an individual is 'universal' if and only if it contains no unrepeatable complex*. Thus all qualia and all sums of several qualia of one kind are universal; all concreta and all sums of concreta are particular; and such individuals as the sum of a concretum and a color foreign to it are neither universal nor particular.

Are there, now, any individuals that are abstract or universal but not both, or any that are concrete or particular but not both? All concrete individuals are particular, but place-times are particular individuals that are nevertheless not concrete but abstract. A place-time is abstract because it has no color as a K-quality; yet it is unrepeatable, for no place-time occurs with two colors or with two qualia of any one kind. Likewise, a sum of place-times is exhaustively divisible into unrepeatable individuals, but contains no concretum as a part, and is accordingly at once particular and abstract. Thus, although all concrete individuals are particular, and all universal individuals are abstract, still certain particular individuals are not concrete but abstract–or, in other words, certain abstract individuals are not universal but particular.

However, although this way of distinguishing between "particular" and "concrete" has theoretical interest, so meticulous a use of these terms is ordinarily impractical even in technical discourse. The noun "particular" is too handy a substitute for "concrete individual" to be reserved for a slightly broader usage. And the reader will recall that I term a system "particularistic" only if all its basic units are concrete.

CHAPTER VIII

SIZE AND SHAPE

1. THE PROBLEM

The sum of two red patches is always red but the sum of two rectangular patches is often not rectangular. An individual and a proper part of it may have the same color but not the same size. As a result of such presystematic differences between size and shape on the one hand and, for example, color and place on the other, we have not counted sizes and shapes among the atomic qualia of our system. In Chapter VI, I explained that the shape term "square", for instance, would be construed otherwise than as naming an atom of our system because all our atoms are qualities of concreta and no concretum is square. More generally and technically, an atomic quale x of our system qualifies a compound y if and only if x qualifies some section of y, while the size or shape of a compound may obviously be quite different from the size and shape of any of the sections. For example, none of the sections of a square individual is square. Is "square", then, rather a name for a compound quality made up of several atomic shape qualia some of which qualify sections of a square concrete individual? Clearly not. "Square" does not name a broad quality like a spatial region, which consists of several definite visual-field locations, or like red, which consists of many specific red colors; "square" is already a completely specific shape term and the shape of a section of a square individual is not a special kind of squareness. Nor is the shape of a section to be construed as an atomic quale according to our criterion; for the shape of the sections of a compound is rarely the shape of the compound itself. No shape term, then, is to be interpreted as naming an atom of our system.

The case of size terms is much the same. Let "is of size a" be applicable to all individuals that occupy exactly one-fifth of the entire visual field. None of the sections of any presentation of size a is also of size a, for a section is a complex that occupies not more than one place in the visual field. Nor is "a" the name of a compound quality made up of the smaller sizes of proper parts of individuals of size a. Smaller sizes are no more contained within larger sizes than a light gray color is contained within a dark gray color; terms for the larger sizes are as specific as terms for the smaller sizes. Nor is even the smallest size, any more than larger ones, to be construed as an atomic quale;

for obviously the fact that all the sections of a compound are of minimal size does not guarantee that the compound itself is of minimal size. Size terms, like shape terms, thus do not name atoms of our system.

From what has already been said, it will be further apparent that a size term or shape term is not to be interpreted as naming any individual of our system at all. Every individual of a system must be a sum of one or more atoms of the system (II,5). Obviously, shapes and sizes are not to be construed as sums of the chosen atoms (colors, places, times, etc.) of our system; and the admission of other atoms of which certain sums might be shapes or sizes would be open to the same sort of objections as the admission of shapes and sizes themselves as atoms.

Thus we must look for some other way of construing size terms and shape terms. They are denied the status of names of atoms or other individuals in our system because, so to speak, there are important ways in which they do not behave like other terms that are taken as naming individuals in the system.

2. SIZE

Let us begin by considering the simple sort of case where a statement about the size of an individual can be interpreted as a statement about the amount of space the individual contains. If we want to base a definition on this interpretation we shall find that one of the predicates needed for the definiens is not yet available in our system. We have not so far defined space, i.e., the category of visual-field places. The problem of defining categories of qualia and distinguishing between them will be left to a later chapter (IX,5) concerned with the construction of quality orders; for the same basic predicate is used for categorizing qualia as for ordering them. Once the predicate "is a place" has been defined, we can tentatively define specific size predicates–"is of size 0", applying to x if and only if x contains no places; "is of size 1", applying to x if and only if x contains just one place, and so on. Translation of the specific numerical terms appearing in these definientia offers no difficulty under either a platonistic or a nominalistic system.

In practice, however, we are almost always concerned with comparative rather than specific size. In a platonistic system, statements to the effect that x has as many places as y, or more or fewer places, are easily translated; for we have number-variables and can speak of "the number" of places in an individual when that number is unspecified. But translation of such indefinite numerical comparisons into a nominalistic system, where numbers are not admitted as values of variables, raises a real problem, as we have already observed in II,3. Since the general apparatus of a nominalistic system does not provide us with the means for translating such statements, we are forced to seek the most economical additional set of predicates that will serve the purpose. As may be seen from the examples of Chapter II, the reduction will be rather a reduction of number to size than of size to number, or better, perhaps, a definition of many size predicates and numerical predicates in terms of a more general predicate that might indifferently be described either as a size predicate or as a numerical predicate. In this book, the general predicate of this kind that I shall employ, without claiming that it enables us to deal with all numerical statements that we may eventually want to translate, is "Z"–for "is of equal aggregate[1] size"–so explained as to apply between two individuals if and only if they contain the same number of qualia. Since it is pretty clearly not definable in terms of the general apparatus of our system together with the predicate "W", the predicate "Z" may for the present be considered a new primitive. However, I shall show (in IX,4) how this predicate and "W" and the basic predicate used for ordering and categorizing qualia can all be reduced to a single predicate as simple as any one of the three.

In terms of "Z" the general predicate "G"–for "is aggregately bigger (or greater) than"–can readily be defined:

D8.021 $\quad \mathrm{G}\,x,y = (\exists t)\,(t \ll x \,.\, \mathrm{Z}\,t,y)$.

For convenience, I shall sometimes place "G" between the expressions for its arguments where these expressions are rather long. No symbol need be defined for "is aggregately smaller than" since x is aggregately smaller than y if and only if $\mathrm{G}\,y,x$.

A calculus of size based on "Z" would be of some interest, especially since we shall need to use certain of its theorems later. I can do no more here, however, than list a few true statements of the calculus that are rather useful. Obviously both "Z" and "G" are transitive; but "Z" is also reflexive and symmetric while "G" is irreflexive and asymmetric. Also:

8.21 $\quad \mathrm{Z}\,x,y \equiv .\, {\sim}\mathrm{G}\,x,y \,.\, {\sim}\mathrm{G}\,y,x$

8.22 $\quad x = y \supset \mathrm{Z}\,x,y$.

Furthermore, since this calculus is introduced within a system having a finite number of individuals, we have:

8.23 $\quad x \ll y \supset \mathrm{G}\,y,x$

[1] The term "aggregate" is inserted to indicate that qualia of all kinds are to be counted together in determining the applicability of this predicate.

8.24 $x \ll y \supset \mathrm{G}\, y, y - x$

8.25 $\sim x < y \supset \mathrm{G}\, x + y, y$

8.26 $x \circ y \,.\, \sim x < y \,.\, \supset \mathrm{G}\, x, x - y.$

Also important sometimes are less simple theorems, such as:

8.27 $\mathrm{G}\, x,y \equiv (\exists t)\, (t < x \,.\, t \wr y \,.\, \mathrm{Z}\, x, y + t)$

8.28 $\mathrm{Z}\, x,y \equiv (t)\, \{\sim(t \wr y \,.\, \mathrm{Z}\, x, y + t) \,.\, \sim(t \wr x \,.\, \mathrm{Z}\, y, t + x)\}.$

But these few examples will have to suffice.

In terms of the comparative aggregate sizes of the spatial regions occupied by two individuals, we may define narrower predicates pertaining to *spatial size*. The predicate "Zs"–for "is of equal spatial size"–may be so defined that it applies between *x* and *y* if and only if (1) neither contains any places or (2) the sum of all the places contained in *x* is of the same aggregate size as the sum of all the places contained in *y*. Similarly, "Gs"–for "is spatially bigger than"–may be so defined that it applies between *x* and *y* if and only if (1) *x* contains at least one place and *y* contains none or (2) the sum of all the places in *x* is aggregately bigger than the sum of all the places in *y*. These definitions are purposely formulated in such a way that the size of a spatial region will be the same as the spatial size of an individual occupying that region; and also in such a way that all individuals containing no places will be of the same spatial size as each other but spatially smaller than any individual containing at least one place. The definitions cannot, of course, be formally introduced into the system until after places have been defined.

In ordinary discourse, we seldom say that one individual is temporally of the same size as or bigger than another; we say rather that one lasts as long as or longer than the other. But duration and spatial size are clearly analogous; and once times have been defined, we may define predicates pertaining to duration or *temporal size* in the same way we define those pertaining to spatial size. We need only supplant the predicate "is a place" by the predicate "is a time". The predicate "Zt", for example, will apply between *x* and *y* if and only if neither contains a time, or the sum of all the times in *x* is of the same aggregate size as the sum of all the times in *y*.

Obviously, individuals may be compared in size with respect to any other category of qualia as well, even though the vocabulary that ordinary language supplies for the purpose is often very meager. For example, just as an individual containing many places is spatially bigger than an individual containing one place, so a multicolored individual is 'bigger in color' than a uniformly colored individual.

I have purposely been oversimplifying the matter up to this point. The systematic predicates so far defined are only the first rudimentary instruments to be used in translating ordinary statements about size; for actually not many such statements are accurately interpreted as speaking solely about the total number of places (or qualia of some other kind) occupied by each of the individuals in question. To take a simple example, the spatial size of an enduring individual is rather the amount of space it occupies at one time than the total amount it covers throughout its entire career. A moving patch of color may in a short time traverse the entire visual field; but the spatial size we ascribe to the whole event is the size of the region it occupies at a single moment. To speak of the spatial size of such an individual is to speak elliptically of the spatial size of its temporal cross sections. If these differ in size, the individual is said to grow or diminish.

What is more important, size usually depends not only upon how many qualia of the given kind are involved but also on how they are distributed. We often regard x as spatially bigger than y, even though y contains more places than x, if x is spread over a larger area than y. In judging the size of a presentation of a celestial constellation, we may consider only those visual-field places that are occupied by the bright patches, or we may consider all those visual-field places that are contained in the smallest continuous convexly bounded region that contains the presentation. To mark such a difference we might introduce the terms "net size" and "gross size", although a precise systematic definition of gross size will obviously not be easy to formulate. From the example given, it may appear that gross size is of comparatively little interest and that in practice we need only make it clear that we shall always speak of net size. But as a matter of fact when the places occupied are close enough together, it is gross rather than net size that we commonly measure. Presentations of two unbroken straight line segments in the visual field are considered equally long if their ends are equally far apart; and their ends may be equally far apart even though the number of intervening places is different. This cannot well be explained in detail here; the further study of size and measure must await the construction of quality orders (see X,5 and X,6).

3. SHAPE

When we speak of the shape of an individual, as when we speak of its size, we are usually saying something about the space it occupies. To say that an individual is of a given spatial shape is to say that the places it contains are related to one another in a given way. The predicate "is square" applies to an individual only if the arrangement of places contained in the individual meets certain requirements. A more general predicate like "is four-sided" applies if the arrangement meets certain less stringent requirements. The 'arrangement' of a set (or sum) of places is, so to speak, the configuration these places mark out in the visual field; and spatial shape thus derives from the order of all places in the array we call the visual field. Obviously I am here again anticipating the construction of quality orders. Only after that problem has been dealt with can the definition of any shape predicate be formulated in the system (see X,12).

Just as we say that an individual is of a certain *spatial* shape if the places it contains are arranged in a certain way, so we may appropriately say that it is of a certain color shape if the colors it contains are arranged in a certain way; and similarly with respect to other kinds of qualia. The color shape of an individual is, so to speak, the configuration that the colors of that individual mark out in the total array of colors sometimes called "the color sphere". Predicates like "is contrastingly colored", "is colored in complementaries", and "is harmoniously colored" are used for describing roughly the color shape of certain individuals. Although ordinary language supplies rather few such predicates, the variation in color shape among phenomenal individuals is even greater than the variation in spatial shape, for the order of colors is more complicated than the order of visual-field places. It must be borne in mind that what I speak of as the color shape of x depends solely upon the relative positions of the colors of x *in the color sphere*; it has nothing to do with the distribution of these colors over the space of x. If a has the colors g, h, j, and k in four stripes while b has the same four colors–or even four other colors related in the color sphere in the same way as g, h, j, and k–in dozens of irregular patches, still a and b are of the same color shape. Thus to say that two individuals have the same color shape is not to say that they have the same (or corresponding) colors at corresponding places.

As with size, what we ordinarily regard as the spatial shape or color shape of an enduring phenomenal individual is not the shape of the sum of all the places or colors the individual contains but rather the shape of the sum of all the colors or places it contains at one time.

Since the order of times is linear, temporally continuous individuals differ rather in temporal size than in temporal shape. But temporally discontinuous individuals–such as auditory presentations of letters in the Morse code–may differ in temporal shape, i.e., in the pattern marked out in the array of time qualia by those times at which the sound occurs. The temporal shape of an individual is determined not by what occurs at the times contained in the individual but by the position of these times relative to one another. Rhythm, when clearly dissociated from such factors as accent in accordance with Prall's acute analysis,[2] is simply temporal shape.

Although individuals may differ in shape with respect to each of the kinds of qualia they contain, far more attention is paid in daily life to spatial shape than to shape in other respects. There are at least three reasons for the superior practical importance of spatial shape. First, the correlation between (visual) spatial shape and tactual spatial shape, and the relevance of objective spatial shape to mechanical laws, make spatial shape vital in the guidance of actual conduct. We may paint a boat in any combination of colors without diminishing its speed, but if we build it with a concave prow we have our troubles. Second, since the number of different colors in the ordinary individual taken as a unit for practical purposes is much smaller than the number of places, we can take inventory of the colors while we incline to note merely the size and shape of the spatial region. Third, the faster and more frequent change of phenomenal place (of an ordinary practical individual) as compared with change of phenomenal color not only focuses practical interest upon objective rather than phenomenal place, as observed earlier, but also gives *relative* phenomenal spatial position a greater importance than specific phenomenal location. With the phenomenal place of a thing varying at every twitch of the eyes, we must in practice concern ourselves rather with relationships obtaining among a set of (simultaneously) presented places; for these relationships–which constitute spatial shape–often remain constant for several different sets of places successively presented by a thing. In the case of color, on the other hand, there is usually less rapid shifting of presented qualia; specific color is less transient, and we are therefore not driven to color shapes to find a comfortable modicum of phenomenal stability.

Nevertheless, color shapes are often important factors in experience. Koehler's experiment demonstrating that a rat learns to go to the lighter-colored of two doors even though the specific colors be greatly varied shows how color relation instead of specific color may become the crucial factor in a practical situation. If the rat were taught to go always to a contrastingly colored in preference to a uniformly colored door, color shape would become the all-important factor. And Prall has argued, in somewhat different words, that the perception of color shape is precisely as integral to aesthetic apprehension in the visual realm as is the perception of rhythms (temporal shapes) to aesthetic apprehension in the auditory realm.

---

[2] D. W. Prall, *Aesthetic Analysis* (New York: T.Y. Crowell Company, 1936) chap. iv.

4. INITIAL AND DERIVATIVE QUALITY TERMS

The specific qualia that an individual contains fix its size and shape with respect to each category of qualia. If we can count and are familiar with the order of qualia in a given category, then we can determine the size and shape of an individual x in that respect if we know exactly what qualia of that kind are contained in x. For the arrangement of a given set of qualia is not variable. Qualia cannot be literally 'moved around'; each has–we might almost say each *is*–a fixed position in the array of the category to which it belongs.

On the other hand, it is not conversely the case that the size and shape of an individual with respect to a given category fix the specific qualia that the individual contains. If we can count and are familiar with the order of qualia in a category, still we cannot in general determine what qualia of that kind are contained in an individual x even if we know the size and shape of x in that respect. Thus, statements ascribing certain qualia to an individual may on this score be regarded as prior to statements concerning shape and size. Conformably, terms for color, place, time, etc. may be called *initial* quality terms, while terms for shape and size are called *derivative* quality terms. In our present system, initial quality terms are construed as names of individuals while derivative quality terms are construed as syncategorematic. In another system, such as a nonplatonistic reformulation of the *Aufbau* system, terms of both sorts may be construed as syncategorematic; while in some still different system, terms of both sorts might perhaps be construed as names of individuals. Yet the distinction between initial and derivative quality terms remains unaffected by such differences in systematic construction; for whichever of these systems is adopted, it remains true that initial quality predications (of color, place, etc.) determine derivative ones (shape, size), but not vice versa.

I am not suggesting that in actual experience we first take inventory of the specific qualia of an individual and then determine its size and shape by counting these qualia and studying out their arrangement. In the case of space especially, as already noted, an ordinary thing moves so rapidly in the visual field and occupies so many least-discernible places at each moment that we are in practice usually concerned not with specific location but rather with more summary characteristics such as shape and size. (Indeed, for much the same reasons we are more often concerned with relative than with actual size). Thus, we commonly observe what derivative quality terms apply to an individual before we consider what initial terms apply–and often we do not consider the latter matter at all. This merely underlines what should already have been clear–that my usage of the terms "initial" and "derivative" is not based on considerations of psychological accessibility or priority.

I have now explained, somewhat more precisely than was possible in earlier chapters, the differences that lead me to construe size terms and shape terms otherwise than as names of individuals; and I have outlined how they are to be construed in our present system. Now it may be argued that we ought to construe color terms syncategorematically, since the color or colors of a concrete visual individual *x* are fixed by the place-times that *x* occupies quite as unambiguously as the shape or size of an individual *y* in any respect is fixed by the qualia of the kind in question that *y* contains. I might use a good many words pointing out several interesting and important ways in which the two cases differ, but this is unnecessary; for the argument mistakes the purpose of construing shape terms and size terms in the way proposed. To define terms syncategorematically in a system is often a difficult task involving the introduction of new predicates. For example, we have seen that to define size terms and shape terms syncategorematically we shall need some predicate for ordering qualia and some such predicate as "is of the same aggregate size". To define color terms syncategorematically as just suggested we should doubtless need some predicate of color identity or color likeness between place-times. The additional complexity of construction and possible sacrifice of economy of basis is justified in the case of size terms and shape terms, where we have to account in our system for the way in which those terms behave differently from other quality terms. It is also justified if for antecedent reasons we are unwilling to countenance any such entities as are ostensibly denoted by given terms; if, for example, nominalistic convictions make us unwilling to countenance entities denoted by so-called 'class-names'. But the trouble, and risk of loss of economy, involved are uncalled for in the interpretation of color terms in our system; for these behave enough like place terms, etc., to make it quite feasible to construe them in the same way, and presumably we are willing to accept colors as individuals if we are willing to accept the equally abstract places and times.

I have repeatedly warned against the notion that our present system is designed to reflect the chronological development of knowledge. I do not suppose that qualia are the 'original givens' and I have expressed doubt whether the question what is originally given in experience can even be made clear. Nevertheless, I may digress briefly to observe one epistemological bearing the system may have, at least according to certain theories of epistemology. Whatever may be the original givens of experience, qualia may still be the elements into which we ordinarily tend to dissect the content of experience in order to comprehend it according to a structural scheme that will be applicable to further experience. This would make it easy to explain, for instance, the ready apprehension of shapes; for while the combination of qualia in a certain presentation might be novel, the qualia themselves and their relations within their several fixed arrays would be familiar. If new content is analyzed as a new combination of familiar and already ordered qualia, its whole structure becomes immediately comprehensible; and this is quite consistent with our earlier observation that the pattern of qualia in a presentation is often noticed before the several qualia themselves. Lewis has argued cogently that knowledge–even "experience" properly so called–depends heavily upon analyzing the presented in an orderly way that will disclose recurrent structural features. Analysis into repeatable qualia obviously serves this purpose, whereas analysis into unrepeatable erlebs does not; and according to Lewis's account, it is something like the former process that is actually performed in cognition. If this is so, then at least one important cognitive process is paralleled more closely by a system like our present one than by a system like that of the *Aufbau*. The reader who remembers my earlier chapters will not mistake this comment for an argument.

In discussing size and shape I have had to refer again and again to the order and categorization of qualia. The latter topics, which are important for other reasons also, thus demand immediate attention.

PART THREE

ON ORDER, MEASURE, AND TIME

CHAPTER IX

THE PROBLEM OF ORDER

1. A NEW PROBLEM

From the venerable problem of the relation between qualities and the concrete, we turn now to a problem that is very young. The problems of abstraction and concretion, though they may appear in modern logical guise, have a clear genealogy running back to the Greeks. The problem of order inherits no such claim to philosophical respect. Only recently have writers like Poincaré, Lewis, and Prall come to emphasize the importance of quality orders, regarding them as integral factors in knowledge. For Lewis, the order of qualities is a necessary condition for the intelligibility of experience; for Prall, it is the basis of aesthetic judgment. The constructionalist's interest is somewhat different, arising in large part from the fact that if the order of qualia in each category can be successfully constructed upon an acceptable basis, a long stride will have been made toward the definition of predicates pertaining to shape and measure.

Since the importance of quality orders is just beginning to be recognized, they have so far been studied very little. Some pioneer work has been done by psychologists, but the theoretical foundations of their experiments hardly meet the constructionalist's demands (see Section 2 below). Philosophers have touched on the problem only rarely and superficially. Carnap recognizes the problem in the *Aufbau*, but does not deal with it in any detail.

The constructionalist thus approaches a formidable task without the help of a formative tradition or of the results of extensive previous research. He has first to crystallize a problem that has hardly been stated, and then to devise effective and legitimate methods for investigating it. Objectives must be set forth, instruments invented, and criteria for testing tentative solutions established. And one must proceed with the chastening awareness that even the most elementary mistakes have yet to be made.

Furthermore, the constructionalist is forced to deal with aspects of the problem that normally do not lie within the philosopher's province. Besides defining elementary ordering predicates in terms of a chosen primitive, he has to discover for himself how to construct complete arrays in terms of those ordering predicates. The latter problem is properly a mathematical one, but so far mathematicians have not provided the wanted treatment of the topology of arrays of finite elements.

Roughly stated, the central problem is to construct, for each category of qualia, a map that will assign to each quale in the category a unique position and that will represent relative likeness of qualia by relative nearness in position. Then by defining a set of co-ordinates, one may develop a systematic nomenclature for the qualia in the category, and also define specific shape, size, and measure predicates. It is the success of these latter definitions that must constitute the ultimate test of any order construction. We can hardly test a constructed order of colors, for example, merely by its correspondence to the familiar color sphere, or test a constructed order of odors by its correspondence to the less familiar odor prism; for the color sphere and the odor prism are themselves the results of attempts at map construction, and are as open to criticism as are any other such constructs. And correspondence of relative nearness of position in a constructed map to relative likeness of qualia is at best a vague criterion, since "likeness" is not so unambiguous that equally good ways of arriving at judgments of likeness may not give divergent results. The proper test of a constructed order is a more remote one–the serviceability of that order as a basis for the further definitions wanted.

Since the arrays that are to be systematically constructed do not lie before us at the beginning, we cannot take them for granted in giving the extrasystematic explanation of a primitive. In explaining the basic[1] concreting predicate "W", we could make free reference to concreta even though the problem was to define concreta in the system. But we cannot explain our basic predicate for order construction as applying between qualia that are related in a certain way in the constructed order, for such an explanation will not be helpful unless the order is already known in detail. We must accordingly make our basic predicate clear in other terms.

2. CHOICE OF A BASIC PREDICATE

If the order to be constructed is to reflect relative likeness of qualia, one naturally thinks first of choosing some similarity predicate as basic. We saw earlier that some similarity is part identity (D7.061), but obviously part identity is not in question here. The predicate we want applies between qualia that, as the indivisible atoms of our system, are discrete from one another. But let us even suppose the qualia we have chosen as atoms are instead construed as divisible–a color, for example, into its hue, chroma, and brightness; and that two colors are construed as similar if and only if they have a

[1] "W" and the predicate chosen for order construction are tentatively taken as primitives but are later to be derived from one primitive and are therefore properly called 'basic' (for the purposes they respectively serve in the system) rather than 'primitive'.

hue or a chroma or a brightness as a common part. Still we could get only a very incomplete order of colors; for so long as each set of components remains unordered within itself, no difference is established between the relative position of two colors made up of very unlike components–i.e., very unlike hues, chromas, and brightnesses–and the relative position of two colors made up of very like components. The order of hues, of chromas, and of brightnesses would have to be constructed before colors could be ordered. Even if we succeed in further analyzing these aspects of colors into other aspects, we are plainly off on an endless chase. As an alternative, someone might propose treating each hue, for example, not as divisible into aspects but as overlapping the immediately preceding and the immediately following hues in the spectrum, or in other words as composed of at least two subphenomenal parts each of which it shares in common with one of the immediately adjacent hues. But there is little antecedent recommendation for this rather Procrustean theory; and since it involves recognizing admittedly subphenomenal particles as atoms, it is patently nonphenomenalistic.

Even when construed as noninternal, "similarity" has marked disadvantages as a basic predicate for our constructions. Normally one asks not whether two qualities are similar but rather whether they are more or less similar than some other pair. Similarity is a comparative matter, and "similarity" thus a 3-place predicate with the complexity-value 5, or a 4-place predicate with the complexity-value 10. (The 3-place predicate–"x is more similar to y than to z"–is irreflexive and asymmetric. The 4-place predicate–"x and y are more similar than w and z"–has an equivalent set of irreflexive predicates consisting of one 4-place predicate symmetric with respect to its first two and also with respect to its last two places, and one 3-place asymmetric predicate that generates all sequences of the original predicate in those not-fully-variegated patterns– x,y,x,z and x,y,z,x and y,x,x,z and y,x,z,x –with respect to which the latter is not regular.) We might, as Carnap does in the *Aufbau*, suppose degree of similarity to be taken as standard, and adopt as basic the similarity predicate that applies between two qualities if and only if they are similar to at least that standard degree. In this way, we get a symmetric 2-place predicate as a basis, but only by resorting to the fiction that a certain degree of similarity is taken as a standard. Since ordinary language provides no ready way of describing definite degrees of similarity, the standard is presumably to be fixed as the interval between two arbitrarily designated qualia–and two qualia must be selected for this purpose from each of the different categories.[2] The predicate is then presumably to be explained, in a

[2] Moreover, in each case care must be taken to choose qualia that are similar enough. →

roundabout way, as applying between two qualia if and only if they are colors more alike than c and c', or sounds more alike than s and s', and so on. This rather forced and devious course hardly recommends itself strongly enough to discourage a search for a better way of achieving equal formal economy.

Before we look further, however, we had better take note of a fact that promises to give us some trouble whatever basis we adopt. Although two qualia q and r exactly match, there may be a third quale s that matches one but not the other. Thus matching qualia are not always identical. Now this is somewhat paradoxical; for since qualia are phenomenal individuals we can hardly say that apparently identical qualia can be objectively distinct. Offhand, it seems that color qualia, for example, that look the same must be the same. Yet if we say that q is identical with r because the two match, then we shall have to say that q does and does not match s. Must we then deny after all that the appearance of identity is a sufficient condition for the identity of appearances, and try to explain how a difference between phenomena can be nonphenomenal? Actually, the fact that some matching qualia are distinct can be accounted for without going beyond appearance; we need only recognize that two qualia are identical *if and only if they match all the same qualia*. Although distinct qualia must indeed be phenomenally distinct, to say they are phenomenally distinct is to say not that they fail to match but that there is some quale that is matched by one but not by the other. Thus the matching of nonidentical qualia does not force us into a contradiction. The principle that two qualia are identical if and only if they match all the same qualia is not a definition in our system; for identity has already been defined in terms of our general apparatus (D2.044). The present discussion has in effect been concerned with the fact that the statement:

$$x \text{ matches } y \supset x = y$$

does not hold, while the following statement:

$$\text{Qu}\, x \supset ((z)\,(z \text{ matches } x \equiv z \text{ matches } y)) \equiv x = y),$$

does hold (see 9.62 below).

A quale, then, not only matches itself but may also match some other qualia. Now may we not, perhaps, have here the basic predicate we want

---

The reason cannot be made entirely clear until later; but roughly speaking, unless the distance between the two qualia chosen as standard is less than half the greatest diameter of the array of the category in question, the relative positions of some qualia will remain undetermined. Thus one who adopts the above approach does well to choose as his referents qualia that are very similar.

for order construction? "Matches", as a 2-place symmetric predicate of qualia, meets the technical requirements above discussed. Furthermore, it may be explained simply as applying between qualia if and only if they are not noticeably different on direct comparison. No other qualia need enter into consideration, either as additional arguments or as referents fixing a standard degree of similarity; for our basic predicate applies between two qualia just in case they are so similar that they match.[3] And to say that two qualia are so similar that they match is merely to say that on direct comparison they appear to be the same.

Against adopting "matches" as basic, the objection may be raised that direct experimental investigation of order in terms of matching will be hampered by the considerable difficulty of handling stimuli that will occasion sensations differing in the slight degrees required, and by the difficulty of securing from observers decisive judgments concerning the matching or non-matching of two very like qualities. These difficulties are admittedly great. But, in the first place, the objection is not really pertinent, for the basic predicate of our systematic construction of order need not be the same as that in terms of which the psychologist frames the questions that provide his initial data. In the second place, moreover, the practical difficulties mentioned are to be blamed on the facts themselves rather than on the choice of a basic predicate. These difficulties cannot be circumvented simply by using some other basic predicate. If, for example, the psychologist proceeds by asking the observer whether two given qualities are more similar than two others, then judgments comparing a pair of matching qualities with a pair of not-quite-matching qualities will be frequently required. Indeed, the delicate decisions called for in the direct experimental application of the calculus of matching are equally called for–usually along with many other judgments–in the use of any other method capable of coping at all adequately with the problem of order. The fact that indiscriminable differences obtain may indeed offer many difficulties to the experimenter, and some to the theoretician; but the choice of "matches" as basic does not create that fact and the choice of a different basis would not destroy it.

The psychologist will naturally propose taking as basic the predicate "is just noticeably different from", often used in psychological studies. But two qualities are just noticeably different if and only if each is the first noticeably different quality encountered in proceeding from the other along the ordering in question. The psychologist here assumes in advance that the order of

[3] Note incidentally, recalling note 2 of the present chapter, that matching qualia are *very* similar.

qualities is the same as some order of stimuli. To set out in this way to discover in what order qualities are arrayed is obviously to beg the question. We could, of course, explain "just-noticeable difference" (without making any reference to stimuli) if we had before us a clear and detailed map of the order to be constructed. But I have already pointed out that we have no such map; and we must not assume that we have one in explaining our basic predicate. Thus, if the predicate "is just noticeably different from" is to appear in our calculus at all, it will have to be introduced by definition.

Certain other predicates used as basic in psychological investigations of order are unsuited to our purposes for other reasons. Sometimes, for example, the psychologist proceeds by asking the observer which of two colors is brighter.[4] This is decidedly uneconomical; for "is (a) brighter (color) than", besides being asymmetric, applies to colors only, so that other predicates would have to be introduced for ordering other kinds of qualities. Still more important, this predicate yields only a partial ordering of colors, since many colors may be equally bright; it is not the case that every two distinct colors have different brightness relationships to some third color. Accordingly, this predicate yields no complete array in which each color occupies a unique position. Furthermore, we can construct many other such partial orders of colors by using predicates like "is more neutral than" or even "is colder than" or "is prettier than"; and there seems to be no limit to the possibilities nor any choice among them unless we are willing to assume that colors are actually composed of certain aspects. Since it is not at all clear just what are the aspects into which colors are properly analyzable, I do not want to make any such assumption; and I expressly avoided doing so by choosing to take colors rather than aspects of them as atoms (VI,2). But if we can first construct a complete order of colors, then we can analyze them by applying a kind of ordinal quasianalysis to be described later (X,13).

In contrast to predicates like those just mentioned, "matches" has many advantages. It applies in all categories of qualia. It distinguishes between every two distinct qualia to which it applies, through the fact that there is always some quale that matches one but not the other of two distinct qualia; and it thus assigns a distinct position to each quale in a category, whether that category is 'unidimensional' or 'multidimensional'. Finally, its use involves no assumption as to how the qualia chosen as atoms are to be analyzed into aspects.

Some may feel, however, that problems of quality order are better dealt

[4] For example, in Harold Gulliksen, "Paired Comparisons and the Logic of Measurement", *Psychological Review*, 53 (1946), pp. 199–213.

with in terms of properties, and that the construction of properties should precede the construction of orders. Two different proposals along this line are worth considering. The first is implicit in a remark that Lewis makes in urging a quite different point. He writes:[5] "Similarity is of two types, partial identity and resemblance proper. ... Resemblance means the possibility of confusion. ... Things are more or less similar according as optimum conditions must be more or less nearly approached for distinction to be made." Now "possibility of confusion", or in other words "difficulty of discrimination", here pertains not to qualia but to properties. Properties are 'easily' discriminable if they involve presentation of discriminable (i.e., nonmatching) qualia under comparatively many different sorts of relevant conditions, that is, under conditions well below the optimum. If similarity is difficulty of discrimination, and if difficulty of discrimination is a matter of the degree of coincidence of property patterns, then after properties are constructed can we perhaps construct a complete order solely on the basis of such coincidence? It turns out that we cannot; for similarity of this sort will not serve to fix the relative positions of three properties that present matching qualia under all relevant conditions and present nonidentical matching qualia under optimum conditions, even though any one of three such properties may belong between the other two or the relationship among the three may be quite symmetric. Accordingly, while coincidence of patterns may offer a helpful auxiliary device for ordering properties, it by no means supplies a substitute for the construction of quale order in terms of matching.

A second proposal is to reject the idea that distinct qualia ever match and to interpret the phenomenon of matching as consisting of the presentation of *identical* qualia, under certain conditions, by different things. For example, take a case where three disks a, b, and c are such that when they are presented under optimum conditions the color of a matches that of b, and the color of b matches that of c, while the color of a does not match that of c. The explanation offered is as follows: disks a and b when compared on a given occasion both present some one color quale q; disks b and c when compared on another occasion present some one color quale r; but disks a and c when compared on a third occasion present two distinct color qualia s and t. It follows that either $s \neq r$ or $r \neq q$ or $q \neq t$; and hence that either a or b or c presents different qualia on two occasions. This means that a single unchanged disk under constant illumination presents different color qualia.

[5] *Mind and the World Order*, p. 364, footnote. The odd thing is that this passage, which stresses the distinction between resemblance and part identity, actually suggests the idea of construing resemblance as partial coincidence of properties.

One might attempt to explain this by saying that although the illumination remains constant, the given disk (say *a*) is compared with different disks on the two occasions–in other words, by treating the difference between the disks with which *a* is compared as constituting a relevant difference in the attendant conditions. But this has very awkward consequences; for it means that whenever two disks *x* and *y* are directly compared they are presented under different conditions, since *x* is being compared with *y* while *y* is being compared with *x*. Hence the fact that *x* and *y* present very different (or identical) color qualia when directly compared under optimum illumination would not be evidence that *x* and *y* have different (or identical) color properties, any more than would be the fact that *x* and *y* present different (or identical) qualia under differing conditions of illumination. And we could not escape this difficulty by saying that *x* and *y* have identical color properties if and only if for every third object *z*–assuming the conditions to remain constant–*x* and *z* present identical qualia on direct comparison if and only if *y* and *z* present identical qualia on direct comparison; for the fact that *x* but not *y* is involved in one case while *y* but not *x* is involved in the other would controvert the assumption that the conditions remain the same. Now I am not at all trying to show that the idea of construing "matches" in terms of coincidence of property patterns cannot somehow be worked out, but only that it offers no easy and obvious alternative to the course I have chosen in approaching the problem of order–even aside from the fact that we are still a long way from having constructed properties.

Having chosen a basic predicate and explained the reasons for my choice, I must remark that the calculus of order to be developed in Chapter X is to a considerable degree independent of this choice. The calculus can readily be adapted to any elements and predicate that meet certain rather meager requirements (see X,1). Accordingly, one who uses some other basic predicate, or who is concerned with ordering elements other than qualia, may nevertheless find the formal calculus applicable.

3. MAPPING AND THE MAPPED

Now qualia obviously do not come to us all neatly labeled with names. We do not have them before us like a set of lettered blocks, which we then proceed to compare and arrange. Rather, facts concerning the matching of qualia may be thought of as first expressed by statements in which the qualia compared are picked out by descriptions; e.g., "the color of the left-hand one of the two round patches now near the center of my visual field matches the color of the right-hand one". On the basis of all such information at our command, we construct a map that assigns a position to each of the described qualia. Quale names may then be treated as indicating positions on this map. Indeed, to order a category of qualia amounts to defining a set of quale names in terms of relative position, and thus eventually (in our system) in terms of matching. When we ask what color a presentation has, we are asking what the name of the color is; and this is to ask what position it has in the order–or in other words to ask which of the ordered qualia it matches. After a map has been considerably used and repeatedly amplified and corrected, we may hardly ever have to alter it again; for although we are constantly having new presentations of qualia, the qualia presented are not by any means always new.

This short account of the nature of mapping raises certain theoretical questions that call for brief informal discussion before we consider methods of construction. In the first place, we never actually compare more than two or three very similar qualia at any one moment. We may note that a great many quite different qualia that are presented at once do not match, but we hardly judge instantaneously whether or not each two of a number of very similar qualia match. Quite plainly any adequate map of a category must thus be constructed on the basis of comparisons made at many different times; and we must therefore often decide whether a quale involved in one comparison is identical with a quale involved in a comparison made at a different time. As was explained earlier (IV,3), identifications of qualia from one moment to another are in a sense decrees but are made and corrected with a view to the difficulties that result from an ill-considered set of decrees. Since two qualia may be distinct even if they are not noticeably different on direct comparison, great caution is obviously needed in making identifications among qualia presented at different times. If, for example, one of two lasting patches that have the same color at the start fades slowly until it contrasts sharply with the other, and if we obey our initial impulses to identify the color of each patch at each moment with its color at the succeeding moment, we shall be at a loss to explain the final difference in color. To avoid trouble of this kind, we often base our identifications upon physical factors: upon the identity of stimuli, of sources of illumination, etc. Any sort of consideration is legitimate that helps us arrive at a trouble-free set of identifications.[6]

[6] Similarly, in a laboratory investigation of matching, the psychologist may well assume identity of qualia wherever he has satisfied himself as to identity of observer, stimulus, and relevant conditions; and he need modify this assumption only if he finds it leading him to anomalous results. He does not here lay himself open to the objections urged in the preceding section against assuming that the *order* of qualia is the same as some *order* of stimuli.

Lest the frequent reliance on physical identities in deciding on the identity of qualia be welcomed by the physicalist as evidence for the epistemological priority of physical things over qualia, I must point out that one can say with equal truth that the identification of a physical object from one occasion to another rests largely upon what qualia are presented on the two occasions. This does not mean that we are doomed to travel in an inescapable circle, but only that identifications of physical objects and identifications of qualia must all be co-ordinated with one another in much the same way that identifications of qualia (or identifications of physical objects) must be co-ordinated among themselves.

In the second place, the actual process of mapping is naturally affected by practical limitations. For one thing, we seldom if ever have the opportunity of starting from a comprehensive set of observations and constructing the map of an entire category. Rather, we begin with observations of what we hope may prove to be a typical set of very similar qualia, seek to determine the pattern exhibited, and then by interpolation, extrapolation, and conjecture construct a tentative complete map that may later be improved and corrected. If we are lucky, our tentative map may take proper care of many qualia not included in our original set of observations. In the second place, it will clearly be difficult to construct a literal pictorial map when, as is probable in the case of color, the array in question turns out to be in more than two dimensions (according to the everyday if not the mathematical usage of "dimension"). Familiar devices will enable us to produce legible diagrams for some of these cases; but pictorial invention will be severely tried if arrays in four or more dimensions are encountered. Here we may do better to look at the problem of ordering in the other, less figurative, way suggested above–as a problem of defining a set of quale names in terms of matching. A different and rather curious situation arises in the mapping of the presumably two-dimensional visual field. Since distinct but matching visual-field places will be represented on a phenomenal map by well-separated dots, a phenomenal map of any segment of the visual field will occupy a region of the field that is larger than that segment. Accordingly, no complete phenomenal map of the visual field can be fitted into the visual field. Or in other words, no complete map of the visual field can be seen in all its detail at any one time; maps of different parts of the field will have to be seen at different moments. All these practical difficulties, however, count less as serious obstacles to the construction of order than as inconveniences along the way.

Finally, the recognition that there are nonnoticeable phenomenal differences reminds us that in ordinary life we seldom deal with single qualia or even single concreta. Each quale occurs together with other qualia, and each concretum is accompanied by many other concreta; and single minimal discernible elements seldom compete successfully with larger wholes for our attention. We normally take experience in larger chunks, and if we try to pulverize it by focusing attention on particles within which no further differences are detected, we usually find ourselves puzzled and uncertain. Paradoxically enough, 'least-discernible' particles are seldom discerned. Thus the data of matching used for the construction of order are perhaps more often inferential and derivative than immediate. This is unobjectionable since we have expressly disavowed any claim to epistemological priority for the basis of our system; but the idea naturally suggests itself that we might easily make the system more faithful to actual experience simply by choosing larger individuals as atoms, and some more rough-and-ready predicate than "matches" as basic. However, to take as indivisible atoms some individuals within which lesser parts are discerned would surely be no less artificial than the course we have chosen. And whatever the size of our atoms, we shall somehow have to cope with non-noticeable differences between them and between other elements that must be defined. We cannot simply ignore the experience of gradually shaded coloring, of phenomenal motion too slow to be momentarily perceived, of phenomenal specks so small that none smaller ever appears. What a faithful system must provide, as I have said more than once before, is not an epistemological diary but a precise, adequate, and integrated description of observed fact. Since it must therefore be capable of formulating the most acute phenomenal distinctions one ever makes, it will inevitably seem overrefined in relation to average daily experience. The anti-intellectualist may point to this as ground for his contention that all systematic analysis distorts fact and should therefore be shunned. One may retort that the standards of fact and distortion implicit here must be clarified and defended before such a complaint can be sustained. Or one may be content to point out that even a good lens distorts somewhat but still often enables us to see better, and that the distortion need not deceive us if we are aware of its nature.

4. REDUCTION OF BASIS

As matters stand now, we have in addition to the general apparatus of the system three undefined predicates: "W", "Z", and "M". Since each of these is a 2-place, reflexively regular ("W" is irreflexive, "Z" totally reflexive, "M" join-reflexive), non-self-complete predicate, they have together a total complexity value of 6. We are naturally prompted to cast about for some way of reducing this basis to a simpler one.

A minor economy could be achieved through replacing "M" by a 1-place predicate (let it be "Ms") explained as applying to all and only those individuals that are sums of an x and a y that match. Then "M" could be defined simply as follows: "$\mathrm{M}\,x,y =_{\mathrm{df}} \mathrm{Ms}\,x+y \,.\, \mathrm{Qu}\,x \,.\, \mathrm{Qu}\,y$". This illustrates a general point of some interest: that once the atoms of a system are defined, any symmetric 2-place primitive predicate that applies solely to atoms can be replaced by a 1-place predicate in this manner. But at the moment we are looking for greater economies.

It turns out that our three special primitives can be reduced to a single 2-place predicate that–being totally reflexive, symmetric, and non-self-complete–has the value 2. This reduction is a good illustration of a non-routine replacement; it depends upon special knowledge concerning the predicates involved, and effects a genuine economy.

Let the new primitive "A" (for "is allied with" or "are allied") be so explained that $\mathrm{A}\,x,y$ if and only if:

either $\mathrm{W}\,x,y$
or $(\exists z)\,(\mathrm{M}\,x,z \,.\, x+z=y \,.\, \mathbf{v} \,.\, \mathrm{M}\,y,z \,.\, y+z=x)$
or $\mathrm{Z}\,x,y \,.\, x \circ y$.

In other words, x and y are said to be allied if and only if they are together, or either is the sum of the other and a quale that matches that other, or they are of equal size and overlap. Just why "A" is explained in this way will become apparent from the way in which the erstwhile primitives "W", "M", and "Z" are defined in terms of it.

Every two individuals that are together, then, are allied, and (by 7.21) discrete. Conversely, every two allied individuals that are discrete are together. Accordingly:

DA-1 $\quad \mathrm{W}\,x,y = \mathrm{A}\,x,y \,.\, x \wr y.$

Qualia are then defined in terms of "W" exactly as in D7.01.

No two distinct matching qualia are allied, but every quale x is allied with the sum of x and any matching quale y. Also, if a quale x is allied with the sum of itself and y, then x matches y; for (1) in no case are x and $x+y$ together, (2) if x is a quale and not identical with y, then x and $x+y$ are not of equal size, and (3) if x is a quale and identical with y, then even though x and $x+y$ are of equal size and overlap, still x matches y. Accordingly:

DA-2 $\quad \mathrm{M}\,x,y = \mathrm{Qu}\,x \,.\, \mathrm{Qu}\,y \,.\, \mathrm{A}\,x,x+y.\ \mathrm{A}\,y,x+y.$

The definition of "Z" is most conveniently reached in two steps. If *x* and *y* are overlapping individuals of equal size, they are allied; and neither is a proper part of the other, since no individual is of equal size with any of its proper parts. Also, if *x* and *y* are allied and overlap and neither is a proper part of the other, they are of equal size. An auxiliary predicate, "Zo", for equality of size *between overlapping individuals* may thus be defined as follows:

DA-3 $\quad \mathrm{Zo}\,x,y = \mathrm{A}\,x,y \,.\, x \circ y \,.\, {\sim}x \ll y \,.\, {\sim}y \ll x.$

Note that "Zo" applies between *x* and *y* if (but not only if) *x* is identical with *y*.

A quale of course is of the same size as itself and as any other quale. Now let *x* and *y* be non-atomic individuals. If they differ in size, there can be no individual *s* that bears the relation Zo to each; but if they are of the same size, there will always be such an *s*. For if *x* and *y* overlap, either will serve as *s*; and if they are discrete and each contains *n* qualia, any individual that contains *n* qualia including at least one from *x* and one from *y* will serve as *s*. Thus equality in size may be defined quite generally as follows:[7]

DA-4 $\quad \mathrm{Z}\,x,y = \mathrm{Qu}\,x \,.\, \mathrm{Qu}\,y \,.\, \mathbf{v}\, (\exists s)\, (\mathrm{Zo}\,s,x \,.\, \mathrm{Zo}\,s,y).$

Note that not all cases where "Z" applies are cases where "A" applies; that "W" and "Z" have common cases, while each has cases not common to both; and that all cases of "M" are cases of "Z".

It is easy enough to object that the above reduction of "M", "W", and "Z" to "A" is trivial, but the grounds for such an objection are not clear. So far as formal simplicity goes, a genuine and appreciable saving has been effected, for a basis having a complexity of 6 has been replaced by one having a complexity of 2. And the new primitive is hardly less clear than whichever is the least clear of the three erstwhile primitives. It is true that the new primitive does not correspond to any simple word used in ordinary discourse; but almost every primitive predicate of a system departs considerably from the nearest everyday predicate and has to be explained at some length in terms of the natural language. The idea of trying to draw a line between permitted and prohibited degrees of departure does not seem very promising. Again, it is true that we are not much interested in our new primitive for its own sake; but for that matter we are quite often less interested in our

[7] This definition, very much simpler than that used in the first edition, is entirely due to Lars Svenonius.

primitive terms than in what we can define from them. I suspect that the impulse to dismiss as trivial a reduction like that carried out above actually stems from the false notion that economy in general is unimportant.

As a matter of fact, there is a technical advantage in having as our primitive a predicate that is in itself of little interest to us. For then we shall want no postulates upon it; and, by the sort of procedure suggested earlier (III,12), we can render demonstrable many of the statements that might otherwise have to be taken as postulates on the erstwhile primitives. For example, theorems affirming that "W" is external and that "M" and "Z" are symmetric all follow directly from definitions DA-1 through DA-4. By complicating these definitions appropriately, we can prove many other statements (e.g., 7.24) that would otherwise probably be postulated.

Thus the replacement of "W", "Z", and "M" as primitives by "A" not only gives us a more economical basis but also conduces to economy of postulates.

5. CATEGORIES AND REALMS

A familiar 'kind' or what I tentatively call 'category' of qualia consists of all colors, or all places, or all sounds, etc. When we compare two concreta with respect, say, to position or color, we are comparing a quale x that is part of one concretum with a quale y that is part of the other and that belongs to the same category as x. And as we noted earlier, comparisons of the size or shape of compounds are not usually comparisons of aggregate size or shape but rather of spatial size or shape, or color size or shape, or size or shape in some other respect, so that the eventual definition of most familiar shape and size predicates will thus require prior definitions of categories of qualia.

Qualia may be sorted into categories by means of "M" before qualia are ordered within the categories. Given any two qualia belonging to the same category, we can trace a path from one to the other by a series of steps, each to a quale matching the preceding one. When two qualia belong to different categories, there is no such path joining them; for example, we cannot go from a color to a time by such a series of steps. This suggests a way of defining categories; for two qualia are joined by an M-path just in case the ancestral of "M" applies between them. However, with our restricted general apparatus, "M*" is not automatically available to us; if we are to use it, it must be defined within our nominalistic language. This may be done by much the same method as was used earlier (II,3) for defining the ancestral of "is a parent of", but with some simplifications resulting from the fact that

"M" is symmetric and reflexive. However, a different and still simpler plan is open: after one preliminary step we can very easily define categories without using "M*", and can also easily define "M*". The technical and psychological simplicity of this set of definitions will perhaps illustrate the point that in restricting ourselves to nominalistic apparatus we are not always condemning ourselves to complex and circuitous ways of achieving results that can be achieved more directly by platonistic methods; sometimes the very paucity of our means leads us to solutions simpler than the customary platonistic ones.

I call an individual a *clan* if it cannot be divided into two parts such that no quale in one matches any quale in the other; or, put another way:

D9.051 $\quad \mathrm{Cln}\, x = (y)\,(y \ll x \supset (\exists r)\,(\exists s)\,(r < y \,.\, s < x - y \,.\, \mathrm{M}\, r,s)).$

A single quale constitutes such a clan, as does the sum of any group of qualia that match each other, the sum of two nonmatching qualia and a third that matches both, and so on. Now it will be seen that a clan contains an M-path between each two of its quale parts; and thus a (quale-) category[8] is simply a most comprehensive clan:

D9.052 $\quad \mathrm{Cg}\, x = \mathrm{Cln}\, x \,.\, (z)\,(\mathrm{Cln}\, z \supset {\sim} x \ll z).$

Accordingly, a clan that fails of being a category lacks one or more qualia of the category containing the clan. We shall see later (X,12) how in some cases a clan that is not a category may so to speak 'spread thinly over' a whole category.

Theoretically, at least, a familiar 'kind' of qualia might, because of a rift somewhere in it, fail to satisfy D9.052. For example, if there happens to be some color that matches no other color, then the sum of all colors will not meet the requirements for a Cg. If any such discrepancy is discovered or suspected, it will have to be rectified when we come eventually to defining each individual familiar kind. In the meantime, however, the categories defined in D9.052 are of greater interest; for each is a most-comprehensive sum of qualia that can be fully ordered on the basis of "M".

We have already noticed that each two qualia in a clan are connected by an M-path; and it is the case, furthermore, that every two qualia connected by an M-path are joint parts of some clan. Thus "M*" is readily defined:

---

[8] In the present context, "category" unqualified is to be taken as short for "category of qualia". Elsewhere, however, (e.g., VII,8) "category" is used as short for the more general "category of complexes". Two complexes belong to the same category if and only if every quality-category overlapping either overlaps the other.

D9.053 $M^* x,y = \text{Qu}\, x \,.\, \text{Qu}\, y \,.\, (\exists z)(\text{Cln}\, z \,.\, x + y < z)$.

Obviously it follows that $M^* x,y$ if and only if x and y are qualia of some one category. Among the alternative readings of "$M^* x,y$" are: "x and y are joined by an M-path"; "the ancestral of 'M' applies between x and y"; "x and y belong to some one clan (or category)". One point must be especially stressed–that there may be an M-path between each two quale parts of an individual x even though x is not a clan; for x is a clan only if x *contains* an M-path between each two quale parts of x. For example, if a and b are two very different colors joined by a long M-path, then clearly each two quale parts of $a + b$ are joined by an M-path; yet $a + b$ is of course not a clan.

Formal introduction of theorems on the predicates above defined is best postponed until after some of the principles of "M" itself have been set forth (see below, 9.64–9.67); but note in passing that a category contains every quale that matches any part of it, and that "M*" is symmetric, reflexive, applies between qualia only, and never applies between qualia that are together.

To define each individual category, we need only discover some peculiarity of it that can be described within the system; for example, a given category might be defined as the largest, or the smallest, or the second smallest, or as the one that is arrayed in the form of a double pyramid. Once each category is thus differentiated from the others, then any familiar 'kind', whether there are rifts in it or not, may be defined as the sum of certain specified categories. Likewise any sense *realm*, such as that of the visual qualities or that of the auditory qualities, is then to be defined as a sum of certain categories; and "realm" as a general term may be defined by enumeration of the several realms. Opportunities for shorter routes of definition may of course be discovered in some cases. For example, if there are no rifts in phenomenal time, then time qualia may be defined as those qualia that are contained in the category that overlaps all concreta; and if there are no rifts in any familiar kind, a realm may be defined as the sum of all the categories that overlap any one concretum. But unless we can assure ourselves of the truth of the assumptions on which such definitions rest we shall have to be content with the slower procedure of differentiating the several categories first, and then combining them into the several familiar kinds and realms.

However, we hardly know enough about the various categories as yet to be able to pick out appropriate distinguishing features with any confidence. In many cases, for example, we shall want to make use of differences in order; but this requires that we have both some such study of order as will be presented in the following chapter and also a much more comprehensive body of pertinent data than is now available. Accordingly, I shall not in this book attempt to define each of the several categories and realms.

Since the categories of qualia are discrete from each other, categorization effects no analysis of qualia. To say that an individual is a color, for instance, is not–even obliquely–to describe an attribute of it, but merely to circumscribe it somewhat more narrowly than when we say it is a quale. The definition of hues (by a method to be suggested in X,13) as certain wholes comprised of colors divides the color category in turn into several mutually discrete parts; and the parallel definitions of chromas and brightnesses do the same. Since the category[9] of colors is exhausted by hues and by chromas and by brightnesses, clearly hues are not discrete, under these definitions, from chromas or from brightnesses, nor are chromas discrete from brightnesses. Indeed each color is a common part of some hue, some chroma, and some brightness; and 'analysis' of a color amounts to specifying the hue, chroma, and brightness that contain it. Qualia being systematically indivisible, the attributes of qualia can be defined only through some such quasianalysis.

6. PRINCIPLES OF MATCHING

Before attempting to define order predicates, we need to have at hand some of the principle governing "M" and the predicates already defined in terms of "M". I shall make no rigid distinction between postulates and theorems here, but merely explain where necessary why the principles set forth are true.

The predicate "M" applies to all qualia and to qualia only:

9.61 $\quad \mathrm{Qu}\,x \equiv (\exists y)\,(\mathrm{M}\,x,y)$.

It is symmetric and reflexive:

9.612 $\quad \mathrm{M}\,x,y \supset \mathrm{M}\,y,x$,

9.613 $\quad \mathrm{Qu}\,x \supset \mathrm{M}\,x,x$,

but not transitive.

The relationship between matching and identity, already discussed at some length, in Section 2, can be summarized in the single theorem:

9.62 $\quad \mathrm{Qu}\,x \supset .\, x = y \equiv (z)\,(\mathrm{M}\,z,x \equiv \mathrm{M}\,z,y)$.

---

[9] I use the word "category" for familiar kinds except where there is reason for emphasizing the fact that a familiar kind may consist of more than one of the individuals satisfying D9.052.

For illustrative purposes here and elsewhere in the text, I speak of colors as describable in terms of hues, chromas, and brightnesses; but I do not mean to suggest that this more or less traditional way of codifying color relationships will prove in the end to be entirely satisfactory.

To ascertain that qualia are identical, we thus must ascertain that *no* quale whatever that matches either fails to match the other; but we are often concerned rather with the question whether *within a given clan* there is any quale that matches one but not both. If x and y are qualia of the clan c, and every quale *of* c matches x if and only if it matches y, then x and y are said to be *indistinguishable* with respect to c.

From 9.62 it is evident that qualia are identical if and only if the sum of all the qualia that match either is identical with the sum of all the qualia that match the other. Since we shall frequently want to speak of the sum of all the qualia that match a given quale, we may call such an individual a 'manor'; and define:

D9.06 The manor of $x = (\iota y)\ \{(z)\ (\mathrm{M}\,x,z \equiv .\ \mathrm{Qu}\,z\ .\ z < y)\}$.

Rather than introduce a symbol for "the manor of", we shall find it convenient to adopt the following notational rule. *Any expression consisting of "the manor of" followed by a lowercase letter may be entirely replaced by the corresponding italicized capital letter.* Thus we may write "X" for "the manor of x", and "A" for "the manor of a", and so on.[10] This gives us as a shorter equivalent of 9.62:

9.621 $\mathrm{Qu}\,x \supset (x = y \equiv X = Y)$.

Every quale, and nothing else, has a manor:

9.63 $\mathrm{Qu}\,x \equiv (\exists y)\ (y = X)$.

Thus unless x is a quale, "X" is an empty description. Every quale is a part of its own manor and of the manor of any matching quale:

9.631 $\mathrm{Qu}\,x \supset x < X$,

9.632 $\mathrm{M}\,x,y \supset x < Y$.

Also, of course, every quale in the manor of x matches x:

9.633 $\mathrm{Qu}\,y\ .\ y < X\ . \supset \mathrm{M}\,x,y$,

but it does *not* follow that every two qualia in the manor of x match each other. Finally, the manors of x and y overlap if and only if something matches both x and y:

[10] Perhaps it would have been satisfactory simply to write "X" as the definiendum of D9.06, and omit this notational rule; but since "x" is not actually a character in "X", the "x" in the definiens might have been mistaken for a free variable, and D9.06 thus misconstrued as the definition of a constant.

9.634 $X \mathbf{o} Y \equiv (\exists z)(\mathrm{M}\, z,x \,.\, \mathrm{M}\, z,y)$.

In the next chapter we shall so often want to refer to the sum of the differences of the manors of two qualia that the following abbreviation is extremely useful:

D9.061 $x \dagger y = (X - Y) + (Y - X)$.

This total difference is also, of course, the sum of the two manors minus their product:

9.635 $x \dagger y = (X + Y) - XY$;

and also the sum of all the qualia that match either of the two qualia but not both:

9.636 $x \dagger y = (\iota z)(t)(\mathrm{Qu}\, t \,.\, t < z \,. \equiv : \mathrm{M}\, t,x \,.\, {\sim}\mathrm{M}\, t,y \,.\, \mathbf{v} \,.\, \mathrm{M}\, t,y \,.\, {\sim}\mathrm{M}\, t,x)$.

The condition of existence, a corollary of 9.62, runs:

9.637 $(\exists z)(z = x \dagger y) \equiv .\, \mathrm{Qu}\, x \,.\, \mathrm{Qu}\, y \,.\, x \neq y$.

Obviously "†" is commutative:

9.638 $x \dagger y = y \dagger x$.

A few theorems on predicates defined in the preceding section may now be set down. The predicate "M*" applies wherever "M" applies:

9.64 $\mathrm{M}\, x,y \supset \mathrm{M}^*\, x,y$;

and "M*", like "M", applies to all qualia and to qualia only:

9.641 $\mathrm{Qu}\, x \equiv (\exists y)(\mathrm{M}^*\, x,y)$;

and is symmetric and reflexive:

9.642 $\mathrm{M}^*\, x,y \supset \mathrm{M}^*\, y,x$,

9.643 $\mathrm{Qu}\, x \supset \mathrm{M}^*\, x,x$.

But "M*", unlike "M", is transitive:

9.644 $\mathrm{M}^*\, x,y \,.\, \mathrm{M}^*\, y,z \,. \supset .\, \mathrm{M}^*\, x,z$.

No two qualia that are together are joined by an M-path:

9.65 $\mathrm{W}\, x,y \supset {\sim}\mathrm{M}^*\, x,y$.

Accordingly, the sum of two distinct qualia joined by an M-path is never a complex. A quale, of course, is a complex (7.34), and also a clan:

9.66 $\text{Qu}\,x \supset \text{Cln}\,x$;

but qualia are the only individuals that are both complexes and clans. Indeed qualia are the only complexes that are parts of any clan, and are the only clans that are parts of any complex.

9.661 $\text{Cm}\,x \,.\, \text{Cln}\,y \,.\supset.\, x < y \supset \text{Qu}\,x \,.\, y < x \supset \text{Qu}\,y$.

One final theorem reminds us that a category, unlike a lesser clan, contains every quale that is joined by an M-path to any part of the category:

9.67 $\text{Cg}\,x \supset (y)\,(z)\,(y < x \,.\, \text{M}^*\, z,y \,.\supset z < x)$.

7. A RULE OF ORDER

Given the data concerning the application of "M" within a given clan, how are we to plot on a map the relative positions of the qualia in the clan? In other words, how shall we proceed to define in terms of matching such explicitly ordering predicates as "is between ... and ..." and "is next to"?

A simple case of transition from matching to betweenness is provided when, from the fact that a quale *b* matches two others *a* and *c* that do not match each other, we conclude that *b* is between *a* and *c*. The considerations supporting this transition seem to be more or less as follows; that two matching qualia are more alike than two nonmatching qualia; that relative likeness represents relative nearness; and that, according to one usage of "between", *b* is between *a* and *c* if *b* is nearer to *a* and to *c* than *a* and *c* are to each other. The point of special interest here is that matching is correlated with distance; for some such correlation could well furnish us with the general rule we need for constructing order in terms of matching. The only question is just what more general principle of correlation is to be adopted.

Obviously, the distances or spans between different matching qualia are not to be taken as uniform; for that would put almost-discriminable[11] qualia as near together as almost-identical ones. Even more obviously, the spans between nonmatching qualia in a clan are not to be taken as uniform, for that would put a strongly contrasting red and green, for example, as near together as two just noticeably different colors. The spans between just noticeably different qualia might be taken as uniform; but this would hardly help us since "just noticeably different" is one of the predicates we have yet to define. What we need may be found in a more relative uniformity. Although

[11] The terms "discriminable" and "noticeably different" are often used in place of "nonmatching".

the spans between matching qualia as well as those between nonmatching qualia vary, there is this much constancy: *the span between any two matching qualia is less than the span between any two nonmatching qualia*.

Quite plainly we did not arrive at this rule altogether arbitrarily, for we found specific reasons for rejecting certain alternative proposals. Yet the rule is arbitrary to some degree. As has already been pointed out, we do not have before us a fully articulated presystematic order against which the details of our constructed order can be checked; we must apply the more remote test of serviceability for further constructions. Our constructed order will of course be unsatisfactory if it does open violence to evident facts of order—if, for example, it puts a red quale directly between two very similar greens. But on the other hand, so long as obvious distortions of this sort are avoided, we need hardly fear that in some very delicate case we shall discover an exception to our rule; for the rule may be taken as decisive in such cases without risk of conflict with any clear presystematic dicta. All this amounts merely to saying that the rule is related to presystematic fact in much the same way that a systematic definition is, but that the proportion of cases that are presystematically undecided and thus open to determination by the rule is comparatively great. *The rule, however, is no part of the system*; unlike a definition, it is not formulated in symbols and given a number. It is an extrasystematic statement of a relationship between matching and order; and it serves as a guide according to which we formulate, test, and use our systematic definitions.

The rule as stated may not seem very powerful, but actually it meets all our needs.[12] At the beginning of our investigation, in fact, a weaker consequence of this rule will suffice, namely, that no span between two nonmatching qualia is enclosed in a span between two matching qualia;[13] or more simply, that every quale between two matching qualia matches both. The full strength of the adopted rule will be needed (X,6) only for regularizing arrays in a way that makes possible the definition of measure predicates and (X,8ff.) for dealing with nonlinear arrays.

Now that a basic predicate has been chosen, certain auxiliary terms

---

[12] Certain cartographical conventions will be adopted from time to time, but these in no way affect the formulation of any definition. See X,7.

[13] This weaker rule was stated, and its use explained, in *A Study of Qualities* (pp. 434ff.). Publication of it ten years later in the first edition of the present book anticipated by five years its adoption by R. Duncan Luce as the fundamental principle of his theory of 'semiorders'. See his article "Semiorders and a Theory of Utility Discrimination", *Econometrica*, 24 (1956) pp. 178–191, especially axioms S3 and S4 and the discussion of them on pp. 181–182.

defined, a few useful theorems set forth, and a rule for constructing order adopted, the aspects of the problem of order that remain to be dealt with are primarily mathematical.

CHAPTER X

TOPOLOGY OF QUALITY

1. THE FORMAL PROBLEM

If we know just which qualia of a given set match, our basic rule will tell us that certain qualia of the set are nearer together than certain others. Our problem then is how to determine which qualia of the set are next to each other. Obviously, this problem is to a large degree independent of the kind of elements to be ordered and of the particular predicate chosen as basic. It may be conceived much more generally as the problem of ordering a finite set of elements of any sort, given certain rather incomplete information about relative distance among them; i.e., that certain pairs are composed of elements that are nearer together than the elements of any remaining pair. For this reason, I have remarked that the problem to be dealt with in this chapter is primarily mathematical; and for this reason also, the calculus of order to be outlined may find uses other than those for which it is introduced here. One might apply it to the problem of ordering properties or any other sort of elements on the basis of an appropriate predicate.

In contrast with standard mathematics, however, our calculus will be designed to apply to *finite* sets of elements. This must be kept constantly in mind, for it explains why some of the most familiar theorems concerning order in a continuum, such as that there is an element between each two distinct elements, will not hold here.

At the same time, our calculus must be adequate for dealing with arrays of any degree of complication. We can by no means take it for granted that all the categories of qualia are linear arrays; indeed, it is clear that the order of colors, for example, is much more complex. Our problem might have been greatly simplified if we had chosen our atoms in such a way as to insure against encountering any but the simplest arrays; but we found (VI,2) that there are good reasons for not choosing our atoms in this way. Among these reasons, which I shall not review here, is the desire to avoid making the formal system narrowly dependent upon a particular choice of atoms. And as a matter of fact, a sufficiently general theory of order will provide us with a principle bearing upon the very question of the level of analysis at which our atoms are best chosen (Section 13 below).

In the first stages of our investigation we may for convenience confine our attention to linear arrays. But although the illustrations and diagrams in these early sections will all be of such arrays, the numbered definitions and theorems are of course subject to no such limitation. Later we must consider how they apply in more complex arrays; and in some cases the reason why a definition is not framed in a simpler way that readily suggests itself, or the reason why a seemingly obvious truth is not a theorem, may not be evident until then. Eventually, of course, the predicate "linear" and predicates for other types of array will have to be defined within the general calculus.

2. BETWIXTNESS

Our first major problem is to discover the conditions under which two qualia of a linear category are juxtaposed, or in other words, so to define the predicate "is beside" that in a linear array it will apply between two elements if and only if they are next to each other. Clearly, matching is a necessary condition for besideness in a category, since if two qualia of a category fail to match there is still an M-path joining them and consequently one or more qualia lying between them. But matching is not a sufficient condition for besideness, since even between two matching qualia there may be a third. This may not be altogether evident until later (see especially Section 3); but, for example, in a linear array of five elements, *b* is between *a* and *c* if all three match and *d* matches *a* alone and *e* matches *c* alone.[1]

Since in a linear array two elements are beside each other just in case there is no element between them, we first need to determine just when a quale is between two matching qualia in such an array. Any quale that lies between matching qualia must match both, since otherwise our rule that no span between nonmatching qualia may be enclosed within a span between mathing qualia will be violated; hence we need now consider only cases of three *matching* qualia. But this leaves us with the question: which of three such qualia is between the other two? Our immediate goal is to answer this question by defining the term "betwixt" in such a way that in a linear array *y* will be betwixt *x* and *z* if and only if all three match and *y* is between *x* and *z*.

Now in a linear array, *y* is between *x* and *z* just in case *y* is nearer to *x* and to *z* than *x* is to *z*. However, our basic rule does not tell us directly that certain matching qualia are nearer together than certain others, but only that

[1] For the purposes of each illustration used in the present chapter, it is to be assumed unless otherwise noted that each of the qualia involved matches only itself and the qualia it is expressly stipulated to match.

matching qualia are nearer together than nonmatching qualia. Accordingly, we must see whether we can derive information of the former kind from information of the latter kind, or more specifically, whether with no other guide than the basic rule we can formulate in terms of "M" the circumstances under which two of three matching qualia are nearer together than another two. It must be noted that the immediate need is not for a general definition of relative nearness–this would indeed be a large initial order–but only for a way of determining relative nearness among three matching elements in a linear array. Let us look for a clue by examining some typical cases.

Suppose that the set to be ordered consists of sixteen qualia designated by the first sixteen letters of the alphabet; and suppose that the quale designated by any letter matches the qualia designated by letters not more than three places removed from that letter in the alphabet–so that the quale *h*, for example, matches the qualia *e, f, g, h, i, j,* and *k*. The first sixteen letters of the alphabet, in alphabetical order:

a b c d e f g h i j k l m n o p,

will then map the sixteen qualia in conformity with the rule that no span between nonmatching elements is to be enclosed within a span between matching elements. This the reader may easily verify for himself. Now the qualia *f, g,* and *h* all match one another; but *f* and *g*–and also *g* and *h*–are nearer together than *f* and *h*. In each case, more qualia match both the qualia that are nearer together than match both the qualia that are farther apart; i.e., six qualia (*d, e, f, g, h,* and *i*) match both *f* and *g*, and six qualia (*e, f, g, h, i,* and *j*) match both *g* and *h*, while only five (*e, f, g, h,* and *i*) match both *f* and *h*. Also *f, g,* and *i* all match one another; but *g* and *i* are nearer together than *f* and *i*. Here likewise the two qualia that are the nearer together are those that match the more qualia in common; for five qualia (*f, g, h, i,* and *j*) match both *g* and *i*, while only four qualia (*f, g, h,* and *i*) match both *f* and *i*. The same principle obviously holds good in many similar cases.

Naturally we are encouraged to inquire whether perhaps in *all* cases two of three matching qualia are nearer together than another two if and only if more qualia match both the former than match both the latter. Unfortunately, this is not true, as we may see by examining another part of the same map. The qualia *a, b,* and *c* all match, and *a* is nearer to *b* than to *c*; yet just four qualia (*a, b, c, d*) match both *a* and *b*, and the same four match both *a* and *c*. This is particularly easy to see from a description of the manors of the qualia in question:

$$A = a+b+c+d,$$
$$B = a+b+c+d+e,$$
$$C = a+b+c+d+e+f.$$

Here the greater distance of c from a than from b is reflected not in fewer qualia matching both a and c than match both a and b, but rather in more elements matching c but not a than match b but not a. To put the matter in just this way, however, will give us the anomalous result that although c is farther from a than from b, still a is no farther from c than from b; for no quale matches a but not c, and none matches a but not b. If instead we compare the total number of qualia that match either a or c but not both with the total number of qualia that match either a or b but not both, it will be immaterial which element of each pair we mention first. This is plainly desirable; for the distance from one element to a second is obviously the same as the distance from the second to the first.

The present proposal, then, is to determine the nearness of x and y as compared to that of x and z–whenever x, y, and z are matching qualia in a linear array–by comparing the number of qualia that match either x or z but not both with the number that match x or y but not both. Speaking nominalistically, we compare the sum of the differences of the manors of x and z with the sum of the differences of the manors of x and y. Briefly, the proposed formula is that for any three matching qualia in a linear array, x and y are nearer together than x and z if and only if

$$x \dagger z \text{ G } x \dagger y.$$

This new formula works properly in other cases where our earlier proposal breaks down, as, for example, in comparing the nearness of a and b with that of a and d or the nearness of n and p with that of m and p. Also, the cases we found to be adequately taken care of by the earlier proposal are equally well taken care of by the new one; for example, $f \dagger i$ contains six qualia (c, d, e, j, k, l), while $f \dagger h$ contains only four (c, d, j, k), and $f \dagger g$ contains only two (c, j). Indeed, the new formula works properly for *all cases in the array illustrated*, as a little further testing will readily show.

Accordingly, for any three qualia in that array, y is *betwixt* x and z if and only if all three match and

$$x \dagger z \text{ G } x \dagger y \,.\, x \dagger z \text{ G } y \dagger z.$$

Furthermore, the same definition is adequate in application to any linear array constructed in accordance with our basic rule. This is unlikely to be immediately evident; but a proof of the completely general adequacy of the definition with respect to linear arrays will be given in the following section.[2] We have arrived, then, at a satisfactory definition of betwixtness. Using "$x/y/z$" as a convenient notation for "y is betwixt x and z", we may set forth the full definition as follows:

D10.02 $x/y/z = \mathrm{M}\, x,y \,.\, \mathrm{M}\, y,z \,.\, \mathrm{M}\, x,z \,.\, x \dagger z \,\mathrm{G}\, x \dagger y \,.\, x \dagger z \,\mathrm{G}\, y \dagger z.$

Since this is numbered among the adopted definitions of the system, its application is not restricted to linear arrays. But for adequacy we require merely that it give intuitively correct results in application to such arrays. Later I shall consider its results and utility in application to nonlinear arrays.

Since D10.02 stipulates that x, y, and z all match, and since by 9.61 only qualia match, "betwixt" applies only among qualia:

10.21 $x/y/z \supset .\, \mathrm{Qu}\, x \,.\, \mathrm{Qu}\, y \,.\, \mathrm{Qu}\, z.$

And it applies only among distinct qualia. No quale is betwixt itself and any other:

10.22 $\sim x/x/y,$

10.23 $\sim y/x/x;$

for if $x/x/y$ or $y/x/x$, then by D10.02 $x \dagger y \,\mathrm{G}\, x \dagger y$; but no individual is greater than itself. Furthermore a quale is never betwixt one and the same quale:

10.24 $\sim x/y/x;$

for "$x/y/x$" implies "$x \dagger x \,\mathrm{G}\, x \dagger y$" which is false because "$x \dagger x$" is an empty description. Thus "betwixt" is thoroughly irreflexive; the theorem

10.25 $x/y/z \supset .\, x \neq y \,.\, x \neq z \,.\, y \neq z$

follows from the preceding three.

"Betwixt" is easily proved to be symmetric with respect to its first and third places:

10.26 $x/y/z \equiv z/y/x;$

for the expansions of the two sides of this theorem according to D10.02 are obviously equivalent in view of the commutativity of conjunction and of "$\dagger$". But 'betwixt" is otherwise asymmetric; only one of three qualia is betwixt

[2] No claim is made that the formula for determining relative nearness, even among three matching qualia, is adequate for all linear arrays. Indeed, we shall see (Section 4) that it breaks down in some cases; but the operation of the definition of betwixtness is not thereby affected, as the proof given in Section 3 will show.

the others. If y is betwixt x and z, then neither is x betwixt y and z nor is z betwixt x and y:

10.27 $x/y/z \supset . \sim x/z/y . \sim y/x/z$.

The proof in outline runs as follows: if $x/y/z$ then $x \dagger z \; G \; x \dagger y$, and therefore $\sim(x \dagger y \; G \; x \dagger z)$, and therefore $\sim x/z/y$; also, if $x/y/z$ then $x \dagger z \; G \; y \dagger z$, and therefore $\sim(y \dagger z \; G \; x \dagger z)$, and therefore $\sim y/x/z$.

In linear arrays, it is also plainly true that one of each of the three non-identical elements lies between the other two, and that one of each of the three distinct matching qualia is betwixt the other two. Nevertheless, we have no theorem to the effect that if x, y, and z are distinct matching qualia, then $x/y/z$ or $y/x/z$ or $x/z/y$; for, as we shall see, this does not hold in nonlinear arrays. All the above theorems, of course hold quite generally.

3. JUSTIFICATION OF THE DEFINITION OF BETWIXTNESS

To justify D10.02 we must show that for any three matching elements in a linear array constructed to conform with our basic rule it is the case that:
I. If y lies between x and z in the array, then y is betwixt x and z according to the definition; and
II. If y is betwixt x and z according to the definition, then y lies between x and z in the array.

The proof of Thesis I will occupy some paragraphs. In the first place, if y lies between x and z, then clearly all three are distinct. We have seen (9.62) that no two distinct qualia match all the same qualia. Accordingly, since by hypothesis x, y, and z all match, there must be in the array other qualia such that for each two of the three qualia x, y, and z there is some quale that matches one but not the other. The minimum number of additional qualia that must be present is two, and these must meet certain requirements. It must not be the case that either matches all or none of the three qualia x, y, and z, or that both match exactly the same qualia from among these three. And it must not be the case that one of the additional qualia matches one of the three qualia x, y, and z while the other additional quale matches the remaining two of the three. Unless these requirements are met, the manors of x, y, and z will not all be different. I shall first show that Thesis I holds for minimum arrays consisting of just five qualia,[3] and then show that it also holds for all larger arrays.

[3] There may be smaller arrays, of course, but none containing three distinct matching qualia (see Section 4, below). By an array I usually mean, as here, a set of elements ordered internally – that is, by their M-relationships to each other only.

Since by hypothesis y lies between x and z, no quale matches x and z without matching y. For suppose such a quale k lies on the opposite side of x from y; then the span between the nonmatching qualia k and y is enclosed within the span between the matching qualia k and z, in violation of our basic rule. Suppose, then, that k lies between x and y; then the span between the discriminables k and y is enclosed within the span between the indiscriminables x and y. Considerations of the same sort show that k can be neither between y and z nor on the opposite side of z from y. Thus there is nowhere in the array any quale that matches x and z but not y.

Likewise, no quale in the array matches y without matching x or z, as can easily be shown by similar proof.

This leaves four possible ways in which the two additional qualia–let us call them "t" and "r"–may be related to the qualia x, y, and z. The four ways are:

(i) r matches x and y only; t matches x only;

(ii) r matches y and z only; t matches z only;

(iii) r matches x only; t matches z only;

(iv) r matches x and y only; t matches y and z only.

All other combinations are excluded by the considerations in the preceding three paragraphs.

In case (i) we have

$$X = x + y + z + r + t,$$
$$Y = x + y + z + r,$$
$$Z = x + y + z.$$

Now since the differences of two individuals never overlap, the size of the sum of the differences varies directly with the arithmetical sum of the numbers of qualia these differences severally contain. Accordingly, as an informal proof for a "G" statement in the present context, I merely indicate the number of elements in each difference of manors. For example, the notation.

"$a \dagger b$
3+4"

indicates that three elements match a but not b while four match b but not a, so that $a \dagger b$ contains seven elements in all. The sign "+" is used here for arithmetical addition. A "G" statement is true if and only if the left-hand arithmetical sum is larger than the right-hand one. Thus in case (i) the proof that $x/y/z$ is indicated as follows:

$$\begin{array}{lll} x \dagger z & \mathrm{G} & x \dagger y \\ 2+0 & & 1+0 \end{array}$$

$$\begin{array}{lll} x \dagger z & \mathrm{G} & y \dagger z \\ 2+0 & & 1+0 \end{array}$$

Since the left-hand side is constant for any two "G" statements supporting a "betwixt" statement, I shall hereafter write the left-hand side but once in each case.

In case (ii) the manors of x, y, and z are as follows,

$$X = x+y+z,$$
$$Y = x+y+z+r,$$
$$Z = x+y+z+r+t.$$

Here also $x/y/z$, for

$$\begin{array}{lll} x \dagger z & \mathrm{G} & x \dagger y \\ 0+2 & & 0+1 \\ & \mathrm{G} & y \dagger z \\ & & 0+1. \end{array}$$

In case (iii) we have

$$X = r+x+y+z,$$
$$Y = x+y+z,$$
$$Z = x+y+z+t.$$

Again $x/y/z$, for

$$\begin{array}{lll} x \dagger z & \mathrm{G} & x \dagger y \\ 1+1 & & 1+0 \\ & \mathrm{G} & y \dagger z \\ & & 0+1. \end{array}$$

Finally, in case (iv) we have

$$X = r+x+y+z,$$
$$Y = r+x+y+z+t,$$
$$Z = x+y+z+t.$$

Here, too, it is the case that $x/y/z$, for

$$\begin{array}{lll} x \dagger z & \mathrm{G}\ x \dagger y \\ 1+1 & \quad 0+1 \\ & \mathrm{G}\ y \dagger z \\ & \quad 1+0. \end{array}$$

Thus for every linear array just big enough to contain three distinct matching qualia x, y, and z, we have shown that if y lies between x and z, then y is betwixt x and z according to D10.02. But we must show further that Thesis I holds as well for all larger linear arrays that contain three distinct matching qualia; that is, we must show that however many other qualia are also in the array, the statements "$x \dagger z$ G $x \dagger y$" and "$x \dagger z$ G $y \dagger z$" will continue to hold. This may be done by showing that all additional qualia that are in $x \dagger y$ or in $y \dagger z$ are also in $x \dagger z$, so that the "G" statements in question will remain true. Now every element in $x \dagger y$ either matches x but not y or matches y but not x. We have shown above that in a linear array in which x, y, and z match, and y lies between x and z, no quale matches x and z without matching y. Thus every quale here that matches x but not y must match x but not z, and hence must be in $x \dagger z$. We have also shown that in the same circumstances no quale matches y without matching either x or z. Thus every quale here that matches y but not x must match z but not x, and hence must be in $x \dagger z$. By exactly similar reasoning, it follows that every quale here that is in $y \dagger z$ is in $x \dagger z$. Consequently, Thesis I holds for all linear arrays whatsoever.

We must now prove Thesis II: for any three matching qualia in a linear array, if $x/y/z$ according to D10.02 then y lies between x and z.

By hypothesis, $x/y/z$; and therefore, by 10.25, x, y, and z are all distinct. It is clear that for three distinct elements in a linear array, if y does *not* lie between x and z, then either x lies between y and z, or z lies between x and y.

Suppose first that x lies between y and z. Then it follows from the proven Thesis I that $y/x/z$. By theorem 10.27, if $y/x/z$ then $\sim x/y/z$. But this contradicts the hypothesis of Thesis II. Hence x cannot lie between y and z.

Suppose, then, that z lies between x and y. Then it follows from Thesis I that $x/z/y$. By 10.27, if $x/z/y$ then $\sim x/y/z$; and this contradicts the hypothesis of Thesis II. Hence z cannot lie between x and y.

Therefore y must lie between x and z, and Thesis II is proved.

By Theses I and II together, then, *for any three matching qualia in a linear array, y is betwixt x and z according to* D10.02 *if and only if y lies between x and z*. In other words, where matching qualia in linear arrays are concerned, the presystematic predicate "between" and the systematic predicate "betwixt" apply in exactly the same cases.

4. BESIDENESS

It will be recalled that we defined "betwixt" as a step toward defining nextness in linear arrays, since linear maps are most conveniently constructed on the basis of information about which elements are next to each other. Now that we have an accurate definition of "betwixt", we can easily so define "beside" that in linear arrays it will apply just between qualia that are next to each other.

D10.04 $\quad \mathrm{B}\,x,y = \mathrm{M}\,x,y \,.\, x \neq y \,.\, (z)\,({\sim}x/z/y)$.

The first clause of the definiens excludes cases where, simply because x and y are too far apart in the array, no z is betwixt them. Obviously "B" is irreflexive:

10.41 $\quad {\sim}\mathrm{B}\,x,x$,

but symmetric:

10.42 $\quad \mathrm{B}\,x,y \supset \mathrm{B}\,y,x$.

"B" is nontransitive.

One might suppose that we could have defined "B" directly, without bothering to define "betwixt" along the way. If the fact that two matching qualia are nearer together than two others in a linear array is reflected by the fact that the sum of the differences of the manors of the latter two is greater in size than the sum of the differences of the manors of the former two, then why not define "beside" simply as applying between two qualia in a linear array just in case they match and the sum of the differences of their manors is no larger than the sum of the differences of the manors of any other two qualia in the array? This plausible suggestion does not work. In the first place, in the example of an array of sixteen elements given in Section 2, g and h would not be beside each other by the alternative definition proposed; for $g\dagger h$ contains two qualia while $a\dagger b$ contains only one. Moreover, there is no guarantee that all linear arrays will be as uniform as the example given, in which all qualia except those near the ends match exactly the same number of qualia. For example, it may perfectly well happen that for the three matching elements r, s, and t in the midst of some other array, r alone also matches p and q, while t alone also matches u, so that the manors of the three qualia are as follows,

$$R = p + q + r + s + t,$$
$$S = \qquad r + s + t,$$
$$T = \qquad r + s + t + u.$$

Here r would not be beside s according to the proposed alternative definition, for $r \dagger s \; \mathrm{G} \; s \dagger t$. The size of the sum of the differences of the manors of two qualia is no generally reliable indication of relative nearness even among matching qualia in the midst of a linear array; but our adopted definition of "betwixt" is neverheless adequate, as was independently shown in Section 3. In the present case, for example, even though $r \dagger s \; \mathrm{G} \; s \dagger t$, still it is not the case that $r \dagger s \; \mathrm{G} \; r \dagger t$ and thus D10.02 does not give us the wrong result that t is betwixt s and r. According to D10.02, s is betwixt r and t; and therefore (by 10.27) neither r nor t is betwixt the other two qualia. But the proposed alternative direct way of defining "beside" is unworkable because, although the idea of defining besideness as maximum nearness is sound enough, relative nearness even among three matching qualia in a linear array cannot be accurately defined as relative size of the sums of the differences of the manors of these qualia.

In terms of "beside", linear orders can be readily and unequivocally constructed; for once we know what elements of a set are next to each other, mapping is a simple, routine process. Moreover, we can define linear arrays in terms of "beside". I use the term "linear" narrowly for what might more explicitly be called "simple open linear" arrays (cf. Section 7 below). A linear array is distinguished from other arrays by the combination of two characteristics: (1) having not more than two qualia beside any given one of its qualia, and (2) having just two ends–i.e., two qualia such that only one other is beside each. Since an array that satisfies condition 1 will have just two ends if it has one end, the definition can be formulated as follows:

D10.041 $\mathrm{Lr}\, x = \mathrm{Cln}\, x \,.\, (\exists a)(\exists b)(a + b < x \,.\, a = (\imath y)(\mathrm{B}\, y,b)) \,.\, (c)(d)$
$(e)(f)(c + d + e + f < x \,.\, \mathrm{B}\, c,d \,.\, \mathrm{B}\, e,d \,.\, \mathrm{B}\, f,d \,.\supset:$
$c = e \,.\mathbf{v}.\, c = f \,.\mathbf{v}.\, e = f)$.

While in linear arrays "beside" applies just in case two qualia are next to each other, in more complex arrays "beside" applies also in certain other cases. For such arrays, then, another predicate for nextness must be defined; and, of course, the more complex types of array must themselves still be defined. But before we turn to these more complex arrays, certain remaining problems concerning linear arrays deserve our attention.

5. JUST NOTICEABLE DIFFERENCE

With a linear array before us it would seem an easy matter to determine which qualia in it are just noticeably different from each other. Roughly speaking, if x is an element in a linear array, then the first element (in either direction) that does not match x may be said to be just noticeably different from x in that array. The trouble is that if we use this as the criterion for just noticeably different pairs, then "is just noticeably different from" unexpectedly turns out to be nonsymmetric. Suppose, for example, we are confronted with a set of elements as follows,

$$\begin{aligned} A &= a + b, \\ B &= a + b + c + d, \\ C &= \phantom{a + {}} b + c + d, \\ D &= \phantom{a + {}} b + c + d + e, \\ E &= \phantom{a + b + c + {}} d + e + f, \\ F &= \phantom{a + b + c + d + {}} e + f. \end{aligned}$$

The resultant linear order is

$$a\ b\ c\ d\ e\ f.$$

The first quale to the right of b that does not match b is e; but the first quale to the left of e that does not match e is c. Thus although e is just noticeably different from b, still c rather than b is just noticeably different from e.

At first glance this paradox may seem to result simply from careless formulation of the test for just noticeably different, or just discriminable, qualia. Surely we can define the predicate in such a way that it will be symmetric. But we soon discover that we get unwelcome results no matter how we frame our definition. Suppose we make c and e just-discriminables and also make b and e just-discriminables; then to the left of e lie two elements, one farther than the other from e, both of which will be termed just discriminable from e. Suppose then we make b and e just-discriminables, but not c and e; then it is not the first but the second element that both is to the left of e and does not match e that will be called just discriminable from e. Let us, then, make c and e just-discriminables but not b and e; then although there are to the right of b many elements that do not match b, none of these is just discriminable from b.

There seems little to choose among these alternatives; whichever way we decide the matter, a conspicuous anomaly results. It is simply the case that in application to certain arrays the predicate "is just noticeably different from" behaves in an intuitively objectionable way. These arrays, however, are all *irregular* ones. Whenever two qualia in a linear set are beside each other and either matches more than one quale that the other does not match, the array constructed by use of our definition of "beside" will be irregular. Such arrays will, as we have seen, satisfy our second, weaker rule (i.e., that no span between nonmatching elements is enclosed within a span between matching elements); but they will not satisfy the primary, stronger rule (i.e., that the span between any two matching elements is less than the span between any two nonmatching elements). We are therefore faced with the problem of constructing maps that satisfy the stronger rule. But if the weaker rule uniquely determines the sequence of elements in every linear set, as our proof of the adequacy of the definition of betwixtness shows, then no rearrangement of this order will work. For if every other order (than, for example *a b c d e f* in the illustration above) violates the weaker rule, then certainly every such order will violate the stronger one. Thus, without changing the order already established by our definitions, we must somehow construct maps that meet the stronger requirement. This is the problem to be dealt with in the following section. A solution is wanted not merely as a means for meeting the difficulties about just-noticeable difference but also as the beginning of a calculus of measure.

Parenthetically, it may be noted that the reason for basing our constructions not upon similarity circles of M, but upon manors, is that the ordering of irregular arrays is thus greatly simplified; for in such arrays there are fewer similarity circles than elements, whereas in every array the manor of any element is different from the manor of any other element (9.621).

6. ADJUSTED LINEAR MAPS

A linear array is irregular just in case there are within it as many elements between some two matching elements as between some two nonmatching elements. N. J. Fine has proved (see X,14) that if and only if a linear array is irregular will it contain two qualia that are beside each other and such that one matches two or more qualia that the other does not. Irregularity may manifest itself in other ways as well; for example, in the array illustrated in the preceding section, the fact that *C* is contained in both *B* and *D*, neither of which is contained in the other, is a sign of irregularity. But somewhere in *every* linear array that satisfies the weaker but not the stronger mapping rule, one of two qualia that are beside each other will match more than one quale that the other does not match.

How are such irregularities to be removed? No technique of mapping will alter the fact that two qualia match *d* that do not match *e*. And if any element of the set is placed between *d* and *e*, then even the weaker rule is violated. Yet in mapping an array, we can compensate for irregularities by making appropriate adjustments in spacing, that is, by including in our map certain positions at which we plot no quale. By thus using more positions than there are qualia to be mapped, and by judiciously selecting the positions to leave vacant, we can satisfy the stronger rule and repair the anomalies pertaining to just-noticeable difference.

Take, for example, the set of six qualia having manors as follows,

$$
\begin{aligned}
X &= x+y+z,\\
Y &= x+y+z+t,\\
Z &= x+y+z+t+r,\\
T &= \phantom{x+{}} y+z+t+r,\\
R &= \phantom{x+y+{}} z+t+r+s,\\
S &= \phantom{x+y+z+t+{}} r+s;
\end{aligned}
$$

and the following map:

$$x\,y\,z\,t\,r\,s.$$

Irregularity is immediately detected from the fact that two qualia match *r* that do not match *s*, which is beside *r*. The stronger rule is clearly violated since there are, for example, as many qualia between the matching elements *z* and *r* as between the nonmatching elements *t* and *s*. Thus we need to revise the map so that there will be more map positions between each two nonmatching qualia than between any two matching qualia. This can be done by leaving a vacant position between *r* and *s*, so that our map becomes:

```
                v
 .   .   .   .   .   .   .
 x   y   z   t   r       s
```

For convenience, I label the vacant position "v"; the other positions are called by the names of the qualia assigned to them.

We have here an *adjusted* map of an irregular set of elements. In this map there are more map positions between any two nonmatching qualia than between any two matching qualia. And now we may be willing to call two nonmatching qualia of the set just noticeably different if all qualia between them match both; for although it will be the case, as in the unadjusted map, that then no quale to the right of *z* is just noticeably different from z, there is now to the right of z a map position, v, such that a and v are the same map distance apart as two just noticeably different qualia.

The same method of adjustment can be applied in the case illustrated in the preceding section. Here two vacant positions must be interpolated; one, v_1, between a and b; the other, v_2, between d and e. But irregularity cannot always be dealt with simply by leaving a vacant position between every two elements that are beside each other and such that one matches two qualia the other does not. Take, for example, the set of eleven elements having manors as follows,

$$
\begin{aligned}
A &= a+b+c+d,\\
B &= a+b+c+d+e,\\
C &= a+b+c+d+e+f,\\
D &= a+b+c+d+e+f+g,\\
E &= b+c+d+e+f+g,\\
F &= c+d+e+f+g+h,\\
G &= d+e+f+g+h+i,\\
H &= f+g+h+i+j,\\
I &= g+h+i+j+k,\\
J &= h+i+j+k,\\
K &= i+j+k;
\end{aligned}
$$

and the unadjusted map:

$$a\,b\,c\,d\,e\,f\,g\,h\,i\,j\,k.$$

We obviously need to interpolate a vacant position, v_1, between g and h; but that alone will not give us a fully adjusted map, for between matching qualia d and g there are as many map positions (two) as between the nonmatching qualia h and k. We therefore need another vacant position, v_2, somewhere between h and k. We cannot put it between h and i, for then we have three map positions (v_1, h, v_2) between the matching qualia g and i. Thus, v_2 must be put between i and j or between j and k. In either case, the stronger rule will be satisfied. Yet there is independent reason for choosing the second alternative. If we put v_2 between i and j, then there will be four map positions (v_1, h, i, v_2) between the just noticeably different qualia g and j, while there will be only three between each other two just noticeably different qualia (e.g., between h and k, or e and h). But if we place v_2 between j and k, then each two just noticeably different qualia in the array will have exactly three map positions between them. Since a fixed span of just-noticeable difference is a convenient feature in a map, v_2 is best placed between j and k. Our fully adjusted map of the above set of qualia is thus as follows:

$$a\,b\,c\,d\,e\,f\,g\,v_1\,h\,i\,j\,v_2\,k.$$

Distance between elements in a linear array is most appropriately measured by the number of map positions between the two on an adjusted map. The distance between just noticeably different qualia in a linear array will then be constant throughout, even though the number of qualia between them varies.

Adjustment in these examples has been by trial-and-error; but formal procedures have been devised (see X, 14). Loose extrasystematic talk of 'map adjustment' and 'vacant positions' must eventually be supplanted by statements in terms of defined predicates of individuals of the system. However, since the whole matter of adjustment has been considered in application to linear arrays alone, I shall make no attempt here to formulate it within the system. For the same reason, the suggested definitions of just-noticeable difference and of distance are left in unofficial verbal form.

For convenience, an array may hereafter be spoken of as having or not having 'gaps' according as it does or does not require interpolation of vacant positions.

7. SOME CARTOGRAPHICAL CONVENTIONS

So far our mapping technique has consisted merely of arranging in a straight line, with vacant postions interpolated where needed, letters standing for the elements of a given set. Since linear sets alone were under consideration, the basic rules could always be satisfied in this way. But not all sets are linear; that is to say, in certain cases no such arrangement in a line will constitute a map conforming to the basic rules. Suppose, for example, we have four qualia with manors as follows,

$$\begin{aligned} A &= d + a + b, \\ B &= \phantom{d + {}} a + b + c, \\ C &= d \phantom{{} + a} + b + c, \\ D &= d + a \phantom{{} + b} + c. \end{aligned}$$

No linear arrangement of the four letters in question will satisfy even the weaker rule. For example, in the map:

a b c d,

the span between the letters for the nonmatching qualia *a* and *c* is enclosed within the span between the letters for the matching qualia *a* and *d*. Every other linear arrangement commits a similar offense; and no interpolation of vacant positions helps at all.

Plainly we need further cartographical conventions to accommodate this and many other cases. Many and widely varied conventions are theoretically adequate, but most of them are impractical. For example, we might distribute the letters in question quite at random along a line or even over a region, and indicate each pair of next elements by joining their letters with a line. This serves well enough in simple cases; but when many nextness relationships obtain among several elements, the resulting tangle of lines is hard to read. We can produce more legible maps by adopting the following conventions:

1. Distribute the letters or other signs for the elements in question in such a way that the actual distance[4] between signs for the next elements is constant and less than the actual distance between the signs for any two non-next elements, except that the fewest vacancy signs needed to satisfy the stronger rule are to be interpolated at appropriate places.

2. For greater clarity, connect the signs for each two next elements by a line.

3. Where the matter under discussion does not require identifying each element on a map, use dots or the meeting points of the lines indicating next-ness (instead of letters) for the anonymous elements.

I must emphasize the fact that the choice of cartographical conventions does not affect the definitions of order predicates or measure predicates in the system. In this, these conventions have a different status from the basic rules. The basic rules directly correlate matching with betweenness, and thus serve as a criterion to be used in testing proposed definitions of "between" and related predicates. But the rules do not prescribe how betweenness shall be represented on a map; we are left to choose our means and this is what we do when we adopt cartographical conventions, either tacitly, as in earlier sections, or explicitly, as in the preceding paragraph. The basic rules set forth a principle on which the whole calculus of order is founded. In contrast,

---

[4] I write "actual distance" to indicate that the distance referred to in formulating this convention is measured straight through the map body itself. Everywhere else in what follows–unless otherwise expressly noted–"distance" and "map distance" are always to be understood as referring to the number of next steps along any minimal connecting path (see Section 6, above). In Section 14, standard graph-theoretical conventions are adopted and the notion of actual distance is dropped.

the cartographical conventions are arbitrarily chosen devices that do not affect the system.

It is true that in describing arrays extrasystematically I shall often use terminology (Section 8, below) suggested by the appearance of maps constructed according to our chosen conventions; but this terminology is to be taken as a set of purely figurative abbreviations for long and complex descriptions in terms of the systematic predicates. Later, certain features of our maps that result accidentally from the particular conventions chosen may even suggest ways of defining certain shape predicates; but satisfactory operation in application to these maps is no sufficient evidence of the adequacy of such definitions.

8. SOME TYPES OF NONLINEAR ARRAY

Linear arrays have already been defined (Section 4, D10.041). Such arrays might have been called 'open linear' arrays, but I use the unmodified term "linear" for them and refer to what might be called 'closed linear' arrays rather as *polygonal* arrays or polygons. A polygon is defined simply as an array in which every element has exactly two others beside it. Thus in polygonal and linear arrays no element has more than two others beside it. Since "beside" and "next" coincide in such arrays, it is also the case that in such arrays no element has more than two others next to it; but "next" has not as yet been generally defined in the system.

We now need terms for some of the other types of array that we shall be discussing in following sections. The terms introduced below are intended for extrasystematic use only, at least for the present, and the explanations are not by any means offered as systematic definitions but only as informal indications of how the terms will be used. Obviously there are many types of array other than those listed.

1. A *square-cell network* is one such that when it is mapped according to our conventions, the lines indicating nextness form a pattern of squares or rhombuses–as for example Figure 4, which maps a set of 132 elements assigned to the meeting points of the lines. The figurative character of the term "square-cell", as of the other tems explained below, is evident from the fact that maps constructed according to other conventions might look quite different. But regardless of the conventions used in mapping, what I call square-cell networks are distinguished from linear arrays and polygons by the occurrence of certain elements having more than two others next to them. On the other hand, they differ from networks of the second type to

be described in that a square-cell network contains no three elements such that each is next to each of the others.

2. *Triangular-cell networks* are such that in our maps the lines for nextness make a pattern of triangles. In these networks there occur sets of three elements each of which is next to each of the others, but no such sets of four. There may be as many as six elements next to a given element in these networks, against a maximum of four in square-cell networks.

3. *Cubical-cell networks* are such that in our maps the lines for nextness mark off a pattern of cubes, or at least of solids with six quadrangular sides and twelve equal edges. Unlike networks of the other types so far considered, these cannot be mapped, by our adopted conventions, in two dimensions. Our maps of these networks will be three-dimensional, and the best we can do on a plane is to make a *picture of* such a map by availing ourselves of familiar conventions of drawing. In a cubical-cell network, as in a triangular-cell network, the maximum number of elements next to any element is six; but, as in a square-cell network, the maximum number of elements each of which is next to each of the others is two.

4. *Tetrahedral-cell networks* are those such that the lines for nextness on our maps mark the edges of tetrahedrons. In such networks as many as twelve elements may be next to a given element; and the maximum number of elements each of which is next to each of the others is four.

For certain other types of network it is obviously easy to invent terms analogous to those here introduced; but many irregular and hybrid networks are less readily labeled.

Since our maps of square-cell and triangular-cell networks happen to be two-dimensional, and our maps of cubical-cell and tetrahedral-cell networks three-dimensional, we may occasionally and tentatively speak of networks of these types as being themselves two- or three-dimensional, but the question of settting up a sound and precise general criterion for classifying arrays by degree of dimensionality is another matter.

9. BESIDENESS IN SQUARE-CELL NETWORKS

Polygons, like linear arrays, can be fully ordered in terms of "beside", for in both types of array two elements are next to each other if and only if they are beside each other.

Square-cell networks, on the other hand, raise new problems that are among the most acute likely to be encountered in developing a calculus of order. We had better begin by seeing how the terms already defined in the system apply in square-cell networks. For the present let us assume that we are dealing with regular square-cell networks that require no adjustment by interpolation of vacant positions, and let us disregard any aberrations that occur as we approach the boundaries of such networks. The problem thus excluded must of course be dealt with eventually; and whether or not I can solve them, I shall return to them later.

We must remember that distance between two elements in an array is measured by the number of next steps along any minimal path connecting the two. Distances measured straight across the map are irrelevant. Thus in any array mapped by Figure 4 the distance between a and b is the same (4 steps) as the distance between a and c but greater than the distance (3)

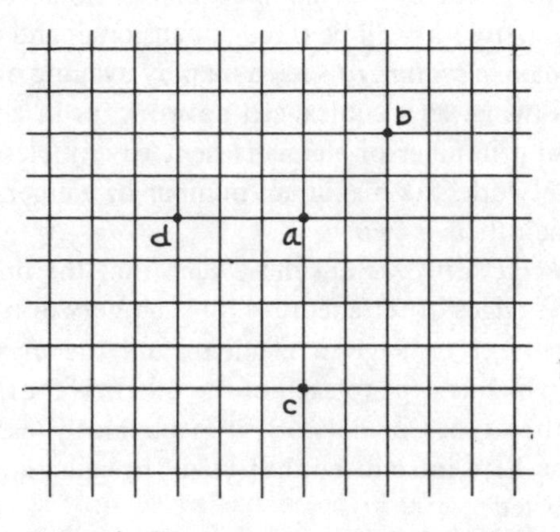

Figure 4. Distance in a square-cell network.

between a and d. The distance between c and d, like that between b and d, is 7; while the distance between b and c is 8. If the M-span for the array is 3–that is, if an element matches all and only those elements that are not more than 3 steps away from it–then a matches d but not b or c.

It will be helpful to become acquainted at once with the pattern marked out by the manor of an element in a square-cell network. In Figure 5, the manor of an element x is indicated (i) where the M-span is 2, and (ii) where the M-span is 5; in each case the intersections within the diamond represent the elements matching x. In the first case the number of elements in the manor is 13, in the second it is 61; in general the maximum number of elements matching a given element in a square-cell network having an M-span of n is $1 + 2n\ (n + 1)$. For the purposes of the present and following sections we may confine our attention to elements having the maximum manor

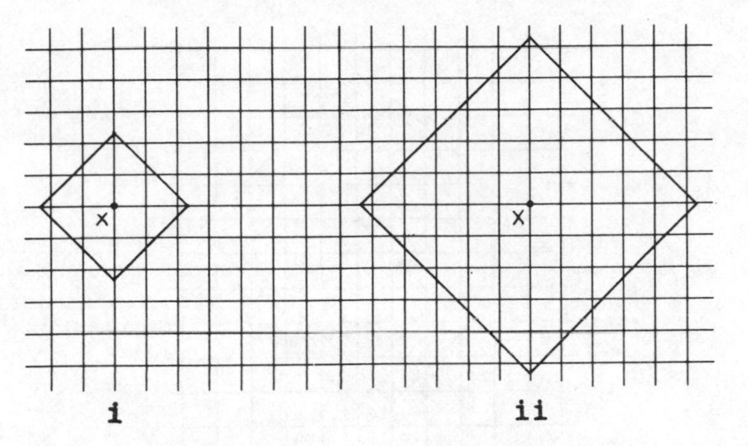

Figure 5. Manors in square-cell networks.

for the M-span in question; that is, we may regard our illustrations as fragments of larger maps that extend indefinitely in each direction.

With arrays in which the manors contain as many as 25 elements when the M-span is only 3, 41 elements where the M-span is 4, and so on, it is clearly impractical to present our original data by tabulating the manors of the elements concerned. I resort therefore to the practice of presenting the data by presenting the map and indicating the M-span. This does not beg the question; for the problem is how to reconstruct the map, or determine what elements are next to each other in the array, given only the information what elements match each other.

Let us now inquire what elements are beside x in the network of which a fragment is mapped in Figure 6 and in which the M-span is 4. First, consider y. Is there any element betwixt x and y? We cannot decide this merely by looking quickly at the map. We have to find out whether the manor of any other element is so related to those of x and y that the element is betwixt x and y according to Definition 10.02. An element such as z would seem to be the most likely candidate. But we find that z is not betwixt x and y; for the sum of the differences of the manors of x and y consists of 18 elements and is therefore not greater than the sum of the differences of the manors of x and z, which also consists of 18 elements. Similar testing will show that no other element is betwixt x and y; and thus x is beside y. Obviously each of the remaining three elements shown by the map as next to x will also be beside x.

We must also test whether any elements other than these four are beside x; and we find that *some elements not next to x are beside it.* Take z for

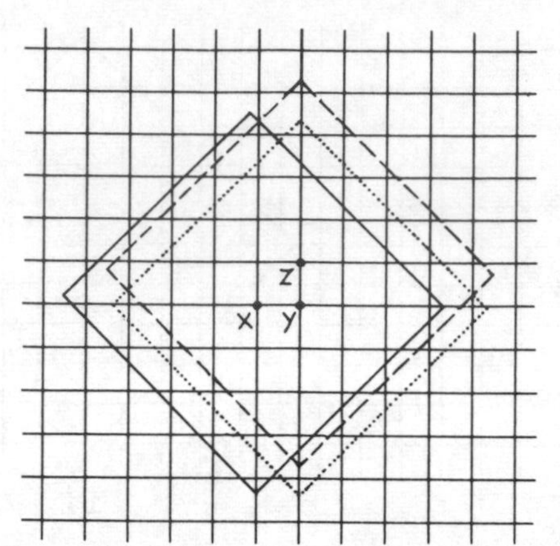

Figure 6. Overlapping of manors. (Solid line: boundary of X; dotted line: boundary of Y; broken line: boundary of Z.)

example. We might expect y to be betwixt x and z, so that z will not be beside x. But y is not betwixt x and z; for $x\dagger z$, which consists of 18 elements, is not greater than $x\dagger y$, which also consists of 18 elements. And there is no other element betwixt x and z. Hence z is beside x, and so is each of the other three elements mapped as diagonally across a square cell from x.

We have here tested for besideness only where the M-span is 4. But if the M-span is of any length greater then 1, exactly the same eight elements and no others will prove to be beside x. The number of elements beside a given element does not increase with the length of the M-span. It is true that if the M-span is 1, then only the four elements mapped as next to x will be beside it; for only these four will match x (since all others are 2 or more steps away), and by D10.04 only elements that match x are beside it. But whether the M-span is 1 or more, an element r in a square-cell network is beside another element s just in case r matches s and is either next to s or next to two elements that are next to s.

Since besideness is thus not generally a sufficient condition for nextness, our immediate problem is how to distinguish formally between those cases of besideness in a square-cell network that are genuine cases of nextness and those that are not. This distinction between elements on a side of a square cell and elements at opposite corners is one of the most delicate we shall be called upon to make in any network; and so if we succeed here, our method may prove to be applicable in networks of other types as well.

10. NEXTNESS

The sum of an element x and all the elements that are beside it I shall call the *barony* of x, and use for it the symbol "$\ddot{x}$". Baronies are defined in terms of "beside" much as manors are defined in terms of "matches". The definition runs:

D10.101 $\ddot{x} = (\imath y)\,\{(z)\,(\mathrm{B}\,z,x \vee z = x \,.\!\equiv\!.\, \mathrm{Qu}\,z \,.\, z < y)\}.$

For the sum of the differences of the baronies of two elements, I place an inverted dagger between their letters; thus

D10.102 $x \mathbin{\rotatebox[origin=c]{180}{$\dagger$}} y = (\ddot{x} - \ddot{y}) + (\ddot{y} - \ddot{x}).$

Now we may define a triadic predicate in terms of such sums of differences just as we define "betwixt" in terms of sums of differences of manors:

D10.103 $x \backslash y \backslash z = \mathrm{B}\,x,y \,.\, \mathrm{B}\,y,z \,.\, \mathrm{B}\,x,z \,.\, x \mathbin{\rotatebox[origin=c]{180}{$\dagger$}} z \;\mathrm{G}\, x \mathbin{\rotatebox[origin=c]{180}{$\dagger$}} y \,.\, x \mathbin{\rotatebox[origin=c]{180}{$\dagger$}} z \;\mathrm{G}\, y \mathbin{\rotatebox[origin=c]{180}{$\dagger$}} z.$

That is, y is *twixt* x and z if each of the three is beside each other, and the sum of the differences of the baronies of x and z is greater in size than the sum of the differences of the baronies of x and y also greater than the sum of the differences of baronies of y and z.

Finally, in terms of twixtness we may define a predicate analogous to "beside". Two elements are said to *flank* each other if they are beside each other and no element is twixt them:

D10.104 $\mathrm{F}\,x,y = \mathrm{B}\,x,y \,.\, (z)\,({\sim}\, x \backslash z \backslash y).$

Clearly, in linear arrays and polygons, flanking coincides with nextness. Again, in square-cell networks with an M-span of 1, only elements next to a given element x flank x, because only such elements are beside x; and every element

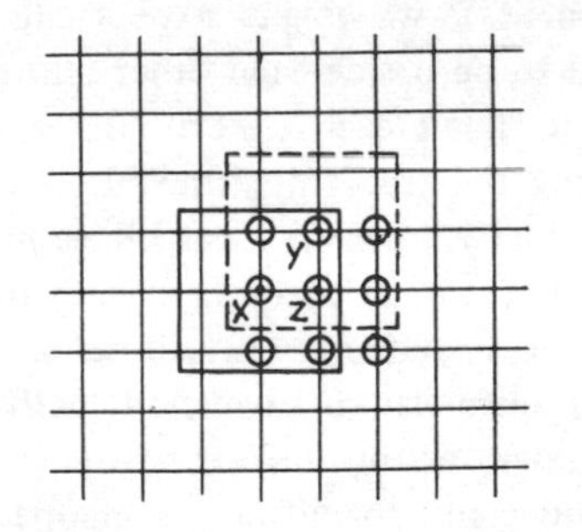

Figure 7. Overlapping of baronies.

next to x flanks x because, since no element is beside both x and an element next to x, no element is twixt the two. The critical case, then, is that where the M-span is 2 or more, so that eight elements are beside a given element x.

In the map fragment shown in Figure 7 the elements comprising the barony of x are represented by intersections within the solid boundary, those comprising the barony of y by intersections within the broken boundary, and those comprising the barony of z by intersections severally enclosed in small circles. Let us see which of the eight elements beside x flank it.

First, does z flank x? It does unless some element such as y which is beside both x and z, is twixt the two. But y is not twixt x and z; for the sum of the differences of the baronies of x and z consists of six elements and is thus not greater than the sum of the differences of the baronies of x and y, which consists of ten. Obviously, then, no other element is twixt x and z; hence z and all the other three elements next to x plainly flank x.

On the other hand, y does not flank x. For z is twixt x and y, since $x \dotplus y$ consists of ten elements and so is greater than $x \dotplus z$ and than $y \dotplus z$, each of which consists of six.

The elements flanking x are thus just the elements that are next to x; and flanking is a necessary and sufficient condition of nextness for the kinds of cases we have been considering.

Further examination shows that the same holds true for other regular networks in two dimensions. In triangular-cell networks, for example, the maximum manor where the M-span is n consists of $1 + 3n\,(n + 1)$ elements; but just the elements–at most six–that are next to a given element are beside it. Here the additional complication in the definition of flanking is superfluous but harmless. This is true as well for hexagonal-cell networks, etc. Square-cell networks seem to provide the acid test.

What, then, of networks in three dimensions? In a cubical-cell network (in which the maximum manor, where the M-span in n, contains $1 + 2n + 2n\,(n + 1)\,(2n + 1)/3$ elements), we might expect all the elements on the corners of a cubical cell to be beside each other and all the elements on one side of a cubical cell to flank each other. This would leave us with the problem of distinguishing the true from the false cases of nextness among the cases of flanking. Actually, however, tests I have made show that only the elements–at most 18–that are on a common side of a cubical cell with a given element x are beside x, and that only those–at most six–that are next to x flank it. Thus here again flanking coincides with nextness. Tetrahedral-cell networks seem to present no new difficulties.

Of course, I make no claim for all arrays mapped in three dimensions. There are a great many different types of such array, and the testing of each with our present techniques is arduous and intricate. But if cubical-cell networks provide as crucial a case for network mapped in three dimensions as square-cell networks for those in two, there is some ground for hope that (under our present assumptions) flanking is a necessary and sufficient condition of nextness for all networks mapped in three dimensions. And if the same criterion of nextness that works for networks mapped in two dimensions does thus work for those mapped in three, there may be ground for the further hope that it works also for networks in four or more dimensions—although I have examined no such networks.

The reason I use the definiens of D10.104 for "flanks" rather than "next" is not solely, however, this lack of certainty concerning arrays in many dimensions. The chief reason is rather that in these two latest sections we have been depending on some very considerable arbitrary restrictions. We have confined ourselves to networks that need no interpolation of vacant positions, and that are of uniform structure; and even within these we have ignored elements so near the boundary as to have truncated manors. As soon as we drop any of these purely artificial restrictions, a host of new problems arises.

An idea of the extent of the task remaining is perhaps best gathered from observing how, even in networks without gaps, difficulties arise with respect to elements so near a boundary as to have less-than-maximum manors. Suppose, for example, that the network mapped in Figure 6 is instead such that the upper edge just cuts off the upper corner of the manor of z. Then z will be betwixt x and y (since $x \dagger y$ will consist of 18 elements, while $x \dagger z$ and $y \dagger z$ will each consist of 17), and thus y will not be beside x. Therefore although x and y are next to each other they will not flank each other. What is needed for dealing with problems like this is not to interpolate vacant positions between the signs for certain elements in the network that are beside each other but rather, in effect, to surround the signs for elements in the network with a fringe of vacant positions that will fill in the mapped manors of these elements. No general rules for accomplishing this are available, however; we must for the present rely upon mere trial and error to reach a satisfactory solution in any given case.

We are also still dependent upon trial and error alone for the proper interpolation of vacant positions in maps of nonlinear arrays with gaps. Only for linear and polygonal arrays do we have anything even approaching adequate rules for adjustment. For nonlinear arrays the whole theory remains to be developed.

Certain clues may, indeed, help us to discover just what supplementation or interpolation is required by one nonlinear array or another. For example, if in a given case we find that the maximum manor consists of 5 or 13 or 25 or 41 (etc.) elements we may well be dealing with a square-cell network having an M-span of, respectively, 1 or 2 or 3 or 4 (etc.). Furthermore, in a square-cell network with an M-span of n, the maximum product of two manors is $2n^2$. Now if two less-than-maximum manors have the maximum product, then the vacant positions to be added to each manor will not be added to their product. Similar clues are easily discovered for other cases, but we are still far from having a general method for dealing with all cases.

What I have offered here, then, constitutes only the beginning of a theory of nonlinear order. The best hope for further progress seems to me to lie in finding ways of making greater use of algebraic techniques.

11. SPURIOUS MAPS

A brief glance at a few spurious maps may serve both to illustrate some of the definitions and rules already presented and also to warn against some time-consuming bogus problems.

Spurious maps are of three sorts. First, there are those figures that are spurious as maps, under our conventions, no matter what M-span is indicated. An example is a single isolated triangle. If the three elements represented by the vertices match each other, then they have the same manors and are not distinct. If they do not match, they are not next; and the sides of the triangle are thus false lines of nextness.

In a second kind of case, it is only the indicated M-span that makes a given figure a spurious map. A single isolated square, for example, may be a genuine map for a set of four elements if the M-span is 1. But if the M-span is given as 2 or more the map is spurious. Obviously any figure is a spurious map with respect to a sufficiently great M-span.

Spurious maps of both these kinds are readily detected by the identity of manors of what are ostensibly two distinct elements; or in other words, by the fact that the map assigns more than one position to a single element. In a spurious map of the third and most interesting kind, all the manors are distinct–i.e., only one position is assigned to each element–but the map does

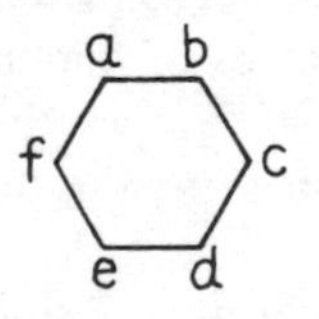

Figure 8. Spurious map.

not consistently represent exactly similar relationships between different elements. Suppose, for example, we are presented with Figure 8, and the M-span is given as 2. If we read off and tabulate the data thus indicated and from which the map was supposedly constructed, we find that each element matches all but one of the others. Every two matching elements here are thus related in the same way as every other two; and there is, for example, no justification for representing *a* as farther from *c* than from *b*. If we work back from the tabulated data, we get a quite different and perfectly satisfactory map of this array: the three dimensional map, with an M-span of 1, depicted in Figure 9.

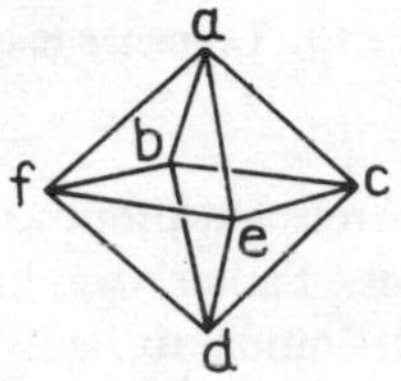

Figure 9. Corrected map.

In Figure 10, with the M-span taken as 2, we have a spurious map exhibiting both faults above explained. First we shall find that *w* and *z* are identical.

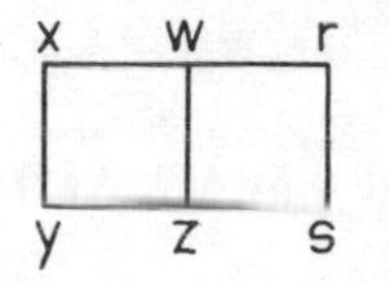

Figure 10. Spurious map.

Let us therefore drop the name "*w*". Then we shall find that an adequate map of the five elements present is that depicted in Figure 11, with the M-span taken as 1.

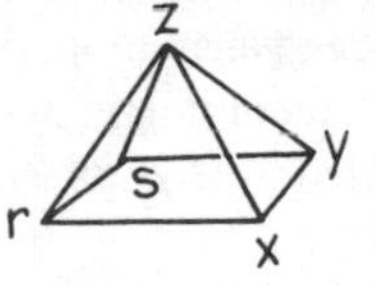

Figure 11. Corrected map.

The maps in Figures 8 and 10 do not show enough nextness relationships. The map in Figure 12, with M-span taken as 1, shows too many; and corrections result in the linear map, with M-span taken as 2, in Figure 13.

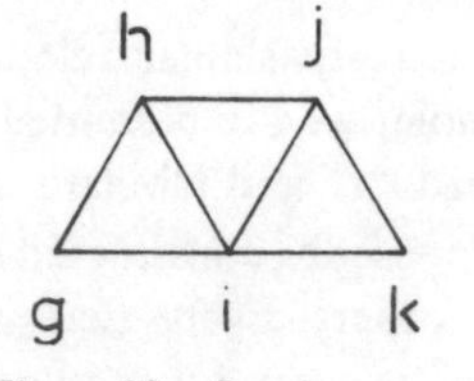

Figure 12. Spurious map.

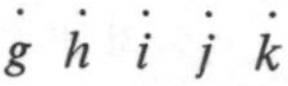

$$\dot{g}\ \ \dot{h}\ \ \dot{i}\ \ \dot{j}\ \ \dot{k}$$

Figure 13. Corrected map.

In all these cases, we have rejected proposed maps in favor of new ones conforming to our definitions. Earlier, we have often considered a discrepancy between map and definition to be evidence of inadequacy in the definition. What makes the difference? A map not conforming to the definitions is discarded as spurious whenever another map with the same manors, for some M-span, conforms to the definitions. Otherwise a way must be found of amending our definitions–as we did in arriving at a satisfactory definition of besideness and still must do before we have a generally adequate definition of nextness.

12. TOWARD SHAPE AND MEASURE

Although the definition of important shape and measure predicates is a major goal of our investigation of order, the investigation has not yet been carried to the point where this goal can be attained. Nevertheless, we may find it worthwhile to see how, for certain arrays, some of these predicates might later be defined. The following definitions contain the predicate "N" ("next"), which has not yet been generally defined, and they are appropriate only for arrays without gaps. Accordingly, these definitions do not belong to our system, but are presented solely for whatever suggestive value they may have for future investigation.

Assuming "next" to have been defined, we may proceed to define a *region* as, in effect, any individual in which every quale is joined with every other by a chain of next elements. The symbolic definition of "region" in terms of "next" is analogous to that (D9.051) of "clan" in terms of "matches".

(i) $$\mathrm{Rg}\,x =_{\mathrm{df}} (y)\,(y \ll x \supset (\exists r)\,(\exists s)\,(r < y \,.\, s < x - y \,.\, \mathrm{N}\,r,s)).$$

All regions are clans but not all clans are regions; for a clan may be thinly spread over a region. For example, if the M-span in Figure 14 is 2 or more, the circled intersections represent elements that make up a clan but not a region.

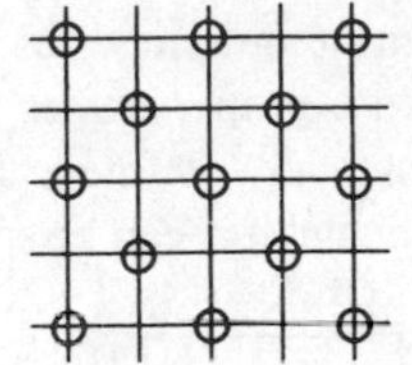

Figure 14. Thinly spread clan.

A *direct* path joining two elements y and z is a minimal region that contains both; i.e.

(ii) $$\text{Dp}\,x,y,z =_{\text{df}} \text{Qu}\,y \,.\, \text{Qu}\,z \,.\, \text{Rg}\,x \,.\, y < x \,.\, z < x \,.\, (t)\,(\text{Rg}\,t \,.\, y < t \,.\, z < t \,.\supset {\sim}\text{G}\,x,t).$$

A direct path is always a line. Between two elements there may be one or several direct paths. For example, between the element mapped at the upper left-hand corner of Figure 14 and the one mapped at the lower right-hand corner there are many direct paths; between the elements mapped at the two lower corners there is only one direct path. Where there is but one it may be called a *rect*:

(iii) $$\text{Rc}\,x,y,z =_{\text{df}} \text{Dp}\,x,y,z \,.\, (t)\,(\text{Dp}\,t,y,z \supset x = t).$$

Here we have a rudimentary shape predicate. Whether "rect" and "straight line" are to be equated will depend upon whether they agree in application to the visual field–to the full category of phenomenal places.

Distances between elements are measured along direct paths. Elements x and y are *farther apart* than z and w, where all belong to the same category, if every direct path joining x and y is greater than every direct path joining z and w:

(iv) $$\text{Ap}\,x,y,z,w =_{\text{df}} \text{Qu}\,x \,.\, \text{Qu}\,y \,.\, \text{Qu}\,z \,.\, \text{Qu}\,w \,.\, (\exists t)\,(\text{Cg}\,t \,.\, x + y + z + w < t) \,.\, (r)\,(s)\,(\text{Dp}\,r,x,y \,.\, \text{Dp}\,s,z,w \,.\supset \text{G}\,r,s).$$

This is obviously a step toward the definition of many shape predicates. Moreover, to have the means for comparing distances is to have the beginning of a theory of measure.

It must be borne in mind, of course, that definientia quite different from those used in (i) to (iv) may be called for when arrays with gaps are taken into account.

13. ORDINAL QUASIANALYSIS

Once a category is ordered it can ordinarily be distinguished from the others by some structural peculiarity. And if the array is sufficiently asymmetric, the position of each quale can be uniquely described in the language of the system, which so far contains no proper names for qualia. In this way, such proper names can be systematically defined. If, for example, an array is mapped as in Figure 15, each quale in it can be so defined.

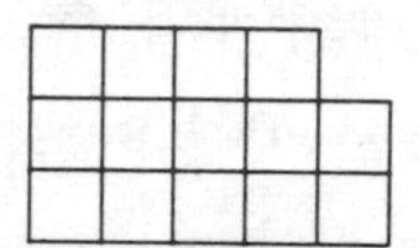

Figure 15. Sufficiently asymmetric array.

Some methods of naming qualia are of special importance. Instead of proceeding directly to define an isolated name for each quale, we might first define a name for each of the straight lines running through the array and then refer to each quale by naming the lines on which it lies. Given an array like that mapped in Figure 15, for example, we might define *"a"*, *"b"*, *"c"*, and *"d"* as names for the horizontal lines (starting at the top) and *"h"*, *"k"*, *"l"*, *"m"*, *"n"*, *"o"* as names for the vertical lines (starting at the left). Each quale in the array may then be assigned a composite name or description, e.g., *"ck"*. This not only provides us with a convenient and informative nomenclature for our atoms but also effects a quasianalysis of them; for in thus naming a quale we are virtually specifying its constituent qualities. This process of *ordinal quasianalysis* makes it possible to deal with visual-field altitudes and azimuths in a system for which visual-field places are not divisible into such attributes, and to deal with hues, chromas, and brightnesses in a system that takes colors as indivisible atoms. The qualities of an atom are represented in the system by lines or planes (etc.) passing through the atom. The fact that a quale in the array mapped in Figure 15 has a certain elevation is systematically reported by saying that the quale is part of a certain horizontal line. The presystematic statement that a visual-field place is comprised of a certain visual-field altitude and a certain visual field azimuth is translated by the systematic statement that the place is the product of a certain horizontal and a certain vertical line.

Now if ordinal quasianalysis works in these applications, why not make more use of it? The idea suggests itself that instead of taking any qualities as atoms we might take concreta, and adopt as primitive a predicate that applies between two concreta just in case they match 'in all respects'; then, by use of this predicate, order all concreta in a single complex array; and finally, by applying ordinal quasianalysis, define qualities of all levels of abstractness. A particularistic system so constructed would be attractive for its closely integrated treatment of the problems of order and abstraction. But a little investigation discloses a serious obstacle; the unified array envisaged would actually be interrupted by many breaks. Every sharp-edged visual event that contrasts in color with all concreta that immediately surround it in visual space-time will be a separate category under this system, for there will be no link between any concretum in the event and any outside. More categories rather than fewer will result, and the hoped-for simplification will not be achieved. Perhaps a completely ordinal particularism can be worked out on a somewhat different basis, but that hardly concerns us here. What does concern us, however, is a paradox that now arises and that bears directly upon our own system.

To begin with, let us recall three facts already noted: (1) that between two just-discriminable places in the visual field there is ordinarily another that matches both; (2) that no place ever occurs without some color; and (3) that two matching places cannot simultaneously present nonmatching colors. Now suppose that at a given time *t* there occurs in the visual field a horizontal line of which the left-hand half is white and the right-hand half black. Let "*a*" be the name of the furthermost right-hand place in the left-hand (white) half of the line, and let "*c*" be the name of the furthermost left-hand place in the right-hand (black) half of the line. Places *a* and *c*, since they simultaneously present contrasting colors, do not match. Let "*b*", then, be the name of a place that matches both. What color is at place *b* at time *t*? Not white, for then place *b* would have to be discriminable from place *c*; not black, for then *b* would have to be discriminable from *a*; finally, not any other color, for since any other color will fail to match either white or black, *b* would have to be discriminable from either *a* or *c*. Thus no color occurs at *b* at time *t*; yet no place ever occurs without some color. We may be tempted to conclude that there is *no* place that matches two discriminable places that simultaneously present contrasting colors; but since any two discriminable places have contrasting colors at some moment or other, this would amount to denying altogether the fact of matching visual-field locations. Thus the paradox remains: at time *t*, place *b* has some color and yet has none.

The solution is simply this: no place that matches two discriminable places is presented *at any time* when contrasting colors appear at these places.[5] We

[5] This is entirely compatible with the fact that no interstice is discerned between the two halves of the line; for the minimal discernible interstice would consist of a place between *a* and *c* that is discriminable from both.

were correct in arguing that at time t no color occurs at b; but this implies neither that there is no such place as b nor that such a place occurs at time t without any color. There may well be one or more such places, but they occur only at other times. The root of the paradox was the mistaken assumption that if a place matches two nonmatching places, it is always presented when they are. We make no like assumption with respect to colors and there is no ground for making it with respect to places. Its paradoxical consequences testify to its falsity.

Now if place b does not occur at time t, then $b + t$ is not a place-time, and there is no place-time that both spatially and temporally matches the place-times $a + t$ and $c + t$. More generally, no place-time thus links simultaneous place-times at which contrasting colors occur. This has an interesting consequence. In trying to find ways of putting ordinal quasianalysis to the fullest use, we thought of taking concreta as atoms and ordering them in a single array by a primitive predicate of matching (in all respects) among them; but we found that each sharp-edged visual event contrasting with its surroundings would then form a separate category. We now see that if our chosen basis is altered to the extent not of taking concreta as atoms but of counting place-times rather than places and times among our atoms and using as primitive a predicate of matching among them, much the same difficulty arises. For the place-times of any such sharp-edged visual event will then form a separate category; we cannot go from any of them to a place-time outside by proceeding always from a place-time to a matching place-time. In both cases, then, the choice of atoms more concrete than those selected for our system fails to achieve the purpose of reducing the number of categories. On the other hand, the *places* of such an event need not form a separate category but may be joined to outside places by an M-path of places that occur at other times; hence on this score there is no reason to reject places as atoms in favor, say, of visual-field altitudes and azimuths.

All this suggests that our official choice of atoms for our system follows consistently what may be called the *Principle of Fewest Categories*. For an obvious reason, a choice of less concrete atoms would seem to give us more categories–e.g., the three categories of hues, chromas, and brightnesses, instead of the category of colors. For the quite different reason explained above, a choice of more concrete atoms would also give us more categories. It cannot be maintained that since there may be rifts in familiar categories (see IX,5), more categories might result under our system than under the alternatives discussed; for all such rifts will have parallels under any system taking more concrete individuals as atoms; e.g., if there is no M-path joining the places x and y, then there will be no matching-path of place-times joining any place-times containing x with any place-times containing y. It is true, however, that although the phenomenon of sharp edges does not by itself indicate rifts in the category of places (as it does in that of place-times), still there may happen to be rifts in this category and yet none in the category of visual-field altitudes or the category of visual-field azimuths. In that case, the choice of these less concrete individuals as atoms will indeed result in fewer categories.

The number of resultant categories is, of course, only one of the factors to be considered in choosing a basis for a system; other criteria, some of them no less important, have been discussed earlier. Moreover, as has already been made clear there is no presumption that one choice of atoms is best for all purposes; and our systematic constructions are in general not narrowly dependent upon a choice of atoms at a given level of analysis.

14. RECENT DEVELOPMENTS[6]

Since the first edition of this book, many new contributions have been made to the theory of M-orders; and only a few of these have been mentioned in the above revised text.

I had conjectured in the first edition that somewhere in every irregular linear array there are two elements beside each other such that one matches two or more elements the other does not. N.J. Fine[7] has since proved this conjecture and shown further that for linear arrays all the following properties are equivalent: *regularity* (conformity with the stronger mapping rule), *goodness* (having no two elements beside each other such that either matches more than one element that the other does not), *symmetry of just-noticeable difference* (being such that for every two elements x and y, if y is the first element to the right of x that does not match x, then x is the first element to the left of y that does not match y), and *uniformity* (having a constant M-span throughout). The equation of uniformity with regularity is especially

[6] Well aware that my fascination with the problems discussed here far outruns both my mathematical equipment and the results achieved, I confine myself to a brief summary, omitting a good deal of supporting and subsidiary material. David Meredith has made a number of valuable criticisms and suggestions; and I have been much helped by the two works of Coxeter (cited in notes 17 and 18 below) that Meredith called to my attention, and by the work of L. Schläfli (1814–95) (see the *Gesammelte Mathematischen Abhandlungen*, 3 vols. [Basel: Verlag Birkhäuser, 1950] referred to by Coxeter).

[7] See his "Proof of a Conjecture of Goodman", *Journal of Symbolic Logic*, 19 (1954), pp. 41–44. The article by Fine and R. Harrop mentioned in the following paragraph above is their "Uniformization of Linear Arrays", *Journal of Symbolic Logic*, 22 (1957), pp. 130–140.

useful, since constancy of M-span thus guarantees and is guaranteed by satisfaction of the stronger mapping rule.

In the first edition, the adjustment of maps was left to trial-and-error. Fine and R. Harrop have since proved the important theorem that every irregular linear map can be adjusted, and have set forth effective procedures for supplying the needed vacant positions. We saw from the example at the end of Section 6 that minimal adjustment is not always unique–that there may be alternative ways of supplying the minimal number of vacant positions to produce a map with constant M-span.[8] But only one of the two ways there indicated produces a map with a constant number of map-positions (vacant or occupied) between every two positions occupied by nonmatching elements such that every element occupying a position between them matches both; and rather clearly no more than one minimal adjustment will satisfy this further requirement. At the moment, I have no proof that, in every case, at least one minimal adjustment will satisfy this requirement; but otherwise the theory of linear arrays is virtually complete.

Progress in the theory of nonlinear arrays has been aided by taking them as graphs of symmetric, irreflexive, connected, nontransitive nextness-relations. In constructing maps, all requirements in terms of background distance are dropped. Elements or *nodes* next to each other are connected by unintersected lines of any length; and all usual topological distortions are permitted. The *order-number o* of a node is the number of lines proceeding from it–that is, the number of nodes next to that node. Where the order-number is the same for all nodes in a graph, the graph itself is said to have that order-number.

A general definition of *cell* has long been wanting in our theory and can now be supplied. Briefly, a cell is a cycle that misses no shortcuts; but this needs some explanation. That a cell misses no shortcuts means that between each two of its nodes it contains a path at least as short (in number of nextness-pairs) as any other path between them in the graph. That a cell is a cycle means that it is finite and of the following structure: a linear cell is simply a next-pair of nodes; in a polygonal cell, each node is common to exactly two linear cells; in a polyhedral cell, each linear cell or edge is common to exactly two polygonal cells; at the next higher level, each polygonal cell or face is common to exactly two polyhedral cells; and so on.

This provides a good basis for dimensional classification. A node is 0-dimensional, a linear cell 1-dimensional, a polygonal cell 2-dimensional, a

[8] In the article cited in note 7 above, Fine and Harrop stated–I am afraid with my full concurrence–that the problem of uniqueness of minimal adjustment was unsolved and difficult. None of us noticed that the question had been definitively settled in the negative, in the first edition of this book, by the example cited above.

polyhedral cell 3-dimensional, and so on. In any n-dimensional cell (for $n > 1$), each $n-2$-dimensional cell is common to exactly two $n-1$-dimensional cells. Since (for $n > 0$) every n-dimensional cell is a cycle of $n-1$-dimensional cells, every n-dimensional cell is made up of cells of each lower dimensionality. A graph is n-dimensional if each of its cells that is not contained in any other is n-dimensional.

All 0-dimensional cells are alike, and so are all 1-dimensional cells. But 2-dimensional cells may have three or more sides; and 3-dimensional cells may have four or more faces, each with three or more sides. Where all the cells not contained in any other in a graph are of a given type, the graph is said to be of that cell-type. Thus in a triangular-cell graph, every cell is contained in a triangular cell but no triangular cell is contained in any polyhedral cell.

The *cell-number* **c** of a node is the number of cells that contain the node but are contained in no other cells in the graph. Where all the nodes have the same cell-number, this is also said to be the cell-number of the graph.

The basic general problem of M-arrays is to define N from M. In earlier sections, we made some examination of how far our present definitions go towards solving this problem. We saw that our definition of besideness (hence also our definition of flanking) successfully defines nextness for all linear graphs, with constant or inconstant M-span, and also for single polygons.

Calling for first consideration after linear graphs are *trees*. A tree is made up of linear cells so strung together as not to form any polygons. For any node in a tree the order-number is the same as the cell-number; but the number may range from 2 on up for interior nodes. End-nodes have the order-number 1; and an end-cell shares only one of its nodes with one or more other cells. Linear graphs may be considered degenerate trees. Sample trees are shown in Figure 16.

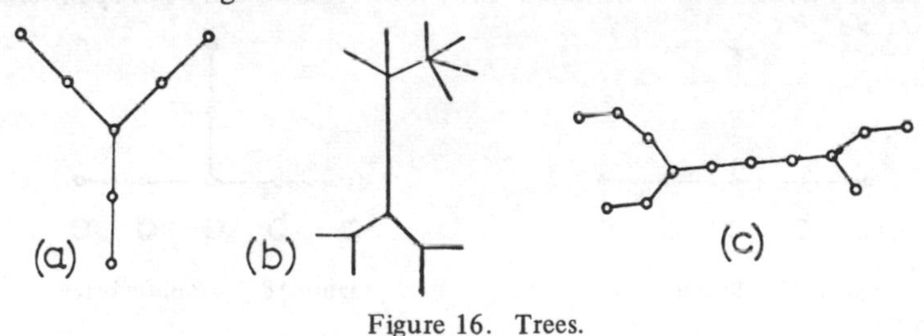

Figure 16. Trees.

Obviously any tree with two end-cells having a node in common (and indeed any graph with two linear end-cells having a node in common) is spurious as a map with M-span greater than 1; and thus for such an M-span, a (finite) tree (e.g. Figure 16*b*) that branches at every interior point is a spurious map.

There is never any problem of deriving N from M for a tree or any other genuine map with an M-span of 1, since then N is M minus identity. The problem of maps with inconstant M-span has not yet been systematically investigated for any non-linear arrays. But what of regular[9] trees with no two end-cells having a node in common?

No finite tree can be without end-cells; for any closure would result in a polygon. Thus since we are concerned solely with finite arrays, we might drop all consideration of trees without end-cells. But the question of such trees, apart from its independent mathematical interest, has this importance: that if flanking coincides with nextness in every such tree, flanking will also coincide with nextness for all pairs of nodes far enough (depending upon M-span) in the interior of any finite tree. For this reason, trees (and other graphs) with infinitely many nodes will in what follows be temporarily admitted for consideration; but local finitude–the finitude of *o*–is still assumed.

I have been able to prove that our definitions work for all regular trees without end-cells.[10] Thus the residual problem for regular trees has to do solely with pairs of nodes near the ends. Our definitions as they stand do not work for all such pairs of nodes. A very simple case of failure is illustrated in Figure 17, with M-span taken as 2. The definitions correctly establish that *a* is next to *b*, *b* next to *c*, *c* next to *d*, and *d* next to *e*; but they make *f* next to *b* and *d* rather than to *c*. Yet if the line from *f* to *c* is deleted, and lines added from *f* to *b* and to *d*, then *f* is falsely shown as matching *a* and *e*. What we have here is not a spurious map like those illustrated in Section 11 that can be readily replaced by a corrected map with a different M-span; for no map of just these six elements and with constant M-span conforms to our definitions. Nevertheless, the remedy in this particular case is easy enough: we need only add a vacant position as shown in Figure 18. The M-span remains

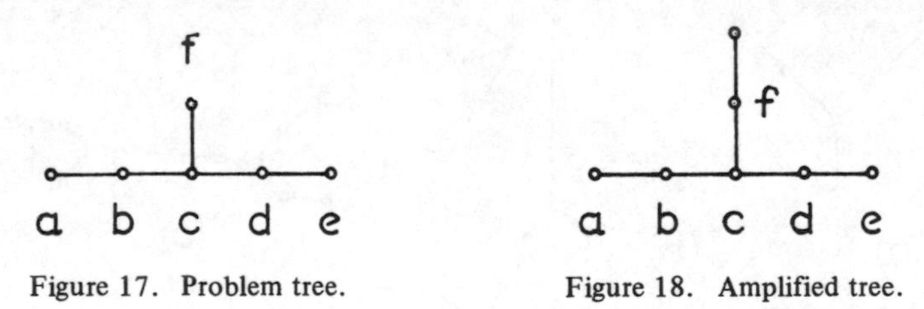

Figure 17. Problem tree. Figure 18. Amplified tree.

[9] A tree or other graph is, of course, regular or irregular with respect to a given M-relation only. To say that a graph is regular is to say elliptically that for some k, two nodes stand in the given M-relation if and only if they are not more than k N-steps apart in the graph.

[10] My proof is outlined in Malcolm W. Pownall's *An Investigation of a Conjecture of Goodman*, doctoral thesis, Department of Mathematics, University of Pennsylvania, January 1960.

at 2. Now all works well; the nextness relationships shown on the map and those called for by the definition are the same.

Amplification of this sort is a very different matter from the adjustment of linear maps explained earlier. There our definitions established correct linear order; vacant positions were interpolated solely for constancy of M-span; and no two of the original elements that were not next by the unsupplemented definitions were next by the final map. Here, on the contrary, our definitions give us neither Figure 17 nor any other tree as a basis; vacant positions are needed not for regularity but for reconciliation of map and definitions; and some original elements (e.g., *f* and *c*) that are not next by the definitions before supplementation are next afterward. Supplementation of our definitions by rules for such extrapolation is thus at once more necessary and more difficult than is mere adjustment for regularity. We are a long way from having such rules for all or for any considerable number of cases.

Suppose the linear cells of a tree so replaced by polygonal cells that no two have more than one node in common. The resultant graph is a *chandelier*.

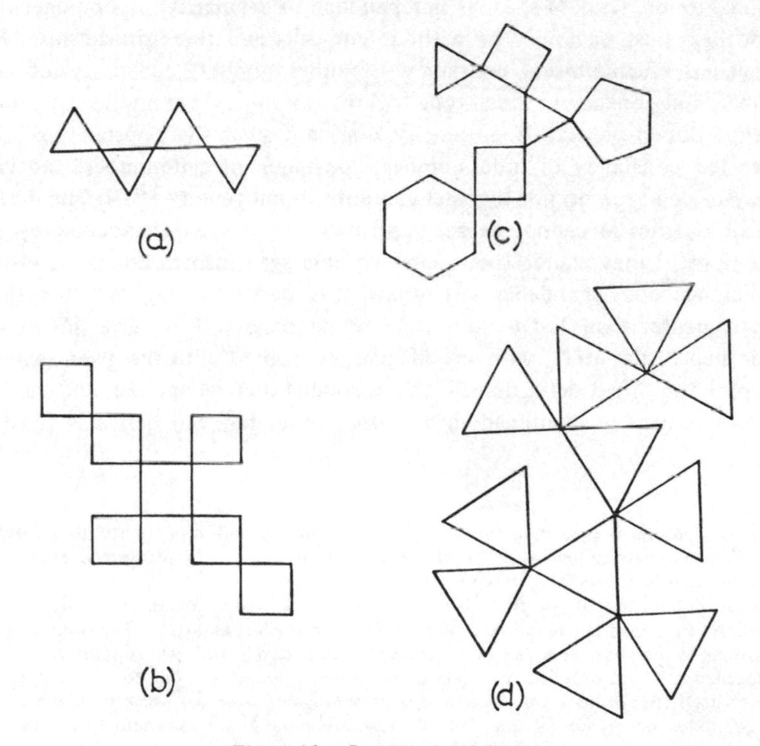

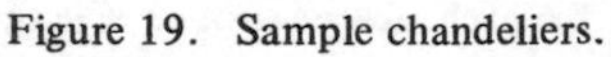
Figure 19. Sample chandeliers.

A chandelier may be of constant or varied cell-type (see Figure 19). The cell-number **c** of a node may be any integer greater than 0; and the order number *o* of a node will always be 2**c** regardless of the type of cells containing that node. A cell that has only one node with *o* greater than 2 is an end-cell. Any chandelier–or other graph–with an end-cell of *k* sides is spurious as a map with M-span greater than $\frac{k-2}{2}$. All finite chandeliers have end-cells; for any closure will result in two polygonal cells having a common side.

Plainly enough, our definitions will not work for all pairs of nodes near the ends of a chandelier. But what is the situation for pairs far enough in the interior, or in other words, for all pairs in chandeliers without end-cells? Malcolm Pownall[11] has studied this question with the following results: (i) N can be defined from M in any regular chandelier that is also homogeneous. Next-pair as well as node-homogeneity is assumed; that is, the relationship of every next-pair to the rest of the graph must be the same as the relationship of every other next-pair to the rest of the graph. Homogeneity, having nothing to do with M-span, is independent of regularity. A homogeneous chandelier must obviously be without end-cells and therefore infinite. For chandeliers, furthermore, constancy of order-number, constancy of cell-number, and constancy of cell-type together imply and are implied by homogeneity. But in general, homogeneity does not guarantee constancy of cell-type; and constancy of order-number, constancy of cell-number, and constancy of cell-type do not together guarantee homogeneity.[12] (ii) The derivation of N from M cannot be accomplished for all such chandeliers by our definitions as they stand. The square-cell case again proves crucial. Consider the homogeneous chandelier part of which is mapped in Figure 19*b* with an M-span greater than 1. Our definition of nextness will not give this or any other map of the array, with any M-span, consonant with the given manors. A slightly modified definition of nextness must then be applied; and the two definitions can be combined into a procedure that will derive N from M

---

[11] See his doctoral thesis, cited in note 10 above. The 'extension postulate' that Pownall adopts is rendered unnecessary by the new definition of *cell* introduced above and the consequently revised classification of types of graph.

[12] For an illustration of the first clause, see Figure 22 below; for an illustration of the second, let Figure 11 above tbe extended endlessly to the left and right. The homogeneity conditions in question here can be more explicitly stated as follows: A graph *G* is node-homogeneous if and only if for each two nodes *x* and *y* there is a one-to-one mapping of *G* onto itself that maps *y* onto *x* and also preserves nextness (i.e., results in every case in two nodes being next if and only if they are next in *G*). A graph *G* is next-pair homogeneous if and only if for each two next-pairs *x*, *y* and *w*, *z* there is a one-to-one mapping of *G* onto itself that maps *w* onto *x* and maps *z* onto *y* and preserves nextness.

whether the array in question happens to be of the kind just described or of one of the kinds for which our original definition works.

Chandeliers are *cut-point* graphs;[13] that is, such that elimination of some node will break them into two or more pieces. Indeed, a chandelier is thus disconnected by the elimination of any node with a cell-number greater than 1. Furthermore, every 2-dimensional homogeneous cut-point graph is a linear graph or tree or chandelier. Thus N has been proved definable from M for all 2-dimensional homogeneous cut-point graphs.

Networks are not cut-point graphs, since each polygonal cell in a network has a side in common with one other such cell. Every interior node is completely surrounded by polygonal cells and (no matter what the type of these cells) has $\mathbf{c} = o$. Networks have no end-cells; but a cell with at least one side that is a side of no other cell is a border-cell.

We saw in Section 10 that our definitions do not work for all pairs of nodes near the boundaries of networks; but we found no failure for cases far enough in the interior of regular square-cell networks with $o\ (= \mathbf{c}) = 4$ or of triangular-cell networks with $o\ (= \mathbf{c}) = 6$. Pownall has now proved that the definitions work for all homogeneous networks of these two types. In a homogeneous network there are no border-cells; each linear cell or side is common to two (and only-two) polygonal cells. A homogeneous square-cell network (or triangular-cell network) must be infinite in all directions; for any closure would result either in a polygonal cell that is not four-sided (or not three-sided) or in a polyhedral cell. Pownall's proof can probably be extended to cover homogeneous square-cell and triangular-cell networks with more cells surrounding each node–that is, with $o\ (= \mathbf{c}) > 4$ in the square-cell case, and $o\ (= \mathbf{c}) > 6$ in the triangular-cell case, and perhaps also extended to cover homogeneous networks with other types of cell; pentagonal-cell networks with $o\ (= \mathbf{c}) > 3$; and (where n is greater than 5) n-agonal-cell networks with $o\ (= \mathbf{c}) > 2$. Obviously for no network is $o < 3$: only for single polygons is o constant at 2, and here $\mathbf{c} = 1$; only for isolated linear cells is $o = 1$; and only for isolated nodes is $o = 0$.

But what of the five remaining cases: square-cell networks with $o\ (= \mathbf{c}) = 3$; triangular-cell networks with $o\ (= \mathbf{c}) = 3$, with $o\ (= \mathbf{c}) = 4$, and with $o\ (= \mathbf{c}) = 5$; and pentagonal-cell networks with $o\ (= \mathbf{c}) = 3$? Here we come upon a rather curious fact.[14]

---

[13] George Schweigert first pointed out to me the importance of this classification.

[14] The substance of most of what follows in this section (but not the discussion of Figure 21) was published in "Graphs for Linguistics" in *Structure of Language and Its Mathematical Aspects*, 12 (1961) of *Proceedings of Symposia in Applied Mathematics* published by the American Mathematical Society, pp. 51–55. The presentation above differs considerably as a result of differences in the use of "cell", "cell-number", "cell-type", and some consequent differences in the classification of graphs.

Although a homogeneous square-cell network with o (= **c**) > 3 runs on infinitely in all directions, we are soon stopped if we try to construct one with o (= **c**) = 3. We cannot add to the graph in Figure 20; for every node already has $o = 3$. Actually what we have here is topologically a single cube.

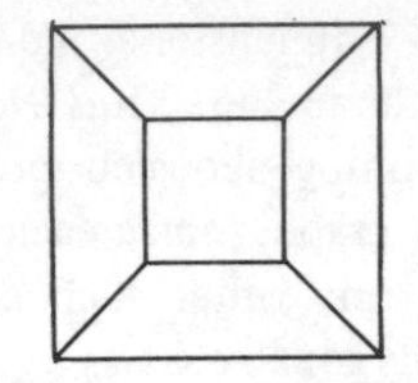

Figure 20. Finite homogeneous graph.

Since all the square-cells are contained in a polyhedron, the graph is strictly not a polygonal-cell or a bidimensional graph at all, but a tridimensional graph consisting of a single cubical cell. The order-number of each node is 3 but the cell-number is 1. Nevertheless the graph is like homogeneous square-cell networks in that the order-number of a node is the same as the number of square cells (counting the one formed by the four outside lines in the figure) that contain a node. And there is no other homogeneous graph for which this number is 3.

The results are similar in the triangular-cell cases. For $o = 3$, we have a homogeneous graph in Figure 21, and nothing can be added. This is the topological equivalent of the tetrahedron; and **c** here is 1. But the order-

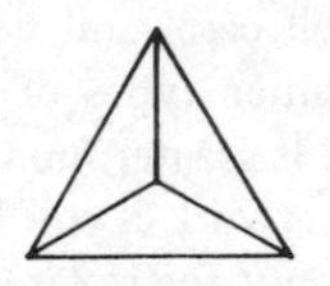

Figure 21. Finite homogeneous graph.

number of a node is the same as the number of triangular cells containing a node; and there is no other homogeneous graph for which this number is 3. Where this number is 4, the unique graph is the topological equivalent of the octahedron;[15] and where the number is 5, the topological equivalent of the icosahedron.

---

[15] In the octahedron, each node is also on two 'interior' square cells; but we are concerned here only with polygonal cells of such a type that the number of these cells is o.

Finally, the only homogeneous graph with $o = 3$ and with each node in three pentagonal cells is the topological equivalent of the dodecahedron.

Thus the graphs resulting from the five cases correspond to the five platonic solids. That for each case the finite graph described is the only one meeting the specifications can be easily proved. Thus uniqueness and overall finitude are here derived from purely local conditions. That there are no other such cases–i.e., no other finite homogeneous graphs made up of polygonal cells of a single type and having the order-number of a node the same as the number of polygonal cells containing that node–follows from the fact that there are only five regular solids.

A network with polygonal cells of more than one type may be called a *lace*. In a homogeneous lace, cells of two different types alternate around each node. As in every homogeneous network, $o = \mathbf{c}$; but in a homogeneous lace o must be even, and each node lies in $o/2$ cells of each of the two types.[16] The simplest case calls for $o = 4$, with two square cells and two triangular cells on each node; but this turns out to yield a finite graph (Figure 22): a single cell topologically equivalent to the cuboctahedron. Nothing can be added and there is no other graph meeting the specifications. The like case with pentagonal instead of the square cells again turns out to be finite: the topological equivalent of the icosidodecahedron.[17] All other homogeneous graphs, with $o/2$ polygonal cells of each of two types alternating around each node, are infinite laces.

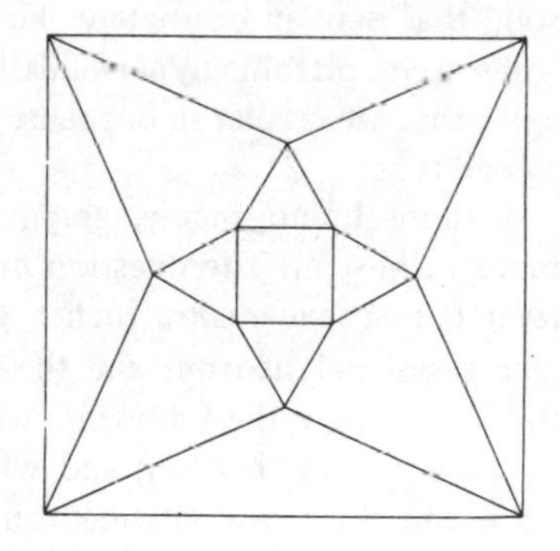

Figure 22. Finite homogeneous graph.

[16] Note that the octahedron does not meet this requirement; for although the number of square cells on each node is $o/2$, the number of triangular cells is o.

[17] See H. S. M. Coxeter, *Regular Polytopes* (London: Methuen & Company, 1948), pp. 18–19 and Plate I, Fig. 10.

Matters are complicated enough for bidimensional graphs to keep us from taking more than the briefest look at tridimensional ones. The tridimensional counterpart of a network is a *stack* of polyhedral cells; one common example consists of cubical cells, with $o = 6$ and $\mathbf{c} = 8$ for each interior node. In a homogeneous stack, every face is common to just two polyhedral cells. Now suppose that the cells are tetrahedral, and that each node lies in just four of these and also has $o = 4$. Under these specifications, nothing can be added to the graph depicted in Figure 23–a finite homogeneous graph with each

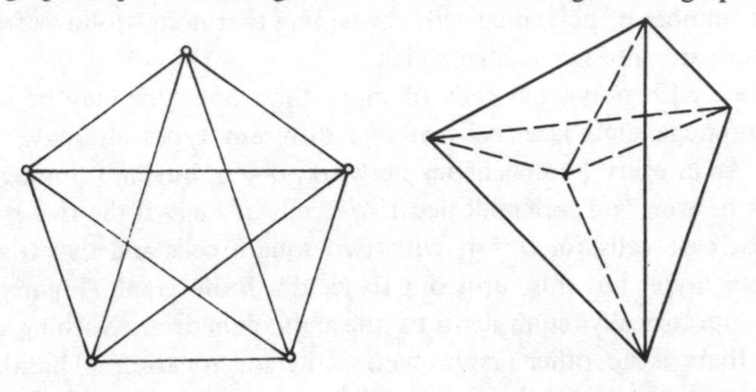

Figure 23. Finite homogeneous graph; two representations.

face common to two tetrahedral cells. Yet this graph is not strictly a stack (any more than Figures 20-22 are strictly networks) but a cycle of polyhedral cells–a platonic hypersolid that may appropriately be called the hypertetrahedron. There are five other such platonic hypersolids;[18] I do not know how many hypersolids there are that are regular in our sense but made up of polyhedral cells of more than one type.

All this discussion of finite homogeneous graphs has been quite independent of any reference to M-span. The question now arises whether our definitions work whenever the M-span makes such a graph a genuine map. The only cases where the seven polyhedrons and the one hypersolid above described constitute genuine maps are the following: with M-span taken as 1, all but the tetrahedron and hypertetrahedron; and with M-span taken as 2 or as 3, the dodecahedron and the icosidodecahedron only. Proof that our definitions work in these cases is not difficult. With greater M-spans, the graphs in question are all spurious.

Thus our procedure works for various types of homogeneous finite and

---

[18] See Coxeter, *Introduction to Geometry* (New York: John Wiley & Sons, 1961), pp. 396–413.

infinite maps with admissible constant M-span; and this gives some ground, far short of proof, for hoping that the same procedure may work for every genuine homogeneous map with constant M-span. We know that it works for all trees, homogeneous or not, without end cells and with admissible constant M-span, and for all genuine linear maps regardless of homogeneity or constancy of M-span. Thus given only M and the information that the map in question belongs to some, unspecified, one among all these kinds, we can define N. Ahead lies the formidable task of covering also trees that have inconstant M-spans and all other nonlinear arrays that either have inconstant M-span or are inhomogeneous.

15. NOTE ADDED IN THIRD EDITION

As this edition goes to press, Ivan Fox is developing a more general way of defining N from M. His method consists of decomposing a non-linear array into certain linear arrays, and then using betwixtness and besideness in these arrays to determine nextness in the nonlinear array in question. This yields what seems a *sufficient* condition for nextness in all nonspurious regular mappings of all types, regardless of homogeneity or finitude, and furthermore a *necessary* condition (and hence a definition) for nextness throughout all arrays where the above definition via flanking succeeds and also many others. The new definition promises to work everywhere in regular nonspurious mappings of all chandeliers with constant cell-type, and of all trees, laces, square-cell and cubical-cell arrays such that every element is at least twice the M-span from some other element along a minimal N-path.

Fox has also offered proof that every linear array can be so adjusted as to preserve just noticeable difference in the way specified above (page 248, lines 9–16), but that minimal adjustment of this sort is not always unique.

CHAPTER XI

OF TIME AND ETERNITY

1. PHENOMENAL TIME

Since phenomenal time is merely one among the several categories of qualia, there might seem to be no more call for a separate chapter on time than for one on, say, place or color. But there are two reasons why time requires special attention. In the first place, ordinary language deals with time by means of special devices that are used and understood with the greatest facility in everyday discourse but misused and misinterpreted with nearly as much facility in most philosophical discourse. Before we can decide with any confidence what are the peculiar problems about time, or whether there are any, we shall have to take precautions against these linguistic pitfalls. In the second place, when linguistic confusions have been cleared away, we shall find that while many supposed problems disappear, there remains indeed one rather important special fact about time.

At the outset it may be well to review and amplify what we have already said about time qualia. Experience obviously has a temporal dimension, and the fact that phenomenal and physical time are different is evident enough from the fact that our temporal acuity is limited: presystematically speaking, a physical event of very short duration (such as the interval of darkness between successive flashes of a fluorescent light) is below the threshold of perception. A time quale is simply a phenomenal moment that has no other as a part. Such moments can be distinguished from one another and from places and colors and other qualia; but no one need start trying to catch one all by itself.[1] As already emphasized (V,2; VI,1) we cannot some-

[1] On the basis of inadequate knowledge of my treatment of time, Bertrand Russell, in P. A. Schilpp, ed., *The Philosophy of Bertrand Russell* (Evanston and Chicago: Northwestern University Press, 1944), p. 716, objects that (1) it involves gratuitously 'inventing entities' since he 'cannot find such things' as time qualia in his experience, and that (2) it 'requires an absolute instead of a relational theory of time'. But Russell himself, in book after book, distinguishes between 'public' and 'private' time; and a time quale is simply a moment of 'private' time. Unless one holds that direct experience has no temporal dimension, the recognition of time qualia no more involves an invention of entities than does the recognition of color qualia or place qualia. Furthermore, the recognition of time qualia no more requires an absolute theory of *physical* (or 'public') time than the recognition of place qualia requires an absolute theory of physical space; and *phenomenal* (or 'private') time seems to me no more 'relational' than phenomenal color.

how literally detach a quale or a concretum or any other part of experience from the rest; but we can analyze the stream of phenomena into elements of any of these kinds for purposes of systematic description.

Every quale that is not itself a time occurs at some time; hence every concretum, no matter of what sense realm, contains a time. The category of times is distinguished from the other categories through thus overlapping all concreta. That is to say, once "Cg" has been satisfactorily defined (see the discussion in IX, 5), a time may be defined as follows:

$$\mathrm{T}x =_{\mathrm{df}} \mathrm{Qu}\, x \,.\, x < (\imath y)\, \{\mathrm{Cg}\, y \,.\, (z)\, (\not{c}\, z \supset y \circ z)\}.$$

Phenomenal time order–to be systematically constructed on the basis of matching by means described in Chapters IX and X–is presumably linear. The problem of defining the earlier-later direction along the line is sometimes mistakenly supposed to have no parallel with respect to qualities other than time. Actually, it has close analogues with respect to all the other categories. If some two individual moments such that one is known to be earlier than the other can be systematically defined, the problem of defining "earlier than" quite generally is readily solved. The problem of defining a direction thus resolves itself into one of defining the requisite individual names. The same holds true for the problem of defining the black-white direction in the series of grays or the left-right direction along a horizontal line of places in the visual field. "Nevertheless", it may be urged, "time flows inevitably in one direction while space, for example, does not go from left to right any more than from right to left". I shall return later to this question of temporal 'flow' (Sections 3, 4).

According to our analysis a concrete phenomenal individual, ordinarily said to be in time, is regarded rather as having time in it; and its temporal size–or duration–depends on how many moments it contains. Consider a colored round patch that appears in the visual field and stays for a while. The total presentation comprises the colors, places, and times involved. It is temporally as well as spatially divisible; and its identity over different times, like its identity over different places, is the identity of a totality of diverse parts. Each of the times Kp-qualifies the whole. What we think of as a phenomenal thing is distinguished from what we think of as a phenomenal event or process only in the pattern of differences among its temporal parts. A thing is a monotonous event; an event is an unstable thing.

In contrast, the sum of color-spots that constitutes the colored patch itself contains no times. Neither it nor any of its parts is temporally divisible. It retains its strict numerical identity throughout the period in question. Every time in that period is *with* some complex in the patch. Such occurrence through a period is thus quite a different matter from the *duration* of a thing or event; and we had better observe the distinction by saying that the patch *persists* through the period. In general:

$$x \textit{ endures } \text{through } y =_{\text{df}} (z)\,(\text{Qu}\, z \,.\, z < y \,.\supset .\, \text{T}z \,.\, \text{Kp}\, z,x);$$

$$x \textit{ persists } \text{through } y =_{\text{df}} (\exists z)\,(z \text{ endures through } y \,.\, (r)\,(\text{Qu}\, r \supset : r < x \equiv . r < z \,.\, \sim \text{T}\, r)).$$

Under these definitions, an individual that endures or persists through a period endures or persists through every part of that period. The endurance or persistence is continuous or discontinuous according as the period is.

The result of adding to a persisting individual the times it occurs at is an enduring individual. The result of extracting from an enduring individual the times it occurs at[2] is a persisting individual. Thus although a color quale, for example, occurs at times and persists through periods, it is nevertheless literally 'out of time', i.e., it is discrete from all times. It is, in a word, *eternal*. This is not to say that it is everpersisting, for nothing occurs at all times. Nor is an eternal individual everlasting; for an everlasting or ever-enduring individual is one–like the total stream of experience or a complete lengthwise strip of it–that contains all times. These distinctions are embodied in the following definitions:

$$x \textit{ is everlasting} =_{\text{df}} (y)\,(\text{T}y \supset \text{Kp}\, y,x);$$

$$x \textit{ is everpersisting} =_{\text{df}} (\exists y)\,(y \text{ is everlasting} \,.\, (z)\,(\text{Qu}\, z \supset : z < x \equiv . z < y \,.\, \sim \text{T}z));$$

$$x \textit{ is eternal} =_{\text{df}} (y)\,(\text{T}y \supset x \rfloor y).$$

Observe that the eternity of an individual is no bar to its occurrence at some times or its failure to occur at others; indeed, only what is eternal is with a time. Theologians have perhaps overlooked something here.

Concerning temporal shape, little need be added to what was said earlier. Because time is linear, the shape of temporally continuous individuals varies directly with their size and need hardly be distinguished from size. It is discontinuous individuals that exhibit, in their different patterns of intermittence, a wide variety of temporal shapes. A dot and a dash sounded in Morse code differ in temporal size; but the letters "k" and "w" sounded in the code, although equal in temporal size, differ in temporal shape. It must be noted that temporal shape depends solely on the order of the occupied moments

---

[2] On the ambiguity of "occurs at", and the resolution of this ambiguity, see VII,5.

relative to the complete temporal array and not upon how these moments are occupied. Two phenomenal events that mark out similar constellations of time qualia have the same temporal shape even though the qualia occurring at corresponding temporal positions in the two events are quite different. Thus enunciations of two very different sentences may have the same temporal size and shape; and a message sounded in code has the same temporal size and shape as the corresponding sequence of dots and dashes flashed in light at the same speed.

The analogy between space and time suggests that time qualia make up a 'present' related to physical time in somewhat the same way that the visual field is related to physical space. But before this matter can be discussed with much hope of avoiding confusion, some features of language as it pertains to time–for example, the usage of terms like "present"–will have to be clarified.

2. TIME AND LANGUAGE

In ordinary discourse, we often indicate the time of events not by explicit description but by such a word as "now", "yesterday", "next week", "past", "later", or by the tense of a verb. As a result, we have quite unequivocal statements that nevertheless seem, paradoxically to change in truth value. For example, when I say "The Red Sox now lead the American league", I am being quite definite; I am not saying that they lead at some unspecified time but am indicating the time unmistakably. What I utter is thus not an open statement like "*x* is yellow" but a closed statement that is either true or false. And yet although it be true when I first utter it, it may be false when I repeat it later.

The point is, of course, that we must be more careful to distinguish between a statement and other statements that resemble it. In the last example given, we have two statements, not one. Each of the utterances is a distinct, definite statement; and the two in fact have different truth values. These utterances may be exactly alike in sound pattern; but it is each utterance and not anything common to the two that constitutes a statement. Similarly, it is each of the utterances of "now"–not anything common to the two–that constitutes a word and refers to a certain time. In platonistic terms, the distinction between the general pattern or *type* of a word or sentence and its particular instances or *tokens* was drawn many years ago by Peirce. Too often, however, those who have noticed the distinction have looked upon it as a matter of isolated academic interest, and assumed that thereafter one need be concerned only with the types. More recently,[3] we have been forced to recognize that often–as in the example above–it is the tokens that function as words or sentences; for we find different tokens of the same type naming and affirming different things.

Indeed, it is the types that we can do without. Actual discourse, after all, is made up of tokens that differ from and resemble each other in various important ways. Some are "now"'s and others "very"'s just as some articles of furniture are desks and others chairs; but the application of a common predicate to several tokens–or to several articles of furniture–does not imply that there is a universal designated by that predicate. And we shall find no case where a word or statement needs to be construed as a type rather than as a token. The exclusion of types not only does away with some excess baggage but also results, I think, in clarifying our immediate problem.

Obviously the term "token" is no longer appropriate. It is both misleading and superfluous; for utterances and inscriptions are no longer to be regarded as mere samples but as the actual words or statements themselves, and the linguistic universals from which they were to be distinguished are no longer to be countenanced at all. Nevertheless, to emphasize the fact that words and statements are utterances or inscriptions–i.e., *events* of shorter or longer duration–I shall sometimes use such terms as "word-events", "noun-events", "'here'-events", "'Paris'-events" and so on, even though the suffix is really redundant in all these cases. "Paris"-events, of course, are not events in Paris but certain utterances and inscriptions–namely, the "Paris"'s. A word-event surrounded by quotes-events is a predicate applicable to utterances and inscriptions; and any

"'Paris' consists of five letters"

is short for any

"Every 'Paris'-inscription consists of five letter-inscriptions".

Although each utterance and inscription is a separate word (or statement or letter, etc.), the difference between two words often has no practical importance. For most purposes, we need not distinguish among the several "Pisa"'s, all of which name the same thing, even though they differ widely in size, shape, color, sound, place, date, etc. On the other hand, we must carefully distinguish a "Pisa" from a "Paris". It is true that a given "Pisa" may be more like a given "Paris" than like some other "Pisa", just as a

---

[3] See, for example, my *A Study of Qualities* (1940), already cited, pp. 594–623; and Reichenbach's *Elements of Symbolic Logic* (New York: The Macmillan Company, 1947), pp. 284–298.

given mushroom may look more like a given toadstool than like some other mushroom; but in both cases we must discern just that overt difference that is correlated with a difference in appropriate use. In the case of "Pisa" 's and "Paris" 's, and in many other cases, some certain difference of shape or sound-pattern is the clue to a difference in what the words name.

Yet by no means every difference of extension is accompanied by a difference in shape or sound-pattern. The nominata of two "Paris" 's that are exactly alike in shape may be as different as those of a "Paris" and a "Pisa"; for some "Paris" 's name a city in France while others name a town in Maine. To note from its shape that a given word is a "Paris" thus is not enough. In order to determine which of two places the particular "Paris" in question names, we must look to the *context*–i.e., to the surrounding words and to certain attendant circumstances. Similarly, the various "this continent" 's name six or more different individuals; the various "John Smith" 's a still greater number; and the various "I" 's name vastly many different individuals.

For convenience, let us speak of words (or letters or statements, etc.) that are catalogued under a single label as *replicas* of one another,[4] so that any "Paris" (or any "I say") is a replica of itself and of any other "Paris" (or "I say"). Roughly speaking, a word is an *indicator* if (but, as will be made clear later in this section, not necessarily only if) it names something not named by some replica of the word. This is admittedly broad, including ambiguous terms as well as what might be regarded as indicators-proper, such as pronouns; but delimitation of the narrower class of indicators-proper is a ticklish business and is not needed for our present purposes.

What has been said above will suggest that almost every name has a replica somewhere that names something different, and that therefore almost every name will be an indicator according to this criterion. But the distinction between indicators and nonindicators becomes effective when applied to a limited discourse. Within such a discourse there will normally be many names that are not indicators–although proper nouns as well as pronouns will still often be indicators, and pronouns will occasionally be nonindicators.

Among the commonest indicators are the personal indicators, the spatial indicators, and the temporal indicators. Of the personal indicators, an "I" or "me" normally refers to its own utterer; a "we" or "us" refers to the utterer and certain others determined by the context; a "you" applies to those addressed by the utterer, and so on. Characteristically, even though there is no

---

[4] This usage differs from that of C. S. Peirce, who speaks of inscriptions or utterances as *replicas* of a word type; see *Collected Papers of Charles Sanders Peirce*, vol. II, (Cambridge, Mass.: Harvard University Press, 1932), p. 143.

variation in what a given personal indicator names, there is wide variation in what several replicas of that indicator name. Much more remains to be said, of course. For one thing, the person in question is sometimes, as in ghostwriting, the ostensible rather than the actual utterer; and the indicators appearing in a copy or transcription relate not to the actual maker of these inscriptions but to the maker of the original inscription or utterance. Furthermore, some indicators, like the "his" 's, not only name but perform a relational function as well. Again, some words of the same shape as indicators are actually not indicators at all but simply variables; a case in point is the "he" in an "If anyone disapproves, he may leave". Finally, an inscription sometimes divides temporarily into several different indicators; for example, if a given placard reading "I hate Hitler" is carried by different persons on three successive days, then the three day-parts of the enduring "I"-inscription name different persons. But all this is by way of subscript to the main point.

The location of a spatial indicator has to be taken into account in much the same way as the producer of a personal indicator. Some spatial indicators like the "here" 's name regions they lie in, while others like the "yonder" 's are discrete from the regions they name. In most cases, just what region a given indicator names depends partly upon its context, including such supplementary aids as pointing. Even among the "here" 's, one may refer to part of a room while others refer severally to a town, a county, a state, a continent, etc. Analogues of all the subsidiary remarks about personal indicators apply to spatial indicators. For example, a "here"-inscription in a personal letter normally refers to the place where it was written; a "here" in a delivered telegram refers rather to the place where the original was written or spoken; and if a "No Parking Here" sign is moved about, certain different temporal parts of the "Here" name different places.

But we are primarily concerned with the temporal indicators. Part of what is to be said concerning them is already evident from our glance at the personal and spatial indicators. The "now" 's, for example, behave much like the "here" 's; each "now" names a period in which it lies, and the periods named by different "now" 's range from a moment to an era. Other terms, like the "yesterday" 's and the "soon" 's, name periods earlier or later than themselves; but in every case the time of a temporal indicator is relevant to what it names. We need hardly review the other points of analogy between temporal indicators and those already discussed, but certain temporal indicators require special attention.

In the first place, the "past" 's, "present" 's, and "future" 's lend themselves to frequent abuse in theoretical discourse. Most "present" 's function exactly like most "now" 's, naming some period they lie in; and the various "presents" 's name many different periods of varying length, some of them remote from others. A "past", however, most often names *all* the time preceding–and a "future" all the time following–a certain period in which it lies. Thus the period named by a given "past" overlaps, and indeed includes or is included in, the period named by any other "past"; and the same holds for "future" 's. This fact, that what is once past is always thereafter past (and that what is once future was always theretofore future), creates an illusion of fixity and leads to treating the "past" 's (or the "future" 's, or even the "present" 's) as if, like the "Eiffel Tower" 's, all named the same thing. Metaphysicians have capitalized on this confusion for some very purple passages on The Past, The Present, and The Future. We must be careful to remember that nonsimultaneous "past" 's (or "present" 's, or "future" 's) commonly name different even if not discrete periods.

Very often, however, temporal indication is accomplished in a sentence not by any word devoted solely or chiefly to that purpose but rather by the tense of the verb. A "Randy ran" tells us not only who did what but also when, i.e., prior to the period of production of the sentence itself. The "ran", besides specifying the action performed, serves also as a temporal indicator; nonsimultaneous "ran" 's ordinarily indicate different periods of time. Incidentally, verbs in some languages may also serve the third purpose of personal indication; for example, a *creo* in Spanish indicates its utterer so definitely that the pronoun *yo* is customarily omitted.

A verb in the present tense normally indicates a period within which the verb is produced, while a verb in the future tense normally indicates a period after its own production. The interpretation of compound tenses and of combinations of tensed verbs with other temporal indicators sometimes requires care but is seldom really difficult. A

"Randy had been running"

tells us that the running took place prior to a moment–presumably further specified in the context–that is in turn prior to the time of production of the sentence itself. An isolated

"World War II was present",

however, tells us simply, as does a

"World War II is past",

that World War II is prior to the sentence in question. The "present" in an "is present" or a "was present" or a "will be present" in no way affects the temporal indication accomplished by the verb alone. On the other hand, an "is past" or an "is future" functions in the same way as, respectively, a "was" or a "will be". No exhaustive survey of such combinations need be attempted here; but it should be noted that some may result in virtually vacuous statements. For instance, a

"World War II was future"

–if unaccompanied by any context determining what prior moment is being affirmed to precede World War II–says only what may be said about any event that did not begin at the first moment of time. Likewise, of any event that does not run to the end of time, we may truly say that it will be past. Of course, a combination such as a "was future" or a "will be past" is usually set within a restrictive context.

In many statements the tense is merely grammatical, the verbs not actually functioning as temporal indicators. This is true more often than not in formal discourse. For one thing, generalizations are usually without effective tense; an

"All men have spines"

refers not only to all men contemporary with the statement but also to all who preceded or will follow it. In many singular statements also the verb, although in the present grammatical tense, is adequately translated by a purely tenseless symbol. For example, where an "*a*" and a "*b*" are proper names, a given

"*a* overlaps *b*"

may speak simply of the overlapping of the two individuals, without indicating anything about the date of their common part; that is, the sentence may just say that *a* **o** *b*. On the other hand, another

"*a* overlaps *b*"

may have effective tense, being used to affirm not just that *a* and *b* have some common part but that they have some common part that is contemporary with the sentence itself. The context makes the difference.

Now one may say that two things overlap [tenseless] if and only if they did or do or will overlap; an "*a* **o** *b*" is implied by an effectively tensed "*a* overlaps *b*" or "*a* overlapped *b*" or "*a* will overlap *b*", while an "*a* **o** *b*" implies none of these but only such a disjunction as an "*a* overlaps *b*, or *a* overlapped *b*, or *a* will overlap *b*". But parallel principles do not hold for all other verbs, indeed, an "*a* ɩ *b*" obviously is not implied by an effectively tensed "*a* is discrete from *b*" or "*a* was discrete from *b*" or "*a* will be discrete from *b*", but implies them all. Moreover, even though each verb that is effectively in, say, the past tense indicates a period preceding the verb, the relationship affirmed to obtain between such a period and other individuals referred to in a sentence varies considerably with different verbs. While an

"*a* overlapped *b*"

places a common part of *a* and *b* within such a period, an

"*a* was earlier than *b*"

places *a* within such a period; a

"color *c* was at place *p*"

places the (color-spot) sum of *c* and *p* within such a period, and a

"color *c* matched color *d*"

seems to place *c* and *d* (but not their sum, of course–VII, 2) at a moment within such a period. These examples will perhaps be sufficient warning against certain kinds of hasty generalization about tense.

Like some verbs, some replicas of other temporal indicators are not themselves indicators. For example, in a

"We can know at a given time only what is past at that time or present at that time, not what is future at that time",

the "past", "present", and "future" name no times. Rather, the "is past at", the "is present at", and the "is future at" are tenseless two-place predicates that may respectively be translated by the tenseless predicates "is earlier than", "is at", and "is later than".

Effective tense does not by itself prevent a string of words from constituting a genuine statement. A tensed statement has as constant a truth value as a tenseless one; and a tenseless statement, no less than a tensed one, is an event in time. The difference is that tensed statements and other statements with indicators are not, so to speak, 'freely repeatable'. Now of course no term or statement is ever repeated in the way a quale is repeated; for a term or statement is a particular event and not a universal. On the other hand, nearly all terms and statements are much repeated in that they have many replicas. But a term or statement is said to be freely repeatable in a given discourse if all its replicas therein are also translations of it. Indicators and statements containing them are not freely repeatable.

Ordinarily, when we want to make continued or renewed use of a given term or sentence that occurs earlier in our discourse, we just repeat it, i.e., introduce a replica to take its place. If the term or sentence is freely repeatable, then for most purposes we need not distinguish between it and its replicas; we proceed as if all were numerically identical. But in the case of an indicator or a sentence containing one, where not all the replicas are translations, this is obviously dangerous. Often, indeed, no available replica of a given term or sentence is a translation of it, so that an inaccessible original cannot, in effect, be brought back into play by repeating it. For this reason, although indicators are of enormous practical utility, they are likely to be awkward for formal discourse. Various remedies may be applied. One lies in supplying a freely repeatable name (or description) of the indicator, or of the sentence containing it, and thereafter, instead of repeating the term or sentence, referring to it by means of a replica of this name. For example, a given "now" might be identified by any

"The 937th word uttered by George Washington in 1776".

A later repetition of that "now" is not a translation of it; but any replica of this descriptive name is a translation of every other, and names just the particular "now" in question. And using such a name, we can readily arrive at a repeatable *translation* of the indicator; e.g., the "now" in question is translated by any

"The period referred to by the 937th word uttered by George Washington in 1776";

or alternatively, if the period is a day, by any

"The day on which George Washington uttered his 937th word in 1776".

Or we may seek a translation that contains no name of the indicator itself, but rather another name for what the indicator names. Thus a certain "here" is translated by any "Philadelphia"; and a certain "ran" is translated by any

"runs [tenseless] on Jan. 7, 1948 at noon E.S.T.".

Against such translations, it is sometimes urged that they do not really convey the content of the originals. A spoken

"Randy is running now"

tells us that the action takes place at the very moment of speaking, while a

"Randy runs [tenseless] on October 17, 1948, at 10 P.M., E.S.T."

does not tell us that the action takes place simultaneously with either utterance unless we know in addition that the time of the utterance is October 17, 1948 at 10 P.M., E.S.T. Since–the argument runs–we recognize the tenseless sentence as a translation of the tensed one only in the light of outside knowledge, we have here no genuine translation at all. But this seems to me no more cogent than would the parallel argument that "L'Angleterre" is not a genuine translation of "England" because we recognize it as a translation only if we know that L'Angleterre is England.

A different question may arise form the auxiliary function of tensed verbs as indicators. If two tensed predicates are coextensive but indicate different times, are they translations of one another? Do we demand that the two agree in what they apply to, or do we demand that they agree also in what they indicate? It is to be noted that ordinarily predicates that indicate different times differ also in extension; for to say that a tensed predicate indicates a time is a convenient way of saying that the application of a tensed predicate is restricted to individuals at that time. Nevertheless in some cases predicates that–according to the looser locution–'indicate different times' may agree in extension. A clear if unimportant example is that of a

"stood still while walking"

and a simultaneous

"will stand still while walking".

Since neither applies to anything, they are coextensive. The question whether they are translations of each other is quite analogous to the question whether

"orders a centaur steak"

and

"orders a unicorn steak"

are translations of each other. Both questions illustrate a general problem concerning the criteria for the use of "translation". That general problem lies outside my province here. I can only remark in passing that I think (1) that the appropriate criteria may vary considerably with the nature and purpose of the discourse and (2) that criteria much more stringent than simple coextensiveness can be formulated within the framework of extensionalism.[5]

[5] For a further discussion, see the articles cited in I,1, note 1.

3. THE PASSAGE OF TIME

We have still to deal with statements that seem most patently to reflect the temporal flow of events. One speaks of time passing, of events moving from the future into the present and on into the past, of things growing steadily older. How is such language to be interpreted?

To say that time passes seems to amount to saying that a moment of time progresses constantly in a future-toward-past direction. Yet obviously a time does not shift its position with respect to other times; it is identified with its position in the temporal series, and if any time moves then all move together. Now we have seen how a

"Time *t* is future",

a later

"Time *t* is present",

and a still later

"Time *t* is past"

may all be true; and how the conjunction of the three might have as a translation any

"Washington's 27th 'future' is earlier than time *t*; his 13th 'present' is at time *t*; and his 49th 'past' is later than time *t*",

(where each "is" is tenseless). The motion of time *t* ostensibly expressed here consists simply of the fact that *t* has different relationships of precedence to different verbal events. Again a

"Time *t* was future, is now present, and will be past"

says merely that this utterance is at time *t*, is later than some earlier time, and earlier than some later time. On the other hand, a statement like

"A time is at first future, then becomes present, then becomes past"

is quite a different matter. The final clause, for example, says neither that a time is earlier than this particular "past" nor that it is earlier than some "past" or other. The clause says rather that a time is past at some time or other; and this, as we have seen, just says that a time is earlier than some time or other. What the clause in question says thus does not depend on the time of its own or any other utterance. Indeed the whole sentence contains no actual indicators at all but is freely repeatable. Of course we may quite understandably want a translation of it free of words having many replicas that are actual indicators; and such a translation is readily provided:

"A time is later than some time x, identical with some time later than x, and earlier than some still later time".

In the case of a

"Time t is past and constantly recedes further into the past",

uttered at time s, the first "past" is an actual indicator while the second is not. The sentence says that time t is earlier than s; and that if q and r are times later than s, and q is later than r, then t precedes q by more than t precedes r. What has been said here of statements concerning times can easily be adapted to the interpretation of parallel statements concerning events.

So far I have not considered statements like

"While it endures, a thing constantly grows older".

This again is normally a tenseless, freely repeatable statement, saying in effect that if two times r and s are within the period of duration of a thing, and r is earlier than s, then a larger part of that period precedes s than precedes r.

Thus are sentences that express the passage of time or the flow of events translated by sentences that merely describe relationships of precedence in the temporal series. The suggestion of flow or of passing or of ageing disappears; and just for this reason, it may be felt that we are missing something important about time. Most efforts to formulate just what is missed end in vague poetry or in hopeless confusion over temporal indicators. Yet I think that underlying these efforts there is a certain peculiarity of time that deserves attention. Strangely enough it turns out not that time is more fluid than (say) space but rather that time is more static.

We saw that the analogy between space and time is indeed close. Duration is comparable to extent. A thing may vary in color in its different spatial or in its different temporal parts. A thing may occupy different places at one time, or the same place at different times, or may vary concomitantly in place and time. The relation between the period of time occupied by a thing during its entire existence and the rest of time is as fixed as the relation between the region the thing covers during its entire existence and the rest of space. And yet there is this difference: two things may approach and then recede from each other in space, may grow more and then less alike in color, shape, etc.; but two things never become nearer and then farther apart in time. The location or the color or the shape of a thing may change, but not its time.

This may seem to depend on a mere verbal accident. Why not simply generalize the use of "change" a little so that a thing changes in a given respect if different parts of the thing have different qualities of the kind in question? Because, it may be fairly answered, this ignores the distinction between a minute mobile thing that travels over a given region, and a spatially large thing that occupies a comparable region at a single instant. Each of the two things has parts that differ from one another in location; but according to ordinary usage, only the former undergoes change. By applying the term "change" in the one case but not the other, ordinary usage marks an important distinction.

In other words, change is concomitant variation in time and some other respect. Since time is always one of the variant factors in change, we speak of *change in* whatever is the other variant factor in the given case. Thus although there is no change that does not involve time, there is no change in time.

4. THE TEMPORAL FIELD

In the past two sections, we have digressed to consider certain features of language likely to confuse any discussion of time. We can now return to the study of phenomenal time.

Quite clearly, there is no change in phenomenal any more than in physical time. A concrete phenomenal individual may change in place or color but not in time. For change, whether physical or phenomenal, is variation concomitant with temporal variation. Thus a minimal spatially changing (moving) compound not merely occupies two places but occupies them at different times, i.e., it comprises two concreta that differ in both place and time. Analogously, a minimal temporally changing compound would have not merely to occupy two times but to occupy them at different times. But no time is at another time, and no concretum contains more than one time. Thus if a compound contains several times, there is no further question whether or not it contains these times at different times. There is no temporal change to be distinguished from temporal size.

Along with the absence of temporal change goes a difference in the relationship between the phenomenal and the physical in the case of time as compared with other qualities. The (spatial) visual field moves freely in relation to physical objects; and by analogy we incline to suppose that the temporal field–the sum of time qualia–moves in relation to physical events. Even if we recognize, as we must, that individuals do not change their temporal relationships to each other as they do their spatial relationships, still we tend to think of time qualia as moving constantly and uniformly in relation to physical events. Each observed event is regarded as presenting one after another the qualia comprised in the temporal field. But in that case the *presentation* of the event would move through phenomenal time, and we have seen that this never happens. In summary, nothing either phenomenal or physical changes its position in relation to either phenomenal or physical time.

Nevertheless, the attempt to complete the analogy between space and time is easier to discredit than to abandon. Our reluctance to drop it derives, I think, from three factors. In the first place, the many genuine points of analogy between time and space naturally lead us to seek analogy in the matter of change as well. But it is precisely here that there is just cause for protest against the spatialization of time. In the second place, we often mistake as evidence for the analogy the obvious truth of a statement like

"An event is at first future, then present, then past".

But our preceding study (Sections 2 and 3 above) has shown that this statement can be translated simply by

"An event is later than some time r, is at some time s that is later than r, and is earlier than some still later time t",

and that other statements that seem to suggest temporal change can be translated by statements concerning temporal precedence. In the third place, I think our error is nourished by a nebulous underlying notion of the self as something that flits through time carrying its specious present along with it. I have no idea of discussing the nature of the self here; but whether it is or is like a thing, event, or quality–or whatever else it may be or may be like–and however many are the times it lights upon, the statement that it lights upon or occupies or is *at different times at different times* will still be absurd.

A consequence of all this is that the temporal field is coextensive with the duration of the stream of experience to which we apply our system; there is no smaller temporal field that moves through this. But how long is this experience that we take as basic? For a phenomenalistic system, surely, it is (in non-phenomenalistic language) confined to the experience of a single subject; but does it comprise the lifetime experience of that subject, or only a moment of that experience, or something in between?

The advocate of 'solipsism of the present moment' confines himself to the experience of one moment. Others may complain that he chooses each different moment as it occurs, and that every time he says that all experience is present he proclaims a different thesis since each "present" refers to a different moment. But for him there is only a single moment; his position depends upon denying that there are any phenomenal temporal distinctions. Now the apparent futility of trying to construct any adequate system based on the experience of a single moment is not by itself a conclusive argument against him. One has no alternative but to strive for such a system if convinced that there is only a single time quale. But it seems to me quite as evident that phenomenal temporal distinctions *are* made–that several time qualia are discerned–as that several place qualia and several color qualia are discerned. Thus I think we are not confined to the experience of a single moment.

If several time qualia are to be admitted, the question exactly how many is a relatively minor one. There is perhaps no good reason for stopping short of the lifetime experience of a subject, but the matter need not be settled with finality and precision. What we call a system is usually rather a schema of systems according to which, given material meeting certain rather broad requirements, we can construct whatever else we want. We saw earlier (VI,2) that it often makes little difference whether qualia at one level of analysis or another are chosen as atoms. In defining a concretum, for example, we did not have to itemize our qualia but merely exhibited a certain general relationship between qualia and concreta, much as logic exhibits certain relationships between unspecified statements. Similarly, whether the category of time qualia for a system contains all the moments in a total stream of experience or only those in some shorter stretch–and whether the choice made be rigidly kept to or constantly changed–the main problem is how, given any such field, we are to deal with the facts of phenomenal and physical time.

The problem of dealing with physical times is not likely to be easy; for even though phenomenal time and physical time do not move with respect to each other, the relationship between them is far from simple. Physical time divides into particles too small to be perceived; some events in physical time lie outside phenomenal time; some physically simultaneous events are not phenomenally simultaneous, and some phenomenally simultaneous events are not physically simultaneous. Indeed, phenomenal time may cut across physical time in amazing ways; at one moment, I may see a leaf fall, hear a word that was spoken a few seconds earlier, and see a stellar explosion that occurred some centuries before that. Moreover, a brief physical event may occupy considerable phenomenal time, while a longer physical event may occupy only an instant of phenomenal time. For example, if an airplane is flying very rapidly away from me, I may hear for an appreciable period the noise made by its motors during a fraction of a second; on the other hand, if the plane is flying very rapidly toward me, I may hear at one instant the noise made by its motors over a longer period. It looks very much as if an adequate account of physical time, even at a more or less common-sense level, might prove to require a rather complex construction of space-time; and difficulties that arise here could easily lead us by degrees into some of the most formidable problems of modern physics. But the problem of physical time lies outside the province of this book.

5. THE PHYSICAL WORLD

Without having considered all the difficulties along the way, we have come to the threshold of the problem of accounting for the physical world upon a phenomenalistic basis. Although this problem is out of bounds for us here, we must glance at it briefly; for it is widely regarded as insoluble, and its insolubility is often taken as sufficient reason for completely abandoning the phenomenalistic approach.

My disagreement on the latter point has already been expressed. In my view, the systematic description of phenomena as such provides genuine answers to important problems; and also the methods developed for accomplishing such a description may often serve as models for the treatment of parallel problems within the objective realm. Thus such a description seems to me to have a value that is quite independent of our ability to solve the problem of giving a phenomenalistic account of the physical world.

Moreover, I am not at all convinced that the problem is insoluble; for I do not see how any sound conclusion concerning its solubility can be reached until we are clearer about just what the problem is. Just what sort of basis are we permitted to use? Just what is to be accounted for or explained? And just what sort of explanation is required? Of course, we are permitted to use any phenomenalistic basis; we are to explain the physical world; and the explanation is to proceed by definition and translation. But in the first place, we have seen that the distinction between phenomenalistic and physicalistic predicates is not altogether sharp. The question is not whether the set of primitives already selected for the system sketched above is adequate for all the wanted constructions. We are free to make needed additions to our basis so long as they do not transgress phenomenalistic restrictions; but it is not very clear just what those restrictions are. In the second place, is the physical world that we are to explain the somewhat inconsistent world of common-sense and stale science or the very abstruse and continually revised world of the latest physical theory? We shall hardly try to accommodate ourselves to both the firmest legend and the newest conjecture any more than we try to reconcile the one with the other. Our explicandum is not so plainly before us that systematic explanation can proceed without a good deal of prior clarification and criticism. In the third place, we saw that traditional criteria of definition are hardly applicable even to some of the most elementary and most obviously satisfactory definitions of a constructional system. The weaker criterion arrived at in Chapter I puts any problem of definition in a new light. Furthermore, when we ask loosely for a translation of certain sentences, it is not always clear whether we are literally demanding a translation or would accept such a syntactical treatment as has recently been outlined for mathematics.[6] Thus 'the problem of accounting for the physical world upon a phenomenalistic basis' is pretty amorphous in more ways than one, and I am suggesting that we decide on its solubility after rather than before the problem has been more precisely formulated. We must avoid the insidious temptation to accept a proposed version because it is insoluble quite as much as the temptation to reject it for the same reason.

What does seem to be fairly clear is that the problem is intimately connected with the problems of distinguishing between laws and nonlaws, of interpreting counterfactual conditionals, and of codifying the principles of confirmation.[7] In recent investigations of these problems some very discouraging difficulties have arisen. But it is worth noting that these difficulties arise primarily in dealing with a natural language and are greatly diminished within a constructional system where control can be maintained over the predicates admitted.

---

[6] In Nelson Goodman and W. V. Quine "Steps Toward a Constructive Nominalism", *Journal of Symbolic Logic*, 12 (1947), pp. 105–122.

[7] These problems are explored in my *Fact, Fiction, and Forecast* (first edition, London: The Athlone Press, 1954; second edition, Indianapolis: The Bobbs-Merrill Company, Inc., 1965; third edition, 1973).

INDEX TO SPECIAL SYMBOLS

Symbols not familiar in standard texts of logic or mathematics and not amenable to placement in the alphabetical index are listed below in the order of their introduction.

INDEX

SYNTHESE LIBRARY

Monographs on Epistemology, Logic, Methodology,
Philosophy of Science, Sociology of Science and of Knowledge, and on the
Mathematical Methods of Social and Behavioral Sciences

1. J. M. Bocheński, *A Precis of Mathematical Logic.* 1959, X + 100 pp.
2. P. L. Guiraud, *Problèmes et méthodes de la statistique linguistique.* 1960, VI + 146 pp.
3. Hans Freudenthal (ed.), *The Concept and the Role of the Model in Mathematics and Natural and Social Sciences, Proceedings of a Colloquium held at Utrecht, The Netherlands, January 1960.* 1961, VI + 194 pp.
4. Evert W. Beth, *Formal Methods. An Introduction to Symbolic Logic and the Study of Effective Operations in Arithmetic and Logic.* 1962, XIV + 170 pp.
5. B. H. Kazemier and D. Vuysje (eds.), *Logic and Language. Studies Dedicated to Professor Rudolf Carnap on the Occasion of His Seventieth Birthday.* 1962, VI + 256 pp.
6. Marx W. Wartofsky (ed.), *Proceedings of the Boston Colloquium for the Philosophy of Science, 1961-1962,* Boston Studies in the Philosophy of Science (ed. by Robert S. Cohen and Marx W. Wartofsky), Volume I. 1973, VIII + 212 pp.
7. A. A. Zinov'ev, *Philosophical Problems of Many-Valued Logic.* 1963, XIV + 155 pp.
8. Georges Gurvitch, *The Spectrum of Social Time.* 1964, XXVI + 152 pp.
9. Paul Lorenzen, *Formal Logic.* 1965, VIII + 123 pp.
10. Robert S. Cohen and Marx W. Wartofsky (eds.), *In Honor of Philipp Frank,* Boston Studies in the Philosophy of Science (ed. by Robert S. Cohen and Marx W. Wartofsky), Volume II. 1965, XXXIV + 475 pp.
11. Evert W. Beth, *Mathematical Thought. An Introduction to the Philosophy of Mathematics.* 1965, XII + 208 pp.
12. Evert W. Beth and Jean Piaget, *Mathematical Epistemology and Psychology.* 1966, XII + 326 pp.
13. Guido Küng, *Ontology and the Logistic Analysis of Language. An Enquiry into the Contemporary Views on Universals.* 1967, XI + 210 pp.
14. Robert S. Cohen and Marx W. Wartofsky (eds.), *Proceedings of the Boston Colloquium for the Philosophy of Science 1964-1966, in Memory of Norwood Russell Hanson,* Boston Studies in the Philosophy of Science (ed. by Robert S. Cohen and Marx W. Wartofsky), Volume III. 1967, XLIX + 489 pp.

15. C. D. Broad, *Induction, Probability, and Causation. Selected Papers.* 1968, XI + 296 pp.
16. Günther Patzig, *Aristotle's Theory of the Syllogism. A Logical-Philosophical Study of Book A of the Prior Analytics.* 1968, XVII + 215 pp.
17. Nicholas Rescher, *Topics in Philosophical Logic.* 1968, XIV + 347 pp.
18. Robert S. Cohen and Marx W. Wartofsky (eds.), *Proceedings of the Boston Colloquium for the Philosophy of Science 1966-1968,* Boston Studies in the Philosophy of Science (ed. by Robert S. Cohen and Marx W. Wartofsky), Volume IV. 1969, VIII + 537 pp.
19. Robert S. Cohen and Marx W. Wartofsky (eds.), *Proceedings of the Boston Colloquium for the Philosophy of Science 1966-1968,* Boston Studies in the Philosophy of Science (ed. by Robert S. Cohen and Marx W. Wartofsky), Volume V. 1969, VIII + 482 pp.
20. J.W. Davis, D. J. Hockney, and W. K. Wilson (eds.), *Philosophical Logic.* 1969, VIII + 277 pp.
21. D. Davidson and J. Hintikka (eds.), *Words and Objections: Essays on the Work of W.V. Quine.* 1969, VIII + 366 pp.
22. Patrick Suppes, *Studies in the Methodology and Foundations of Science. Selected Papers from 1911 to 1969.* 1969, XII + 473 pp.
23. Jaakko Hintikka, *Models for Modalities. Selected Essays.* 1969, IX + 220 pp.
24. Nicholas Rescher *et al.* (eds.), *Essays in Honor of Carl G. Hempel. A Tribute on the Occasion of His Sixty-Fifth Birthday.* 1969, VII + 272 pp.
25. P. V. Tavanec (ed.), *Problems of the Logic of Scientific Knowledge.* 1969, XII + 429 pp.
26. Marshall Swain (ed.), *Induction, Acceptance, and Rational Belief.* 1970, VII + 232 pp.
27. Robert S. Cohen and Raymond J. Seeger (eds.), *Ernst Mach: Physicist and Philosopher,* Boston Studies in the Philosophy of Science (ed. by Robert S. Cohen and Marx W. Wartofsky), Volume VI. 1970, VIII + 295 pp.
28. Jaakko Hintikka and Patrick Suppes, *Information and Inference.* 1970, X + 336 pp.
29. Karel Lambert, *Philosophical Problems in Logic. Some Recent Developments.* 1970, VII + 176 pp.
30. Rolf A. Eberle, *Nominalistic Systems.* 1970, IX + 217 pp.
31. Paul Weingartner and Gerhard Zecha (eds.), *Induction, Physics, and Ethics: Proceedings and Discussions of the 1968 Salzburg Colloquium in the Philosophy of Science.* 1970, X + 382 pp.
32. Evert W. Beth, *Aspects of Modern Logic.* 1970, XI + 176 pp.
33. Risto Hilpinen (ed.), *Deontic Logic: Introductory and Systematic Readings.* 1971, VII + 182 pp.
34. Jean-Louis Krivine, *Introduction to Axiomatic Set Theory.* 1971, VII + 98 pp.
35. Joseph D. Sneed, *The Logical Structure of Mathematical Physics.* 1971, XV + 311 pp.
36. Carl R. Kordig, *The Justification of Scientific Change.* 1971, XIV + 119 pp.
37. Milič Čapek, *Bergson and Modern Physics,* Boston Studies in the Philosophy of Science (ed. by Robert S. Cohen and Marx W. Wartofsky), Volume VII. 1971, XV + 414 pp.

38. Norwood Russell Hanson, *What I Do Not Believe, and Other Essays* (ed. by Stephen Toulmin and Harry Woolf), 1971, XII + 390 pp.
39. Roger C. Buck and Robert S. Cohen (eds.), *PSA 1970. In Memory of Rudolf Carnap,* Boston Studies in the Philosophy of Science (ed. by Robert S. Cohen and Marx W. Wartofsky), Volume VIII. 1971, LXVI + 615 pp. Also available as paperback.
40. Donald Davidson and Gilbert Harman (eds.), *Semantics of Natural Language.* 1972, X + 769 pp. Also available as paperback.
41. Yehoshua Bar-Hillel (ed.), *Pragmatics of Natural Languages.* 1971, VII + 231 pp.
42. Sören Stenlund, *Combinators, λ-Terms and Proof Theory.* 1972, 184 pp.
43. Martin Strauss, *Modern Physics and Its Philosophy. Selected Papers in the Logic, History, and Philosophy of Science.* 1972, X + 297 pp.
44. Mario Bunge, *Method, Model and Matter.* 1973, VII + 196 pp.
45. Mario Bunge, *Philosophy of Physics.* 1973, IX + 248 pp.
46. A. A. Zinov'ev, *Foundations of the Logical Theory of Scientific Knowledge (Complex Logic),* Boston Studies in the Philosophy of Science (ed. by Robert S. Cohen and Marx W. Wartofsky), Volume IX. Revised and enlarged English edition with an appendix, by G. A. Smirnov, E. A. Sidorenka, A. M. Fedina, and L. A. Bobrova. 1973, XXII + 301 pp. Also available as paperback.
47. Ladislav Tondl, *Scientific Procedures,* Boston Studies in the Philosophy of Science (ed. by Robert S. Cohen and Marx W. Wartofsky), Volume X. 1973, XII + 268 pp. Also available as paperback.
48. Norwood Russell Hanson, *Constellations and Conjectures* (ed. by Willard C. Humphreys, Jr.). 1973, X + 282 pp.
49. K. J. J. Hintikka, J. M. E. Moravcsik, and P. Suppes (eds.), *Approaches to Natural Language. Proceedings of the 1970 Stanford Workshop on Grammar and Semantics.* 1973, VIII + 526 pp. Also available as paperback.
50. Mario Bunge (ed.), *Exact Philosophy – Problems, Tools, and Goals.* 1973, X + 214 pp.
51. Radu J. Bogdan and Ilkka Niiniluoto (eds.), *Logic, Language, and Probability. A Selection of Papers Contributed to Sections IV, VI, and XI of the Fourth International Congress for Logic, Methodology, and Philosophy of Science, Bucharest, September 1971.* 1973, X + 323 pp.
52. Glenn Pearce and Patrick Maynard (eds.), *Conceptual Chance.* 1973, XII + 282 pp.
53. Ilkka Niiniluoto and Raimo Tuomela, *Theoretical Concepts and Hypothetico-Inductive Inference.* 1973, VII + 264 pp.
54. Roland Fraïssé, *Course of Mathematical Logic* – Volume 1: *Relation and Logical Formula.* 1973, XVI + 186 pp. Also available as paperback.
55. Adolf Grünbaum, *Philosophical Problems of Space and Time.* Second, enlarged edition, Boston Studies in the Philosophy of Science (ed. by Robert S. Cohen and Marx W. Wartofsky), Volume XII. 1973, XXIII + 884 pp. Also available as paperback.
56. Patrick Suppes (ed.), *Space, Time, and Geometry.* 1973, XI + 424 pp.
57. Hans Kelsen, *Essays in Legal and Moral Philosophy,* selected and introduced by Ota Weinberger. 1973, XXVIII + 300 pp.
58. R. J. Seeger and Robert S. Cohen (eds.), *Philosophical Foundations of Science. Proceedings of an AAAS Program, 1969,* Boston Studies in the Philosophy of

Science (ed. by Robert S. Cohen and Marx W. Wartofsky), Volume XI. 1974, X + 545 pp. Also available as paperback.

59. Robert S. Cohen and Marx W. Wartofsky (eds.), *Logical and Epistemological Studies in Contemporary Physics,* Boston Studies in the Philosophy of Science (ed. by Robert S. Cohen and Marx W. Wartofsky), Volume XIII. 1973, VIII + 462 pp. Also available as paperback.
60. Robert S. Cohen and Marx W. Wartofsky (eds.), *Methodological and Historical Essays in the Natural and Social Sciences. Proceedings of the Boston Colloquium for the Philosophy of Science, 1969-1972,* Boston Studies in the Philosophy of Science (ed. by Robert S. Cohen and Marx W. Wartofsky), Volume XIV. 1974, VIII + 405 pp. Also available as paperback.
61. Robert S. Cohen, J. J. Stachel and Marx W. Wartofsky (eds.), *For Dirk Struik. Scientific, Historical and Political Essays in Honor of Dirk J. Struik,* Boston Studies in the Philosophy of Science (ed. by Robert S. Cohen and Marx W. Wartofsky), Volume XV. 1974, XXVII + 652 pp. Also available as paperback.
62. Kazimierz Ajdukiewicz, *Pragmatic Logic,* transl. from the Polish by Olgierd Wojtasiewicz. 1974, XV + 460 pp.
63. Sören Stenlund (ed.), *Logical Theory and Semantic Analysis. Essays Dedicated to Stig Kanger on His Fiftieth Birthday.* 1974, V + 217 pp.
64. Kenneth F. Schaffner and Robert S. Cohen (eds.), *Proceedings of the 1972 Biennial Meeting, Philosophy of Science Association,* Boston Studies in the Philosophy of Science (ed. by Robert S. Cohen and Marx W. Wartofsky), Volume XX. 1974, IX + 444 pp. Also available as paperback.
65. Henry E. Kyburg, Jr., *The Logical Foundations of Statistical Inference.* 1974, IX + 421 pp.
66. Marjorie Grene, *The Understanding of Nature: Essays in the Philosophy of Biology,* Boston Studies in the Philosophy of Science (ed. by Robert S. Cohen and Marx W. Wartofsky), Volume XXIII. 1974, XII + 360 pp. Also available as paperback.
67. Jan M. Broekman, *Structuralism: Moscow, Prague, Paris.* 1974, IX + 117 pp.
68. Norman Geschwind, *Selected Papers on Language and the Brain,* Boston Studies in the Philosophy of Science (ed. by Robert S. Cohen and Marx W. Wartofsky), Volume XVI. 1974, XII + 549 pp. Also available as paperback.
69. Roland Fraïssé, *Course of Mathematical Logic* – Volume 2: *Model Theory.* 1974, XIX + 192 pp.
70. Andrzej Grzegorczyk, *An Outline of Mathematical Logic. Fundamental Results and Notions Explained with All Details.* 1974, X + 596 pp.
71. Franz von Kutschera, *Philosophy of Language.* 1975, VII + 305 pp.
72. Juha Manninen and Raimo Tuomela (eds.), *Essays on Explanation and Understanding. Studies in the Foundations of Humanities and Social Sciences.* 1976, VII + 440 pp.
73. Jaakko Hintikka (ed.), *Rudolf Carnap, Logical Empiricist. Materials and Perspectives.* 1975, LXVIII + 400 pp.
74. Milič Čapek (ed.), *The Concepts of Space and Time. Their Structure and Their Development,* Boston Studies in the Philosophy of Science (ed. by Robert S. Cohen and Marx W. Wartofsky), Volume XXII. 1976, LVI + 570 pp. Also available as paperback.

75. Jaakko Hintikka and Unto Remes, *The Method of Analysis. Its Geometrical Origin and Its General Significance*, Boston Studies in the Philosophy of Science (ed. by Robert S. Cohen and Marx W. Wartofsky), Volume XXV. 1974, XVIII + 144 pp. Also available as paperback.
76. John Emery Murdoch and Edith Dudley Sylla, *The Cultural Context of Medieval Learning. Proceedings of the First International Colloquium on Philosophy, Science, and Theology in the Middle Ages – September 1973*, Boston Studies in the Philosophy of Science (ed. by Robert S. Cohen and Marx W. Wartofsky), Volume XXVI. 1975, X + 566 pp. Also available as paperback.
77. Stefan Amsterdamski, *Between Experience and Metaphysics. Philosophical Problems of the Evolution of Science*, Boston Studies in the Philosophy of Science (ed. by Robert S. Cohen and Marx W. Wartofsky), Volume XXXV. 1975, XVIII + 193 pp. Also available as paperback.
78. Patrick Suppes (ed.), *Logic and Probability in Quantum Mechanics*. 1976, XV + 541 pp.
79. H. von Helmholtz, *Epistemological Writings*. (A New Selection Based upon the 1921 Volume edited by Paul Hertz and Moritz Schlick, Newly Translated and Edited by R. S. Cohen and Y. Elkana), Boston Studies in the Philosophy of Science, Volume XXXVII. 1977 (forthcoming).
80. Joseph Agassi, *Science in Flux*, Boston Studies in the Philosophy of Science (ed. by Robert S. Cohen and Marx W. Wartofsky), Volume XXVIII. 1975, XXVI + 553 pp. Also available as paperback.
81. Sandra G. Harding (ed.), *Can Theories Be Refuted? Essays on the Duhem-Quine Thesis*. 1976, XXI + 318 pp. Also available as paperback.
82. Stefan Nowak, *Methodology of Sociological Research: General Problems*. 1977, XVIII + 504 pp. (forthcoming).
83. Jean Piaget, Jean-Blaise Grize, Alina Szeminska, and Vinh Bang, *Epistemology and Psychology of Functions*. 1977 (forthcoming).
84. Marjorie Grene and Everett Mendelsohn (eds.), *Topics in the Philosophy of Biology*, Boston Studies in the Philosophy of Science (ed. by Robert S. Cohen and Marx W. Wartofsky), Volume XXVII. 1976, XIII + 454 pp. Also available as paperback.
85. E. Fischbein, *The Intuitive Sources of Probabilistic Thinking in Children*. 1975, XIII + 204 pp.
86. Ernest W. Adams, *The Logic of Conditionals. An Application of Probability to Deductive Logic*. 1975, XIII + 156 pp.
87. Marian Przełęcki and Ryszard Wójcicki (eds.), *Twenty-Five Years of Logical Methodology in Poland*. 1977, VIII + 803 pp. (forthcoming).
88. J. Topolski, *The Methodology of History*. 1976, X + 673 pp.
89. A. Kasher (ed.), *Language in Focus: Foundations, Methods and Systems. Essays Dedicated to Yehoshua Bar-Hillel*, Boston Studies in the Philosophy of Science (ed. by Robert S. Cohen and Marx W. Wartofsky), Volume XLIII. 1976, XXVIII + 679 pp. Also available as paperback.
90. Jaakko Hintikka, *The Intentions of Intentionality and Other New Models for Modalities*. 1975, XVIII + 262 pp. Also available as paperback.
91. Wolfgang Stegmüller, *Collected Papers on Epistemology, Philosophy of Science and History of Philosophy*, 2 Volumes, 1977 (forthcoming).

92. Dov M. Gabbay, *Investigations in Modal and Tense Logics with Applications to Problems in Philosophy and Linguistics.* 1976, XI + 306 pp.
93. Radu J. Bogdan, *Local Induction.* 1976, XIV + 340 pp.
94. Stefan Nowak, *Understanding and Prediction: Essays in the Methodology of Social and Behavioral Theories.* 1976, XIX + 482 pp.
95. Peter Mittelstaedt, *Philosophical Problems of Modern Physics,* Boston Studies in the Philosophy of Science (ed. by Robert S. Cohen and Marx W. Wartofsky), Volume XVIII. 1976, X + 211 pp. Also available as paperback.
96. Gerald Holton and William Blanpied (eds.), *Science and Its Public: The Changing Relationship,* Boston Studies in the Philosophy of Science (ed. by Robert S. Cohen and Marx W. Wartofsky), Volume XXXIII. 1976, XXV + 289 pp. Also available as paperback.
97. Myles Brand and Douglas Walton (eds.), *Action Theory. Proceedings of the Winnipeg Conference on Human Action, Held at Winnipeg, Manitoba, Canada, 9-11 May 1975.* 1976, VI + 345 pp.
98. Risto Hilpinen, *Knowledge and Rational Belief.* 1978 (forthcoming).
99. R. S. Cohen, P. K. Feyerabend, and M. W. Wartofsky (eds.), *Essays in Memory of Imre Lakatos,* Boston Studies in the Philosophy of Science (ed. by Robert S. Cohen and Marx W. Wartofsky), Volume XXXIX. 1976, XI + 762 pp. Also available as paperback.
100. R. S. Cohen and J. Stachel (eds.), *Leon Rosenfeld, Selected Papers.* Boston Studies in the Philosophy of Science (ed. by Robert S. Cohen and Marx W. Wartofsky), Volume XXI. 1977 (forthcoming).
101. R. S. Cohen, C. A. Hooker, A. C. Michalos, and J. W. van Evra (eds.), *PSA 1974: Proceedings of the 1974 Biennial Meeting of the Philosophy of Science Association,* Boston Studies in the Philosophy of Science (ed. by Robert S. Cohen and Marx W. Wartofsky), Volume XXXII. 1976, XIII + 734 pp. Also available as paperback.
102. Yehuda Fried and Joseph Agassi, *Paranoia: A Study in Diagnosis,* Boston Studies in the Philosophy of Science (ed. by Robert S. Cohen and Marx W. Wartofsky), Volume L. 1976, XV + 212 pp. Also available as paperback.
103. Marian Przełęcki, Klemens Szaniawski, and Ryszard Wójcicki (eds.), *Formal Methods in the Methodology of Empirical Sciences.* 1976, 455 pp.
104. John M. Vickers, *Belief and Probability.* 1976, VIII + 202 pp.
105. Kurt H. Wolff, *Surrender and Catch: Experience and Inquiry Today,* Boston Studies in the Philosophy of Science (ed. by Robert S. Cohen and Marx W. Wartofsky), Volume LI. 1976, XII + 410 pp. Also available as paperback.
106. Karel Kosík, *Dialectics of the Concrete,* Boston Studies in the Philosophy of Science (ed. by Robert S. Cohen and Marx W. Wartofsky), Volume LII. 1976, VIII + 158 pp. Also available as paperback.
107. Nelson Goodman, *The Structure of Appearance,* Boston Studies in the Philosophy of Science (ed. by Robert S. Cohen and Marx W. Wartofsky), Volume LIII. 1977 (forthcoming).
108. Jerzy Giedymin (ed.), *Kazimierz Ajdukiewicz: Scientific World-Perspective and Other Essays, 1931–1963.* 1977 (forthcoming).
109. Robert L. Causey, *Unity of Science.* 1977, VIII+185 pp.
110. Richard Grandy, *Advanced Logic for Applications.* 1977 (forthcoming).

111. Robert P. McArthur, *Tense Logic.* 1976, VII + 84 pp.
112. Lars Lindahl, *Position and Change: A Study in Law and Logic.* 1977, IX + 299 pp.
113. Raimo Tuomela, *Dispositions.* 1977 (forthcoming).
114. Herbert A. Simon, *Models of Discovery and Other Topics in the Methods of Science,* Boston Studies in the Philosophy of Science (ed. by Robert S. Cohen and Marx W. Wartofsky), Volume LIV. 1977 (forthcoming).
115. Roger D. Rosenkrantz, *Inference, Method and Decision.* 1977 (forthcoming).
116. Raimo Tuomela, *Human Action and Its Explanation. A Study on the Philosophical Foundations of Psychology.* 1977 (forthcoming).
117. Morris Lazerowitz, *The Language of Philosophy,* Boston Studies in the Philosophy of Science (ed. by Robert S. Cohen and Marx W. Wartofsky), Volume LV. 1977 (forthcoming).
118. Tran Duc Thao, *Origins of Language and Consciousness,* Boston Studies in the Philosophy of Science (ed. by Robert S. Cohen and Marx. W. Wartofsky), Volume LVI. 1977 (forthcoming).
119. Jerzy Pelc, *Polish Semiotic Studies, 1894–1969.* 1977 (forthcoming).
120. Ingmar Pörn, *Action Theory and Social Science. Some Formal Models.* 1977 (forthcoming).
121. Joseph Margolis, *Persons and Minds,* Boston Studies in the Philosophy of Science (ed. by Robert S. Cohen and Marx W. Wartofsky), Volume LVII. 1977 (forthcoming).

SYNTHESE HISTORICAL LIBRARY

Texts and Studies
in the History of Logic and Philosophy

Editors:

1. M. T. Beonio-Brocchieri Fumagalli, *The Logic of Abelard.* Translated from the Italian. 1969, IX + 101 pp.
2. Gottfried Wilhelm Leibniz, *Philosophical Papers and Letters.* A selection translated and edited, with an introduction, by Leroy E. Loemker. 1969, XII + 736 pp.
3. Ernst Mally, *Logische Schriften,* ed. by Karl Wolf and Paul Weingartner. 1971, X + 340 pp.
4. Lewis White Beck (ed.), *Proceedings of the Third International Kant Congress.* 1972, XI + 718 pp.
5. Bernard Bolzano, *Theory of Science,* ed. by Jan Berg. 1973, XV + 398 pp.
6. J. M. E. Moravcsik (ed.), *Patterns in Plato's Thought. Papers Arising Out of the 1971 West Coast Greek Philosophy Conference.* 1973, VIII + 212 pp.
7. Nabil Shehaby, *The Propositional Logic of Avicenna: A Translation from al-Shifā: al-Qiyās,* with Introduction, Commentary and Glossary. 1973, XIII + 296 pp.
8. Desmond Paul Henry, *Commentary on De Grammatico: The Historical-Logical Dimensions of a Dialogue of St. Anselm's.* 1974, IX + 345 pp.
9. John Corcoran, *Ancient Logic and Its Modern Interpretations.* 1974, X + 208 pp.
10. E. M. Barth, *The Logic of the Articles in Traditional Philosophy.* 1974, XXVII + 533 pp.
11. Jaakko Hintikka, *Knowledge and the Known. Historical Perspectives in Epistemology.* 1974, XII + 243 pp.
12. E. J. Ashworth, *Language and Logic in the Post-Medieval Period.* 1974, XIII + 304 pp.
13. Aristotle, *The Nicomachean Ethics.* Translated with Commentaries and Glossary by Hypocrates G. Apostle. 1975, XXI + 372 pp.
14. R. M. Dancy, *Sense and Contradiction: A Study in Aristotle.* 1975, XII + 184 pp.
15. Wilbur Richard Knorr, *The Evolution of the Euclidean Elements. A Study of the Theory of Incommensurable Magnitudes and Its Significance for Early Greek Geometry.* 1975, IX + 374 pp.
16. Augustine, *De Dialectica.* Translated with Introduction and Notes by B. Darrell Jackson. 1975, XI + 151 pp.